INSTRUCTOR'S MANUAL

J. RICHARD CHRISTMAN
United States Coast Guard Academy

With the Assistance of
Stanley A. Williams
Iowa State University

Walter Eppenstein
Rensselaer Polytechnic Institute

to accompany

FUNDAMENTALS OF
PHYSICS

SIXTH EDITION

DAVID HALLIDAY
University of Pittsburgh

ROBERT RESNICK
Rensselaer Polytechnic Institute

JEARL WALKER
Cleveland State University

JOHN WILEY & SONS, INC.
New York • Chichester • Weinheim • Brisbane • Singapore • Toronto

To order books or for customer service call 1-800-CALL-WILEY (225-5945).

ISBN 0-471-37870-4

Printed in the United States of America

10 9 8 7 6 5 4 3 2 1

Printed and bound by Victor Graphics, Inc.

PREFACE

This manual contains material we hope will be useful in the design of an introductory physics course based on the text *FUNDAMENTALS OF PHYSICS*, sixth edition, by David Halliday, Robert Resnick, and Jearl Walker. It may be used with either the extended or regular versions of the text. In Section One, we have included material to help instructors choose topics and design courses. In Section Two, we have also provided lecture notes, outlining the important topics of each chapter, and have suggested demonstration and laboratory experiments, computer software, video cassettes, and video disks.

Because some instructors avoid assigning problems that are discussed on the website associated with *A Student's Companion* and in the *Student Solution Manual*, while others desire to include a few of these in many assignments, Section Three of the manual contains a list of problems in the two student supplements. Sections Four, Five, and Six contain answers to checkpoints, answers to end-of-chapter questions, and answers to end-of-chapter problems. To help ease the transition from the fifth to the sixth edition of the text, Section Seven of the manual cross references end-of-chapter problems between the two editions.

Some instructors include in their courses computational exercises to be carried out by students on computers, using either spreadsheet software or their own programs. To aid these instructors, the final section of the manual contains a selection of computer projects.

The principal author is grateful to Stanley Williams, who co-authored the first edition of the instructor manual for *Fundamentals of Physics*. Much of his material has been retained in this manual. He is also grateful to Walter Eppenstein, who helped with suggestions for demonstration and laboratory experiments. Jearl Walker helped significantly by supplying answers to checkpoint questions, end-of-chapter questions, and new and revised end-of-chapter problems.

The author is indebted to Joan Kalkut who provided much good advice and managed many aspects of the production side of this project. Many good people at Wiley helped in a host of ways. Special thanks go to Stuart Johnson, Cliff Mills, Catherine Donovan, and Tom Hempstead. Karen Christman carefully read earlier editions of the manuscript and made many useful suggestions. Her fine work is gratefully noted. The unfailing support of Mary Ellen Christman is joyfully acknowledged.

J. Richard Christman
U.S. Coast Guard Academy
New London, Connecticut 06320

TABLE OF CONTENTS

SECTION ONE
PATHWAYS

Fundamentals of Physics, sixth edition, follows the sequence of topics found in most introductory courses. In fact, earlier editions of this text were instrumental in establishing the sequence. It is, however, extremely flexible in regard to both the range of topics and the depth of coverage. As a result, it can be used for a two, three, or four term course along traditional lines. It can also be used with many of the innovative courses that are presently being designed and taught. In many instances sections that discuss fundamental principles and give applications are followed by other sections that go deeper into the physics. Some instructors prefer to cover fewer topics than others but treat the topics they do cover in great depth. Others prefer to cover more topics with less depth. Courses of both types can easily be accommodated by selecting appropriate sections of the text.

By carefully choosing sections of the text to be included, your course might be a two-term, in-depth study of the fundamentals of classical mechanics and electromagnetism. With the addition of another term you might include more applications and the thermodynamics and optics chapters. In a three-term course, you might also forgo thermodynamics and optics but include Chapter 38 (Relativity) and some of the quantum mechanics chapters added in the extended version.

When designing the course, some care must be taken in the selection of topics because many discussions in later chapters presume coverage of prior material. Here are some comments you might find useful in designing your course. Also refer to the *Lecture Notes* section of this manual.

Mechanics. The central concepts of classical mechanics are covered in Chapters 1 through 12. Some minor changes that are possible, chiefly in the nature of postponements, are mentioned in the Lecture Notes. For example, the scalar product can be postponed until the discussion of work in Chapter 7 and the vector product can be postponed until the discussion of torque in Chapter 12.

Coverage of Chapter 5 can be shortened to two lectures or elongated to over four, depending on the time spent on applications. Other sections in the first twelve chapters that can be used to adjust the length of the course are 2–7, 3–6, 4–8, 4–9, 6–3, 9–7, 9–8, and 11–3. However, you should consider covering section 9–8 if you include the thermodynamics chapters in your course. Coverage of Chapter 10 (Collisions) can be shortened significantly with safety. Section 11–7, which deals with the calculation of the rotational inertias of extended bodies, can be covered in detail or can be shortened by simply stating results once the definition as a sum over particles has been discussed. The parallel axis theorem is needed to solve some end-of-chapter problems in this chapter and in Chapter 16 and it should be covered if those problems are assigned.

The order of the chapters should be retained. For example, difficulties arise if you precede dynamics with statics as is sometimes done in other texts. To do so, you would need to discuss torque, introduced in Chapter 11, and explain its relation to angular acceleration. This involves considerable effort and is of questionable value.

Chapters 13 through 18 apply the fundamental principles of the first 12 chapters to special systems and, in many cases, lay the groundwork for what is to come. Many courses omit one or more of Chapters 13 (Equilibrium and Elasticity), 14 (Gravitation), 15 (Fluids), and 18 (Waves — II). There is some peril in these omissions, however. Chapter 14, for example, is pedagogically important. The central idea of the chapter is a force law and the discussions of many of its ramifications show by example how physics works. Since the chapter brings together many previously discussed ideas it can be used as a review. In addition, Newton's law of gravity is used later to introduce Coulomb's law and the proof that the electrostatic force is conservative relies on the analogy. The basis of Gauss' law is laid in Chapter 14 and inclusion of this chapter makes teaching of the law easier.

The idea of a velocity field is first discussed in Chapter 15 and is used to introduce electric flux in Chapter 24 (Gauss' Law). The concepts of pressure and density are explained in Chapter 15 and are used again in the thermodynamics chapters. If Chapter 15 is omitted, you should be prepared to make up for the loss of material by presenting definitions and discussions of velocity field, pressure, and density when they are first used in your course.

Chapter 13 (Equilibrium of Rigid Bodies) can be safely omitted. If it is, a brief description of the equilibrium conditions might be included in the discussion of Chapter 11 or 12. The few problems in later chapters that depend on material in this chapter can be passed over. If Chapter 13 is included, be sure you have already covered torque and have explained its relation to angular acceleration.

Chapters 16 (Oscillations) and 17 (Waves — I) are important parts of an introductory course and should be covered except when time constraints are severe. Chapter 16 is required for Chapter 17 and both are required for Chapter 18 (Waves — II). Chapter 16 is also required for Chapter 33 (Electromagnetic Oscillations and Alternating Current) and parts of Chapter 17 are required for Chapters 34 (Electromagnetic Waves), 36 (Interference), 37 (Diffraction), 39 (Photons and Matter Waves), and 40 (More About Matter Waves). Chapters 16 and 17 may be covered in the mechanics part of the course or may be delayed until electromagnetic waves are covered.

Sections of Chapters 13 through 18 that can be used to adjust the length of the course are 13–5, 13–6, 14–8, and 14–9.

Thermodynamics. Chapters 19 through 21 cover the ideas of thermodynamics. Most two-term courses and some three-term courses omit these chapters entirely. If they are covered, they can be placed as a unit almost anywhere after the mechanics chapters. The idea of temperature is used in Chapter 27 (Current and Resistance) and in some of the modern physics chapters, as well as in the other thermodynamics chapters. If Chapter 19 is not covered prior to Chapter 27, you should plan to discuss the idea of temperature in connection with that chapter or else omit the section that deals with the temperature dependence of the resistivity.

Electromagnetism. The fundamentals of electricity and magnetism are covered in Chapters 22 through 34. Chapter 34 (Electromagnetic Waves) may be considered a capstone to the electromagnetism chapters or as an introduction to the optics chapters. Sections that might be omitted to adjust the length of the course are 26–6, 26–7, 26–8, 27–6, 27–8, 27–9, 28–7, 28–8, 29–6, 31–9, 31–12, 32–3, 32–4, 32–5, 32–6, 32–7, and 32–8.

Sections 26–6, 26–7, and 26–8, on dielectrics, should be included in an in-depth course but may be omitted in other courses to make room for other topics. Similarly, coverage of Chapters 28 (Circuits) and 33 (Electromagnetic Oscillations and Alternating Currents) may be adjusted considerably, depending on the extent to which the course emphasizes practical applications. Section 27–6 is required if Chapter 42 is covered although the material can be shorted and presented in conjunction with Chapter 42 rather than at an earlier time.

Section 32–2 contains a discussion of Gauss' law for magnetism, one of Maxwell's equations, and should be included in every course, as should sections 32–9, 32–10, and 32–11, on the displacement current and the Ampere-Maxwell law. For continuity, the first section of the chapter should also be included. The central portion of the chapter deals with magnetic properties of materials and some of ramifications of those properties. It nicely complements the previous sections on dielectrics. These parts of the chapter might be omitted or passed over swiftly to gain time for other sections. On the other hand, they should be included if you intend to emphasize properties of materials.

Optics. Chapters 35 through 37 are the optics chapters. You might wish to precede them with Chapter 34 (Electromagnetic Waves) or you might wish to replace Chapter 34 with a short qualitative discussion. You can be somewhat selective in your coverage of Chapter 35 (Images).

It can be covered as lightly or as deeply as desired. Much of the material in this chapter can be covered as laboratory exercises.

Chapters 36 (Interference) and 37 (Diffraction) are important in their own right and are quite useful for the discussion of photons and matter waves in Chapter 39. Chapter 37 cannot be included without Chapter 36 but coverage of both chapters can be reduced somewhat to make room for other topics. The fundamentals of interference and diffraction are contained in Sections 36–1 through 36–6 and 37–1 through 37–4. Other sections of these chapters can be included or excluded, as desired.

Modern Physics. Chapter 38 (Relativity) may be used as a capstone to the mechanics section of the course, as a capstone to the entire course, or as an introduction to the modern physics included in the extended version of the text. Some results of relativity theory are needed for the chapters that follow. If you do not wish to cover Chapter 38 in detail you can describe these results as they are needed. However, it is probably more satisfying to present a more complete and logically connected description of relativity theory. If you plan to cover some of the other modern physics chapters you should consider including Chapter 38.

The fundamentals of the quantum theory are presented in Chapters 39 (Photons and Matter Waves) and 40 (More About Matter Waves). This material should be treated as a unit and must follow in the order written. If you include these chapters, be sure earlier parts of the course include discussions of uniform circular motion, angular momentum, Coulomb's law, electrostatic potential energy, electromagnetic waves, and diffraction. $E = mc^2$ and $E^2 = (pc)^2 + (mc^2)^2$, from relativity theory, are used in discussions of the Compton effect.

The introductory modern physics chapters are followed by application chapters: Chapters 41 (All About Atoms), 42 (Conduction of Electricity in Solids), 43 (Nuclear Physics), 44 (Energy from the Nucleus), and 45 (Quarks, Leptons, and the Big Bang). You may choose to end the course with Chapter 40 or you may choose to include one or more of the application chapters.

The ideas of temperature and the Kelvin scale are used in several places in the modern physics chapters: Sections 41–12 (How a Laser Works), 42–5 (Metals), 42–6 (Semiconductors), 44–6 (Thermonuclear Fusion), and 45–12 (The Microwave Background Radiation). With a little supplementary material, these sections can be covered even if Chapter 19 is not.

Chapter 44 (Energy from the Nucleus) requires Chapter 43 (Nuclear Physics) for background material, but Chapter 43 need not be followed by Chapter 44. $E = mc^2$ and $E^2 = (pc)^2 + (mc^2)^2$ from relativity theory are also used. The discussion of thermonuclear fusion uses some of the ideas of kinetic theory, chiefly the distribution of molecular speeds. Either Chapter 20 (particularly Section 20–7) should be covered first or you should be prepared to supply a little supplementary material here.

Chapter 45 includes an introduction to high energy particle physics and tells how the ideas of physics are applied to cosmology. Both these topics fascinate many students. In addition, the chapter provides a nice overview of physics.

Some knowledge of the Pauli exclusion principle (from Chapter 41) and spin angular momentum (from Chapters 32 and 41) is required. Knowledge of the strong nuclear force (discussed in Chapters 43 and 44) is also required. In addition, beta decay (discussed in Chapter 43) is used several times as an illustrative example. Nevertheless, the chapter can be made to stand alone with the addition of only a small amount of supplementary material.

SUGGESTED COURSES

A bare bones two-semester course (about 90 meetings) can be constructed around Chapters 1 through 12, 16, 17, and 22 through 34, with the omission of sections 32–3 through 32–8. The course can be adjusted to the proper length by the inclusion or omission of supplementary material and optional topics. If four to eight additional meetings are available each term, Chapter 14 or 15 (or perhaps both) can be inserted after Chapter 12 and one or more of the optics chapters can be inserted after Chapter 34. As an alternative, you might consider including sections on dielectrics, magnetic properties, semiconductors, and superconductors to emphasize properties of materials.

A three-term course (about 135 meetings) can be constructed by adding the thermodynamics chapters (19 through 21) and some or all of the modern physics chapters (38 through 45) to those mentioned above. If the needs of the class dictate a section on alternating current, some modern physics material can be replaced by Chapter 33.

ESTIMATES OF TIME

The following chart gives estimates of the time required to cover all of each chapter, in units of 50 minute periods. The second and fifth columns of the chart contain estimates of the number of lecture periods needed and includes the time needed to perform demonstrations and discuss the main points of the chapter. The third and sixth columns contain estimates of the number of recitation periods required and includes the time needed to go over problem solutions, answers to end-of-chapter questions, and points raised by students. If your course is organized differently, you may wish to add the two numbers to obtain the total estimated time for each chapter.

Use the chart as a rough guide when planning the syllabus for a semester, quarter, or year course. If you omit parts of chapters, reduce the estimated time accordingly.

Text Chapter	Number of Lectures	Number of Recitations	Text Chapter	Number of Lectures	Number of Recitations
1	0.3	0.2	24	1.8	1.8
2	2.0	2.0	25	1.8	1.8
3	1.0	1.0	26	1.5	2.0
4	2.0	2.5	27	1.0	1.0
5	2.0	2.0	28	2.0	2.3
6	2.0	2.0	29	2.0	1.8
7	1.8	1.5	30	2.0	1.2
8	2.0	2.0	31	2.5	2.5
9	2.0	1.5	32	1.5	1.8
10	2.0	1.6	33	2.5	2.5
11	2.0	1.5	34	2.9	2.7
12	2.0	2.0	35	2.5	2.5
13	1.0	2.0	36	2.0	2.0
14	2.3	2.0	37	2.0	2.0
15	2.0	2.0	38	2.5	2.0
16	2.5	1.8	39	2.0	2.0
17	2.5	2.0	40	2.0	2.0
18	2.5	2.0	41	2.2	2.0
19	2.5	2.5	42	2.0	2.0
20	2.0	1.4	43	1.8	2.0
21	1.5	1.6	44	2.0	2.0
22	1.0	1.0	45	2.0	2.0
23	1.6	1.3			

SECTION TWO
LECTURE NOTES

Lecture notes for each chapter of the text are grouped under the headings BASIC TOPICS and SUGGESTIONS.

BASIC TOPICS contains the main points of the chapter in outline form. In addition, one or two demonstrations are recommended to show the main theme of the chapter. You may wish to pattern your lectures after the notes, suitably modified, or simply use them as a check on the completeness of your own notes.

The SUGGESTIONS sections recommend end-of-chapter questions and problems, video cassettes, video disks, computer software, computer projects, alternate demonstrations, and other material that might be useful for the course. Many of the questions concentrate on points that seem to give students trouble, and it is worthwhile dealing with some of them before students tackle a problem assignment. Some questions and problems might be incorporated into the lectures while some might be assigned and used to generate discussion by students in small recitation sections. Answers to the questions appear in Section Five of this manual and answers to the problems appear in Section Six.

Each chapter contains several semi-quantitative questions, called checkpoints. Encourage students to use them to check their understanding of the concepts and relationships discussed in the chapter. Go over some or all of them in recitation classes or lectures. Answers to the checkpoint questions are given in Section Four of this manual.

If funds are available, consider setting up an interactive class room or lecture hall in which students can be polled remotely. The checkpoints and end-of-chapter questions are excellent for this purpose. A system called Classtalk, which utilizes student input via individual computers, is available from Better Education, Inc., 4824 George Washington Memorial Highway, Yorktown, Va 23692.

General Two excellent books that deal with teaching the introductory calculus-based course are

Teaching Introductory Physics; Arnold B. Arons; John Wiley (1997).

Teaching Introductory Physics (A Sourcebook); Clifford E. Swartz and Thomas Miner; Springer-Verlag (1998).

Both of these provide well thought-out explanations of some of the concepts that perplex students and give help with teaching those concepts. They are also excellent sources of demonstration and laboratory experiments that illuminate the important ideas of the introductory physics course.

Over the past ten years or so the field of physics education research has grown tremendously. Many research projects focus on the troubles students have in learning physics and analyze proposed remedies. Lillian McDermot and Edward Redish have compiled an extensive resource letter that lists books and journal articles in the field. It appeared in the September 1999 issue of the American Journal of Physics and is highly recommended as a source of material for improvement of the course.

Video. All of the video cassette and video disk items listed in the SUGGESTIONS sections are short, well done, and highly pertinent to the chapter. It is not possible to review all available material and there are undoubtedly many other fine video cassettes and disks that are not listed. Video might be incorporated into the lectures, shown during laboratory periods, or set up in a special room for more informal viewing.

An excellent set of video cassettes and disks, *THE MECHANICAL UNIVERSE*, can be obtained from The Annenberg CPB Collection, PO Box 2345, South Burlington, VT 05407–2345. The set consists of 52 half-hour segments dealing with nearly all the important concepts of introductory physics. Historical information and animated graphics are used to present the concepts in an imaginative and engaging fashion. Some physics departments run appropriate segments throughout the course in special viewing rooms. Accompanying textbooks, teacher manuals, and study guides are also available.

Many time-tested films originally from Encyclopaedia Britannica, PSSC, Project Physics, and elsewhere have been transferred to video disk by the AAPT Instructional Media Center and are available under the title *Physics: Cinema Classics* from Ztek Co., PO Box 11768, Lexington, KY 40577–1768. The films cover a host of topics in mechanics, thermodynamics, electricity and magnetism, optics, and modern physics. Other short films that have been transferred to video are the AAPT Collections 1 and 2 and the Miller Collection. These and many other video tapes and disks are referenced in the SUGGESTIONS section of the Lecture Notes.

Computer Software. Computers have made significant contributions to the teaching of physics. They are widely used in lectures to provide animated illustrations, with parameters under the control of the lecturer; they also provide tutorials and drills that students can work through on their own. Specialized programs are listed in appropriate SUGGESTION sections of the Lecture Notes in this manual. In addition, several available software packages cover large portions of an introductory course. Four of them are:

Core Concepts in Physics; CD-ROM; Macintosh, Windows; Saunders College Publishing, The Public Ledger Building, Suite 1250, 150 South Independence Mall West, Philadelphia, PA 19106–3412. A great many animations and live videos, laboratory demonstrations, and graphics. Most are interactive. Many step-by-step solutions are given to example problems. Covers a large portion of the introductory calculus-based course.

Interactive Journey Through Physics; CD-Rom; Macintosh, Windows; Cindy Schwartz and Bob Beicher; Prentice-Hall, 240 Frisch Ct., Paramus, NJ 07652-5240. A large number of animations and simulations, many with audio descriptions. Self-check quizzes are associated with the simulations. Over fifty narrated video segments. Covers most of the topics of the introductory course.

Multimedia Enhanced Physics Instruction; CD ROM; Macintosh, Windows; Maha Ashour-Abdalla; McGraw-Hill, P.O. Box 182604, Columbus, OH 43272–303143004. Extensive illustrated concepts sections, which give the important ideas of the introductory course, along with applications and examples, often by way of video. Numerous problems with interactive help. Each concept module also has a self-check quiz.

Interactive Physics; Macintosh, Windows; Knowledge Revolution, 15 Bush Pl., San Francisco, CA 94103). Animations and graphs for a wide variety of mechanical phenomena. The user can set up "experiments" with massive objects, strings, springs, dampers, and constant forces. Parameters can easily be changed. For Macintosh and Windows computers. Reviewed in The Physics Teacher, September 1991.

You might consider setting aside a room or portion of a lab, equip it with several computers, and make tutorial, drill, and simulation programs available to students. If you have sufficient hardware (and software), you might base some assignments on computer materials.

Computers and top-of-the line graphing calculators might also be used by students to perform calculations. Properly selected computer projects can add greatly to the students' understanding of physics. Projects involving the investigation of some physical system of interest might be assigned to individuals or might be carried out by a laboratory class. Some computer projects are outlined in Section Eight of this manual and are referenced in the Lecture Notes of this manual. A large number of suitable problems and projects can also be found in the book *Introduction to Computational Physics* by Marvin L. De Jong (Addison-Wesley, 1991).

Commercial spreadsheet programs can facilitate problem solving. PSI-Plot (Windows; Poly Software International, P.O. Box 1457, Sandy, UT 84091) and $f(g)$ Scholar (Macintosh, Windows; Future Graph, Inc., 75 James Way, Southampton, PA 18966) are high-end spreadsheet programs that incorporate many science and engineering problem-solving and graphing capabilities. Commercial problem-solving programs such as *MathCAD* (Windows; MathSoft, Inc., 101 Main Street, Cambridge, MA 02142-1521), *DERIVE* (Windows; Soft Warehouse; 3660 Waialae Avenue, Suite 304, Honolulu, HI 96815), *MAPLE* (Macintosh, Windows; Waterloo Maple, 57 Erb Street W., Waterloo, Ontario, Canada N2L 6C2), *Pro Solve* (Macintosh, Windows; Problem Solving Concepts, Inc., 1980 E. 116th Street, Ste 220, Carmel, IN 46032), and *Mathematica* (Macintosh, Windows; Wolfram Research, Inc., 100 Trade Center Drive, Champaign, IL 61820–7237) can easily be used by students to solve problems and graph results. All these programs allow students to set up a problem generically, then view solutions for various values of input parameters. For example, the range or maximum height of a projectile can be found as a function of initial speed or firing angle, even if air resistance is taken into account.

A number of computer programs allow you to view digitized video on a computer monitor and mark the position of an object in each frame. The coordinates of the object can be listed and plotted. They can then be used to find the velocity and acceleration of the object, either within the program itself or by exporting the data to a spreadsheet. Three of these are: *Videopoint* (Windows, CD-ROM; Pasco Scientific, 10101 Foothills Blvd., Roseville, CA 95678), *VideoGraph* (Macintosh; Physics Academic Software, Box 8202, North Carolina State University, Raleigh, NC 27695–8202), and *World in Motion* (Windows; Physics Curriculum and Instruction, 22585 Woodhill Drive, Lakeville, MN 55044). All of these come with an assortment of video clips. Home-made videos can also be used. The capabilities of the programs are different. Check carefully before purchasing.

Demonstrations. Notes for most of the chapters are developed around demonstration experiments. Generally speaking, these use relatively inexpensive, readily available equipment, yet clearly demonstrate the main ideas of the chapter. The choice of demonstrations, however, is highly personal and you may wish to substitute others for those suggested here or you may wish to present the same ideas using chalkboard diagrams. Several excellent books give many other examples of demonstration experiments. The following are available from the American Association of Physics Teachers, One Physics Ellipse, College Park MD 20740–3845:

A Demonstration Handbook for Physics, G.D. Freier and F.J. Anderson, 320 pages (1981). Contains over 800 demonstrations, including many that use everyday materials and that can be constructed with minimal expense. Line drawings are used to illustrate the demonstrations.

String and Sticky Tape Experiments, Ronald Edge, 448 pages (1987). Contains a large number of illuminating experiments that can be constructed from inexpensive, readily available materials.

Apparatus for Teaching Physics, edited by Karl C. Mamola. A collection of articles from The Physics Teacher that describe laboratory and demonstration apparatus.

How Things Work, H. Richard Crane, 114 pages, 1992. A collection of 20 articles from The Physics Teacher.

The following is currently out of print but is available in many college libraries and physics departments:

Physics Demonstration Experiments, H.F. Meiners, ed. An excellent source of ideas, information, and construction details on a large number of experiments, with over 2000 line drawings and photographs. It also contains some excellent articles on the philosophical aspects of lecture demonstrations, the use of shadow projectors, TV, films, overhead projectors, and stroboscopes.

Appropriate demonstrations described in Freier and Anderson are listed in the SUGGESTIONS sections of the notes. This book does not give any construction details, but more information about most demonstrations can be obtained from the book edited by Meiners.

The *Physics InfoMall* CD-ROM (The Learning Team, 84 Business Park Drive, Armonk, NY 10504), a searchable database of over 1000 demonstrations, is another excellent source. There are both Windows and Macintosh versions. The CD also contains articles and abstracts, problems with solutions, whole reference books, and a physics calendar. The following three books, all available from the AAPT, are also sources of ideas for demonstrations and examples:

Physics of Sports; edited by C. Frohlich. Contains reprints and a resource letter.

Amusement Park Physics; edited by Carole Escobar. In workbook form. The activities described are perhaps more appropriate for a high school class but some can be used in college level lectures as examples.

Potpourri of Physics Teaching Ideas; edited by Donna Berry; reprints of articles on apparatus from The Physics Teacher.

A computer can also be used for data acquisition during demonstrations. Photogate timers, temperature probes, strain gauges, voltage probes, and other devices can be input directly into the computer and results can be displayed as tables or graphs. The screen can be shown to a large class by using a large monitor, a TV projection system, or an overhead projector adapter. Inexpensive software and hardware can be purchased from Vernier Software (8565 SW Beaverton-Hillsdale Hwy., Portland, OR 97225-2429). PASCO Scientific (PO Box 619011, 10101 Foothills Boulevard, Roseville, CA 95661–9011) has data acquisition software and an extensive variety of probes for both Macintosh and Windows computers. If more sophisticated software is desired, consider the commercial package Labview (National Instruments Corporation, 11500 N. Mopac Expwy., Austin, TX 78759–3504). The monograph *Photodetectors* by Jon W. McWane, J. Edward Neighbor, and Robert F. Tinker (available from the AAPT) is a good source of technical information about photodetectors.

Laboratories. Hands-on experience with actual equipment is an extremely important element of an introductory physics course. There are many different views as to the objectives of the physics laboratory and the final decision on the types of experiments to be used has to be made by the individual instructor or department. This decision is usually based on financial and personnel considerations as well as on the pedagogical objectives of the laboratory.

Existing laboratories vary widely. Some use strictly cookbook type experiments while others allow the students to experiment freely, with practically no instructions. The equipment ranges from very simple apparatus to rather complex and sophisticated equipment. Physical phenomena may be observed directly or simulated on a computer. Data may be taken by the students or fed into a computer.

The equipment described above can be used for data acquisition in a student lab. Even if data acquisition software is not used, consider having students use computers and spreadsheet programs to analyze and graph data.

Many physics departments have written their own notes or laboratory manuals and relatively few physics laboratory texts are on the market. Two such books, both available from John Wiley & Sons, are

Laboratory Physics, second edition, H.F. Meiners, W. Eppenstein, R.A. Oliva, and T. Shannon. (1987).

Laboratory Experiments in College Physics, seventh edition, C.H. Bernard and C.D. Epp. (1994).

Experiments from these books are listed in the SUGGESTIONS section of the Lecture Notes. Meiners is used to designate the Meiners, Eppenstein, Oliva, and Shannon book while Bernard is used to designate the Bernard and Epp book. Both books contain excellent introductory sections explaining laboratory procedures to students. Meiners also contains a large amount of material on the use of microprocessors in the lab.

Student supplements. Several supplements, all available from Wiley, might be recommended to the students:

A Student's Companion to Fundamentals of Physics. A study guide. The basic concepts of each chapter are reviewed in a workbook format that helps students focus their attention on the important ideas and their relationships to each other. Each copy contains a password for an internet site that provides the students with hints for about 650 end-of-chapter exercises and problems. The list of exercises and problems on the site are given in Section Three of this manual.

Student Solution Manual. Contains fully worked solutions to about 880 end-of-chapter problems and exercises.

CD Physics. A CD ROM version of the text and supplements for Windows and Macintosh. This contains the complete text, the *Student Solution Manual*, the interactive tutorials, the interactive simulations, and a glossary. It is extensively hyperlinked.

Instructor aids. In addition to this *Instructor Manual* Wiley provides several other aids for instructors:

Instructor's Solution Manual Contains fully worked solutions to all the end-of-chapter problems.

Test Bank. Contains over 2800 multiple choice questions (with answers) for use on exams and quizzes. Both quantitative and qualitative questions are included. In each chapter, some of the questions are modeled after the checkpoints and end-of-chapter questions, as well as after the end-of-chapter problems and exercises. There is also a computerized version of the *Test bank* on a CD ROM, which also contains a program for constructing exams (along with answer keys).

Instructor's Resource CD. A CD ROM for Windows. Contains the *Instructor's Solution Manual* (as both LaTeX(2e) source files and in PDF form), reproductions of illustrations from the text (in JPEG form), and the *Test Bank*. There is a computer program that allows instructors to generate exams from the test bank questions.

WebCT. A collections instructor resources that can be downloaded from the Wiley server. It includes the set of text illustrations, the interactive simulations, and test questions.

Chapter 1 MEASUREMENT

BASIC TOPICS

I. Base and derived units.

  A. Explain that standards are associated with base units and that measurement of a physical quantity takes place by means of comparison with a standard. Discuss qualitatively the SI standards for time, length, and mass. Show a 1 kg mass and a meter stick. Show the simple well-known procedure for measuring length with a meter stick.

  B. Explain that derived units are combinations of base units. Emphasize that the speed of light is now a defined unit and the meter is a derived unit. Discuss an experiment in which the time taken for light to travel a certain distance is measured. Example: the reflection of a light signal from the Moon. Use a clock and a meter stick to find your walking speed in m/s.

  C. This is a good place to review area, volume, and mass density. Use simple geometric figures (circle, rectangle, triangle, cube, sphere, cylinder, etc.) as examples.

II. Systems of units.

  A. Explain what a system of units is. Give the 1971 SI base units (Table 1–1). Stress that they will be used extensively.

  B. Point out the SI prefixes (Table 1–2). The important ones for this course are mega, kilo, centi, milli, micro, nano, and pico. Discuss powers of ten arithmetic and stress the simplicity of the notation. This might be a good place to say something about significant digits.

  C. Discuss unit arithmetic and unit conversion.

  D. Most of the students' experience is with the British system. Relate the inch to the centimeter and the slug to the kilogram. Discuss unit conversion. Use speed as an example: convert 50 mph and 3 mph to km/h and m/s. Point out the conversion tables in Appendix D.

III. Properties of standards.

  A. Discuss accessibility and invariability as desirable properties of standards.

  B. Discuss secondary standards such as the meter stick used earlier.

IV. Measurements.

  A. Stress the wide range of magnitudes measured. See Tables 1–3, 1–4, and 1–5. Explain the atomic mass unit. One atom of ^{12}C has a mass of exactly 12 u. 1 u is approximately 1.661×10^{-27} kg.

  B. Discuss indirect measurements.

SUGGESTIONS

1. Assignments

  a. To emphasize SI prefixes assign problems 1 and 5.

  b. Unit conversion is covered in many problems. Choose some, such as 2, 3, 4, that deal with unfamiliar units.

  c. According to the needs of the class, assign one or more problems that deal with area and volume calculations, such as 5, 7, and 8.

2. Demonstrations

  Examples of "standards" and measuring instruments: Freier and Anderson Ma1 — 3.

3. Books and Monographs

  a. *Frequency and Time Measurements*, edited by Christine Hackman and Donald B. Sullivan; available from AAPT, One Physics Ellipse, College Park MD 20740–3845.

b. *SI: The International System of Units*; edited by Robert A. Nelson; available from AAPT (see above for address).
c. *Connecting Time and Space*; edited by Harry E. Bates; available from AAPT (see above for address). Reprints that discuss measurements of the speed of light and the redefinition of the meter. Students will not be able to understand much of this material at this stage of the course but it is nevertheless useful for background.

4. Audio/Visual
a. *Time and Place*, *Measuring Short Distances*; Side A: Mechanics (I) of Cinema Classics; video disk; available from Ztek Co., PO Box 11768, Lexington, KY 40577–1768.
b. *Powers of Ten* from the Films of Charles and Ray Eames; video disk; available from Ztek Co. (see above for address).

5. Laboratory
a. Meiners Experiment 7-1: *Measurement of Length, Area, and Volume.* Gives students experience using the vernier caliper, micrometer, and polar planimeter. Good introduction to the determination of error limits (random and least count) and calculation of errors in derived quantities (volume and area).
b. Bernard Experiments 1 and 2: *Determination of Length, Mass, and Density* and *Measurements, Measurement Errors, and Graphical Analysis.* Roughly the same as the Meiners experiment, but a laboratory balance is added to the group of instruments and the polar planimeter is not included. Graphs of mass versus radius and radius squared for a collection of disks made of the same material, with the same thickness, are used to establish the quadratic dependence of mass on radius.
c. Meiners Experiment 7-3: *The Simple Pendulum* and Bernard Experiment 3: *The Period of a Pendulum — An Application of the Experimental Method.* Students time simple pendulums of different lengths, then use the data and graphs (including a logarithmic plot) to determine the relationship between length and period. They calculate the acceleration due to gravity. This is an exercise in finding functional relationships and does not require knowledge of dynamics.

Chapter 2 MOTION ALONG A STRAIGHT LINE

BASIC TOPICS

I. Position and displacement.
A. Move a toy cart with constant velocity along a table top. Select an origin, place a meter stick and clock on the table, and demonstrate how x(t) is measured in principle. Emphasize that x is always measured from the origin; it is not the cart's displacement during any time interval.
B. Draw a graph of $x(t)$ and point out that it is a straight line. Show what the graph looks like if the cart is not moving. Point out that the line has a greater slope if the cart is going faster. Move the cart so its speed increases with time and show what the curve $x(t)$ looks like. Do with same for a cart that is slowing down.
C. Some students think of a coordinate as distance. Distinguish between these concepts. Point out that a coordinate defines a position on an axis and can be positive or negative. Demonstrate a negative velocity, both with the cart and on a graph. As another example, throw a ball into the air, pick a coordinate axis (positive in the upward direction, say), and point out when the velocity is positive and when it is negative. Draw the graph of the coordinate as a function of time. Repeat with the positive direction upward.

D. Define the displacement of an object during a time interval. Emphasize that only the initial and final coordinates enter and that an object may have many different motions between these while still having the same displacement. Point out that the displacement is zero if the initial and final coordinates are the same.

II. Velocity.

A. Define average velocity over an interval. Stress the meaning of the sign. Go over Sample Problem 2–1. Draw a graph of x versus t for an object that is accelerating. Pick an interval and draw the line between the end points on the graph. Observe that the average velocity in the interval is the slope of the line. Figs. 2–3 and 2–4 may also be used. Show how to calculate average velocity if the function $x(t)$ is given in algebraic form.

B. Define instantaneous velocity. Demonstrate the limiting process. Use a graph of x versus t for an accelerating cart to demonstrate that the line used to find the average velocity becomes tangent to the curve in the limit as Δt vanishes. Remark that the slope of the tangent line gives the instantaneous velocity. Show a plot of v versus t that corresponds to the x versus t graph used previously. Show how to calculate the instantaneous velocity if the function $x(t)$ is given in algebraic form. See Sample Problem 2–4. Stress that a value of the instantaneous velocity is associated with each instant of time. Some students think of velocity as being associated with a time interval rather than an instant of time.

C. Define instantaneous speed as the magnitude of the velocity. Compare to the average speed in an interval, which is the total path length divided by the time. Remark that the average speed is not necessarily the same as the magnitude of the average velocity.

D. Note that many calculus texts use a prime to denote a derivative. They also define the derivative of x with respect to time by the limit of $[x(t + \Delta t) - x(t)]/\Delta t$ rather than by the limit of $\Delta x/\Delta t$. Mention the different notations in class so students can relate their physics and calculus texts.

III. Acceleration.

A. Define average and instantaneous acceleration. Show the previous v versus t graph and point out the lines used to find the average acceleration in an interval and the instantaneous acceleration at a given time. Show how to calculate the average and instantaneous acceleration if either $x(t)$ or $v(t)$ is given in algebraic form.

B. Interpret the sign of the acceleration. Give examples of objects with acceleration in the same direction as the velocity (speeding up) and in the opposite direction (slowing down). Be sure to include both directions of velocity. Emphasize that a positive acceleration does not necessarily imply speeding up and a negative acceleration does not necessarily imply slowing down.

C. Use graphs of $x(t)$ and $v(t)$ to point out that an object may simultaneously have zero velocity and non-zero acceleration. Explain that if the direction of motion reverses the object must have zero velocity at some instant. Give the position as a function of time as $x(t) = At^2$, for example, and show that the velocity is 0 at $t = 0$ but the acceleration is not 0. Illustrate the function with a graph.

IV. Motion in one dimension with constant acceleration.

A. Derive the kinematic equations for $x(t)$ and $v(t)$. If students know about integration, use methods of the integral calculus (as in Section 2–7). In any event, show that $v(t)$ is the derivative of $x(t)$ and that a is the derivative of $v(t)$.

B. Discuss kinematics problems in terms of a set of simultaneous equations to be solved. Examples: use equations for $x(t)$ and $v(t)$ to algebraically eliminate the time and to algebraically eliminate the acceleration. The equations of constant acceleration motion are listed in Table 2–1. Some instructors teach students to use the table. Others ask students to always start with Eqs. 2–11 and 2–15, then use algebra to obtain the equations needed for a particular problem. See Sample Problem 2–5.

C. To help students see the influence of the initial conditions, sketch graphs of $v(t)$ and $x(t)$ for various initial conditions but the same acceleration. Include both positive and negative initial velocities. Draw a different set of graphs for positive and negative acceleration. Point out where the particle has zero velocity and when it returns to its initial position.

V. Free fall.

A. Give the value for g in SI units. Point out that the free-fall acceleration is due to gravity and that it is directed toward the center of Earth. Say that locally Earth's surface is essentially flat and the free-fall acceleration may be taken to be in the same direction at slightly different points. Explain that $a = +g$ if down is taken to be the positive direction and $a = -g$ if up is the positive direction. Do examples using both choices. Throw a ball into the air and emphasize that its acceleration is g throughout its motion, even at the top of its trajectory.

B. Drop a small ball through two photogates, one near the top to turn on a timer and one further down to turn it off. Repeat for various distances and plot the position of the ball as a function of time. Explain that the curve is parabolic and indicates a constant acceleration.

C. Explain that all objects at the same place have the same free-fall acceleration. In reality, different objects may have different accelerations because air influences their motions differently. This can be demonstrated by placing a coin and a wad of cotton in a glass cylinder about 1 m long. Turn the cylinder over and note that the coin reaches the bottom first. Now use a vacuum pump to partially evacuate the cylinder and repeat the experiment. Repeat again with as much air as possible pumped out.

D. Point out that free-fall problems are special cases of constant acceleration kinematics and the methods described earlier can be used. Work a few examples. For an object thrown into the air, calculate the time to reach the highest point, the height of the highest point, the time to return to the initial height, and its velocity when it returns, all in terms of the initial velocity.

SUGGESTIONS

1. Assignments

   a. To help students obtain some qualitative understanding of velocity and acceleration, ask them to discuss questions 1, 2, 3, and 4. Some aspects of motion with constant acceleration are covered in questions 5 and 6. Free fall is covered in questions 8 and 9.

   b. To make more use of the calculus assign some of problems 7, 11, 12, 15, 19, and 21. Problems 12 and 19 can also be used to discuss differences between average and instantaneous velocity.

   c. To emphasize the interpretation of graphs assign a few of problems 6, 7, 10, 12, 13, 16, and 18. Some of these require students to draw graphs after performing calculations.

   d. Ask students to solve a few problems dealing with motion with constant acceleration. Consider problems 25, 26, 34, and 35. For a little more challenge, consider 37.

   e. Problems 42, 45, 46, 50, and 55 are good problems to test understanding of free-fall motion. Problem 62 is more challenging.

2. Demonstrations

   Uniform velocity and acceleration, velocity as a limiting process: Freier and Anderson Mb10 — 13, 15, 18, 21, 22.

3. Audio/Visual
   a. *Acceleration due to Gravity*; from AAPT collection 1 of single-concept films; video tape; available from Ztek Co., PO Box 11768, Lexington, KY 40577–1768.
   b. *One Dimensional Motion*; *Distance, Time & Speed*; *One Dimensional Acceleration*; *Constant Velocity & Uniform Acceleration*; from AAPT collection 2 of single-concept films; video tape; available from Ztek Co. (see above for address).
   c. *Uniform Motion*, *Free Fall*; Side A: Mechanics (I) of Cinema Classics; video disk; available from Ztek Co. (see above for address).
   d. *Acceleration*; VHS video tape (8 min); Films for the Humanities & Sciences, PO Box 2053, Princeton, NJ 08543–2053.
   e. *Falling Motion*; VHS video tape (8 min); Films for the Humanities & Sciences (see above for address).
   f. *Uniform and Accelerated Motion*; *Gravitational Acceleration*; Physics Demonstrations in Mechanics, Part I; VHS video tape; ≈3 min each; Physics Curriculum & Instruction, 22585 Woodhill Drive, Lakeville, MN 55044.
   g. *Graphical Analysis of Motion*; Demonstrations in Mechanics, Part III; VHS video tape or video disk; ≈3 min; Physics Curriculum & Instruction (see above for address).
4. Computer Software
   a. *Graphs and Tracks*; David Trowbridge; Windows, Macintosh; available from Physics Academic Software, North Carolina State University, PO Box 8202, Raleigh, NC 27690–0739. A ball rolls on a series of connected inclines. In one part the student is given graphs of the position, velocity, and acceleration and is asked to adjust the tracks to produce the graphed motion. In a second part the student is shown the motion and asked to sketch the graphs. Complements lab experiments with a sonic ranger.
   b. *Newtonian Sandbox*; Judah Schwartz; Windows, Macintosh; available from Physics Academic Software (see above for address). Generates the motion of a point particle in one and two dimensions. Plots trajectories, coordinates, velocity components, radial and angular positions, and phase space trajectories.
   c. *Objects in Motion*; Peter Cramer; Windows, Macintosh; available from Physics Academic Software (see above for address). Simulates the motion of an object under various conditions and plots graphs of the position, velocity and accelerations. Situations considered are: uniform acceleration along a straight line, projectile motion, relative motion, circular motion, planetary motion, and elastic collisions.
   d. *Physics Demonstrations*; Julien C. Sprott; Windows; available from Physics Academic Software (see above for address). Ten simulations of motion and sound demonstrations. Includes "the monkey and the coconut", "ballistics cat", "flame pipe", "Doppler effect".
   e. *Motion*; Macintosh, Windows; Cross Educational Software, Inc., 508 E. Kentucky Avenue, PO Box 1536, Ruston, LA 71270. Contains sections on graphing motion and motion with constant acceleration.
   f. *Accelerated Motion*; Windows; Cross Educational Software, Inc. (see above for address).
   g. *Conceptual Kinematics*; Frank Griffin and Louis Turner; Macintosh, Windows; available from Physics Academic Software (see above for address). An interactive, animated tutorial, with quiz questions for self-testing.
   h. *Dynamic Analyzer*; Roger F. Sipson; available from Physics Academic Software (see above for address).
   i. *Motion in One Dimension*; Socrates Software, 6187 Rosecommon Drive, Norcross, GA 30092. Uses animation and graphical analysis to demonstrate the relationships between displacement, velocity, and acceleration.

5. Computer Projects
   a. Use a spreadsheet or your own computer program to demonstrate the limiting processes used to define velocity and acceleration. Given the functional form of $x(t)$, have the computer calculate and display the coordinate for some time t and a succession of later times, closer and closer to t. For each interval, have it calculate and display the average velocity. Be careful to refrain from displaying non-significant figures and be sure to stop the process before all significance is lost.
   b. Have students use the root finding capability of a commercial math program or their own computer programs to solve kinematic problems for which $x(t)$ and $v(t)$ are given functions. Nearly all of them can be set up as problems that involve finding the root of either the coordinate or velocity as a function of time, followed perhaps by substitution of the root into another kinematic equation. Problems need not be limited to those involving constant acceleration. Air resistance, for example, can be taken into account. The same program can be used to solve rotational kinematic problems in Chapter 11.
6. Laboratory
   a. Motion detectors. As a student moves toward and away from a sonar-like ranging device, his position, velocity, or acceleration is plotted on a computer monitor. A graph can be designed by the instructor and the student can be asked to duplicate it by moving in front of the ranger. Several sonic rangers are reviewed in The Physics Teacher of January 1988. An extremely popular model is available from Vernier Software, 8565 SW Beaverton-Hillsdale Hwy., Portland, OR 97225-2429.
   b. Meiners Experiment 7–5: *Analysis of Rectilinear Motion.* Students measure the position as a function of time for various objects rolling down an incline, then use the data to plot speeds and accelerations as functions of time. No knowledge of rotational motion is required. This experiment emphasizes the definitions of velocity and acceleration as differences over a time interval.
   c. Meiners Experiment 8–1: *Motion in One Dimension* (omit the part dealing with conservation of energy). Essentially the same experiment except pucks sliding on a nearly frictionless surface are used. This experiment may be done with dry ice pucks or on an air table or air track.
   d. Bernard Experiment 7: *Uniformly Accelerated Motion.* The same technique as the Meiners experiments but a variety of setups are described: the standard free fall apparatus, the free fall apparatus with an Atwood attachment, an inclined plane, an inclined air track, and a horizontal air track with a pulley attachment.

Chapter 3 VECTORS

BASIC TOPICS

I. Definition.
   A. Explain that vectors have magnitude and direction, and that they obey certain rules of addition.
   B. Example of a vector: displacement. Give the definition of displacement and point out that a displacement does not describe the path of the object. Give the definition and physical interpretation of the sum of two displacements. Demonstrate vector addition by walking along two sides of the room. Point out the two displacements and their sum. Note that the distance traveled is not the magnitude of the displacement. Go back to your original position and point out that the displacement is now zero.
   C. Compare vectors with scalars and present a list of each.

D. Go over vector notation and insist that students use it to identify vectors clearly. In this text a vector is indicated by placing an arrow over an algebraic symbol. The italic version of the symbol, without the arrow, indicates the magnitude of the vector. Point out that many other texts use boldface type to indicate vectors.

II. Vector addition and subtraction by the graphical method.
   A. Draw two vectors tail to head, draw the resultant, and point out its direction. Explain how the magnitude of the resultant can be measured with a ruler and the orientation can be measured with a protractor. Explain how a scale is used to draw the original vectors and find the magnitude of the resultant.
   B. Define the negative of a vector and define vector subtraction as $\vec{a}-\vec{b} = \vec{a}+(-\vec{b})$. Graphically show that if $\vec{a}+\vec{b}=\vec{c}$ then $\vec{a}=\vec{c}-\vec{b}$.
   C. Show that vector addition is both commutative and associative.

III. Vector addition and subtraction by the analytic method.
   A. Derive expressions for the components of a vector, given its magnitude and the angles it makes with the coordinate axes. In preparation for the analysis of forces, find the x component of a vector in the xy plane in terms of the angles it makes with the positive and negative x axis and also in terms of the angles it makes with the positive and negative y axis.
   B. Point out that the components depend on the choice of coordinate system, and compare the behavior of vector components with the behavior of a scalar when the orientation of the coordinate system is changed. Find the components of a vector using two differently oriented coordinate systems. Point out that it is possible to orient the coordinate system so that only one component of a given vector is not zero. Remark that a pure translation of a vector (or coordinate system) does not change the components.
   C. Define the unit vectors along the coordinate axes. Give the form used to write a vector in terms of its components and the unit vectors. Explain that unit vectors are unitless so they can be used to write any vector quantity.
   D. Vector addition. Give the expressions for the components of the resultant in terms of the components of the addends. Demonstrate the equivalence of the graphical and analytic methods of finding a vector sum. See the diagram to the right.

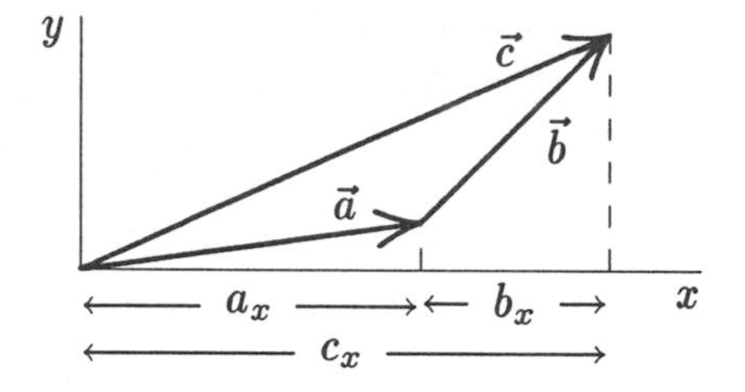

   E. Give the expression for vector subtraction in terms of components. You may also wish to demonstrate the equivalence of the graphical and analytical methods of vector subtraction.
   F. Show how to find the magnitude and angles with the coordinate axes, given the components. Explain that calculators give only one of the two possible values for the inverse tangent and show how to determine the correct angle for a given situation.
   G. State that two vectors are equal only if their corresponding components are equal. State that many physical laws are written in terms of vectors and that many take the form of an equality between two vectors.

IV. Multiplication involving vectors.
   A. Multiplication by a scalar. Give examples of both positive and negative scalars multiplying a vector. Give the components of the resulting vector as well as its magnitude and direction. Remark that division of a vector by a scalar is equivalent to multiplication by the reciprocal of the scalar.
   B. (May be postponed until Chapter 7.) Scalar product of two vectors. Emphasize that the product is a scalar. Give the expression for the product in terms of the magnitudes of the

vectors and the angle between them. To determine the angle, the vectors must be drawn with their tails at the same point. Point out that $\vec{a} \cdot \vec{b}$ is the magnitude of $\vec{a}$ multiplied by the component of $\vec{b}$ along an axis in the direction of $\vec{a}$. Explain that $\vec{a} \cdot \vec{b} = 0$ if $\vec{a}$ is perpendicular to $\vec{b}$.

C. Either derive the expression for a scalar product in terms of cartesian components or else assign Problem 30. See the discussion leading to Eq. 3–23. Specialize the expression to show that $\vec{a} \cdot \vec{a} = a^2$. Show how to use the scalar product to calculate the angle between two vectors if their components are known. Consider problem 31.

D. Vector product of two vectors (may be postponed until Chapter 12). Emphasize that the product is a vector. Give the expression for the magnitude of the product and the right hand rule for determining the direction. Explain that $\vec{a} \times \vec{b} = 0$ if $\vec{a}$ and $\vec{b}$ are parallel. Point out that $|\vec{a} \times \vec{b}|$ is the magnitude of $\vec{a}$ multiplied by the component of $\vec{b}$ along an axis perpendicular to $\vec{a}$ and in the plane of $\vec{a}$ and $\vec{b}$. Show that $\vec{b} \times \vec{a} = -\vec{a} \times \vec{b}$.

E. Either derive the expression for a vector product in terms of cartesian components or else assign Problem 32. See the discussion leading to Eq. 3–30. Give students the useful mnemonic for the vector products of the unit vectors $\hat{\imath}$, $\hat{\jmath}$, and $\hat{k}$, written in that order clockwise around a circle. One starts with the first named vector in the vector product and goes around the circle toward the second named vector. If the direction of travel is clockwise the result, is the third vector. If it is counterclockwise, the result is the negative of the third vector.

SUGGESTIONS

1. Assignments
   a. Use questions 1, 2, and 3 to discuss properties of vectors. Questions 4 and 5 deal with vector addition and subtraction. Question 6 deals with the signs of components.
   b. Ask students to use graphical representations of vectors to think about problems such as 1 and 2.
   c. Problems 3, 5, and 6 cover the fundamentals of vector components. Problems 7 and 8 stress the physical meaning of vector components. Some good problems to test understanding of analytic vector addition and subtraction are 12, 13, 15, 18, and 19.

2. Demonstrations

   Vector addition: Freier and Anderson Mb2, 3.

3. Audio/Visual
   a. *Vector Addition — Velocity of a Boat*; from AAPT collection 1 of single-concept films; video tape; available from Ztek Co., PO Box 11768, Lexington, KY 40577–1768.
   b. *Vectors*; Side A: Mechanics (I) of Cinema Classics; video disk; available from Ztek Co. (see above for address).
   c. *Vector Addition*; Physics Demonstrations in Mechanics, Part III; VHS video tape or video disk; ≈3 min; Physics Curriculum & Instruction, 22585 Woodhill Drive, Lakeville, MN 55044.

4. Computer Software
   a. *Math Methods in Physics*; Macintosh, Windows; Cross Educational Software, Inc., 508 E. Kentucky Avenue, PO Box 1536, Ruston, LA 71270. Contains a section on vector components, vector addition, vector and scalar products. Also contains sections on numbers and units, algebra, geometry, trigonometry, and calculus.
   b. *Vectors*; Richard R. Silbar; available from Physics Academic Software, Box 8202, North Carolina State University, Raleigh, NC 27695–8202.

5. Computer Project
   Have students use a commercial math program or write their own computer programs to carry out conversions between polar and cartesian forms of vectors, vector addition, scalar and vector products.
6. Laboratory
   Bernard Experiment 4: *Composition and Resolution of Coplanar Concurrent Forces.* Students mathematically determine a force that balances 2 or 3 given forces, then check the calculation using a commercial force table. They need not know the definition of a force, only that the forces in the experiment are vectors along the strings used, with magnitudes proportional to the weights hung on the strings. The focus is on resolving vectors into components and finding the magnitude and direction of a vector, given its components.

Chapter 4 MOTION IN TWO AND THREE DIMENSIONS

BASIC TOPICS

I. Definitions.
   A. Draw a curved particle path. Show the position vector for several times and the displacement vector for several intervals. Define average velocity over an interval. Write the definition in both vector and component form.
   B. Define velocity as $d\vec{r}/dt$. Write the definition in both vector and component form. Point out that the velocity vector is tangent to the path. Define speed of the magnitude of the velocity.
   C. Define acceleration as $d\vec{v}/dt$. Write the definition in both vector and component form. Point out that $\vec{a}$ is not zero if either the magnitude or direction of $\vec{v}$ changes with time.
   D. Show that the particle is speeding up only if $\vec{a}\cdot\vec{v}$ is positive. If $\vec{a}\cdot\vec{v}$ is negative, the particle is slowing down, and if $\vec{a}\cdot\vec{v} = 0$, its speed is not changing.
   E. Remark that sometimes the magnitude and direction of the acceleration are given, rather than its components. Remind students how to find the components if such is the case.
   F. Go over Sample Problem 4–4 or a similar problem of your own devising. It shows how to find and use the components of the acceleration.

II. Projectile motion.
   A. Demonstrate projectile motion by using a spring gun to fire a ball onto a surface at the firing height. Use various firing angles, including 45°, and point out that the maximum range occurs for a firing angle of 45°. Remark on the symmetry of the range as a function of firing angle. Mention that the maximum range occurs for a different angle when the ball is fired onto a surface at a different height.
   B. Draw the trajectory of a projectile, show the direction of the initial velocity, and derive its components in terms of the initial speed and firing angle.
   C. Write down the kinematic equations for $x(t)$, $y(t)$, $v_x(t)$, and $v_y(t)$. At first, include both a_x and a_y but then specialize to $a_x = 0$ and $a_y = -g$ for positive y up. Stress that these form two sets of one dimensional equations, linked by the common variable t and are to be solved simultaneously. Note that a_x affects only v_x, not v_y or v_z. Make similar statements about the other components. Throw a ball vertically, then catch it. Repeat while walking with constant velocity across the room. Ask students to observe the motion of the ball relative to the chalkboard and to describe its motion relative to your hand.
   D. Point out that the acceleration is the same at all points of the trajectory, even the highest point. Also point out that the horizontal component of the velocity is constant.

E. Work examples. Use punted footballs, hit baseballs, or thrown basketballs according to season.
   1. Find the time for the projectile to reach its highest point, then find the coordinates of the highest point.
   2. Find the time for the projectile to hit the ground, at the same level as the firing point. Then, find the horizontal range and the velocity components just before landing.
   3. Show that maximum range over level ground is achieved when the firing angle is 45°.
   4. Show how to work problems for which the landing point is not at the same level as the firing point.

F. Point out that all projectiles follow some piece of the full parabolic trajectory. For example, A to D could be the trajectory of a ball thrown at an upward angle from a roof to the street; B to D could be the trajectory of a ball thrown horizontally; C to D could be the trajectory of a ball thrown downward.

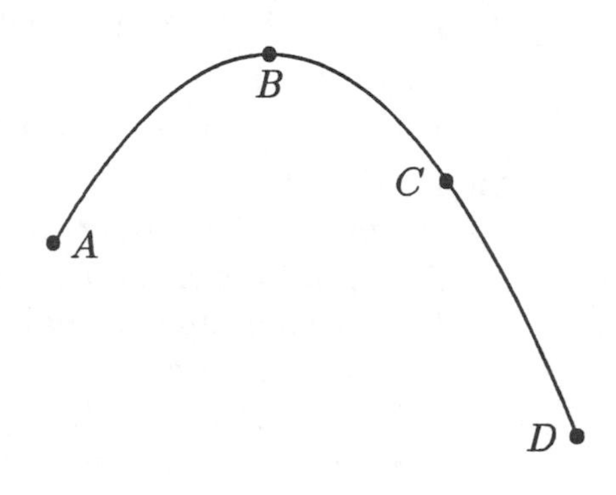

G. Explain how to find the speed and direction of travel for any time. Specialize to the time of impact on level ground and show that the speed is the same as the firing speed but that the vertical component of the velocity has changed sign. Remark that this result is true only because air resistance has been neglected.

H. Work some sample problems. Consider Sample Problems 4–7 and 4–8 or others of your own devising.

III. Circular motion.

A. Draw the path and describe uniform circular motion, emphasizing that the speed remains constant. Remind students that the acceleration must be perpendicular to the velocity. By drawing the velocity vector at two times, argue that it must be directed inward. On the diagram show the velocity and acceleration vectors for several positions of the particle.

B. Derive $a = v^2/r$. As an alternative to the derivation given in the text, write the equations for the particle coordinates as functions of time, then differentiate twice.

C. Example: calculate the speed of an Earth satellite, given the orbit radius and the acceleration to due to gravity at the orbit. Emphasize that the acceleration is toward Earth.

IV. Relative motion.

A. Material in this section is used in Chapter 5 to discuss inertial frames and in Chapter 12 to discuss rolling without slipping. It is also useful as a prelude to relativity.

B. Relate the position of a particle as given in coordinate system A to the position as given in coordinate system B by $\vec{r}_{\rm PA} = \vec{r}_{\rm PB} + \vec{r}_{\rm BA}$, where $\vec{r}_{\rm BA}$ is the position of the origin of B relative to the origin of A. Differentiate to show that $\vec{v}_{\rm PA} = \vec{v}_{\rm PB} + \vec{v}_{\rm BA}$ and $\vec{a}_{\rm PA} = \vec{a}_{\rm PB} + \vec{a}_{\rm BA}$, where $\vec{v}_{\rm BA}$ and $\vec{a}_{\rm BA}$ are the velocity and acceleration, respectively, of B relative to A.

C. Discuss examples of a ball thrown or rolled in accelerating and non-accelerating trains. The discussion may be carried out for either one- or two-dimensional motion.

D. Remark that $\vec{a}_{\rm PB} = \vec{a}_{\rm PA}$ if the two coordinate systems are not accelerating with respect to each other. This is an important point for the discussion of inertial reference frames in Chapter 5.

E. Work several problems dealing with airplanes flying in the wind and boats sailing in moving water. Emphasize that relative motion problems are chiefly exercises in vector addition. To help students understand some of the problems explain that a boat's "heading" is its direction of motion in a frame attached to the water, while its direction of travel is its direction of motion in a frame attached to the ground.

SUGGESTIONS

1. Assignments
   a. Assign problems 1, 10, and 15 to have students think about the analysis of motion in two dimensions.
   b. Use questions 3 through 10 to generate discussions of ideal projectile motion.
   c. Ask questions 12 and 13 in connection with centripetal acceleration.
   d. Have students work several of problems 17 through 26 and 28 through 41. Some of these deal with sports. See, for example, problems 39, 40, and 41.
   e. Assign problems 46 and 48 in connection with uniform circular motion.
   f. Assign one or two problems that deal with relative motion. Good examples are 61 and 64.
2. Demonstrations
   Projectile motion: Freier and Anderson Mb14, 16, 17, 19, 20, 23, 24, 28.
3. Audio/Visual
   a. *A Matter of Relative Motion*, *Galilean Relativity — Ball Dropped from Mast of Ship*; *Object Dropped from Aircraft*, *Projectile Fired Vertically*; from AAPT collection 1 of single-concept films; video tape; available from Ztek Co., PO Box 11768, Lexington, KY 40577–1768.
   b. *Projectile Motion*, *Circular Motion*; Side B: Mechanics (II) of Cinema Classics; video disk; available from Ztek Co. (see above for address).
   c. *Projectile Motion*; VHS video tape (15 min); Films for the Humanities & Sciences, PO Box 2053, Princeton, NJ 08543–2053.
   d. *Circular Motion*; VHS video tape (10 min); Films for the Humanities & Sciences (see above for address).
   e. *Introduction to Relative Motion*; VHS video tape (8 min); Films for the Humanities & Sciences (see above for address).
   f. *The Ape in the Tree*; Demonstrations of Physics: Motion; VHS video tape (3:13); Media Design Associates, Inc., Box 3189, Boulder, CO 80307–3190.
   g. *Reference Frames* from Skylab Physics Videodisc; video disk; available from Ztek Co. (see above for address).
   h. *Projectile Motion*; Physics Demonstrations in Mechanics, Part I; VHS video tape; ≈3 min; Physics Curriculum & Instruction, 22585 Woodhill Drive, Lakeville, MN 55044.
   i. *Circular Motion*; Physics Demonstrations in Mechanics, Part I; VHS video tape; ≈3 min; Physics Curriculum & Instruction (see above for address).
   j. *Velocity and Acceleration Vectors*; *Frame of Reference*; Physics Demonstrations in Mechanics, Part III; VHS video tape or video disk; ≈3 min each; Physics Curriculum & Instruction (see above for address).
   k. *Projectile Motion*; *Circular Motion*; Physics Demonstrations in Mechanics, Part V; VHS video tape or video disk; ≈3 min; Physics Curriculum & Instruction (see above for address).
4. Computer Software
   a. *Adding Velocities*; Windows; Cross Educational Software, Inc., 508 E. Kentucky Avenue, PO Box 1536, Ruston, LA 71270. Relative velocity is considered in both one and two dimensions.
   b. *Projectiles*; Vernier Software, 8565 S.W. Beaverton-Hillside Hwy., Portland, OR 97225–2429. A simulation program that allows students to experiment with projectile motion.
   c. *Mechanics in Motion*; Stephen Saxon; available from Physics Academic Software, Box 8202, North Carolina State University, Raleigh, NC 27695–8202. Contains projectile, pendulum, and collision simulators. Can also be used to demonstrate conservation of energy and rotational motion.
   d. *Newtonian Sandbox*. See Chapter 2 SUGGESTIONS.

e. *Objects in Motion.* See Chapter 2 SUGGESTIONS.
f. *Physics Demonstrations.* See Chapter 2 SUGGESTIONS.
g. *Dynamic Analyzer.* See Chapter 2 SUGGESTIONS.

5. Computer Projects
   a. Have students use a commercial math program or their own root finding programs to solve projectile motion problems.
   b. Have students use a spreadsheet or write a computer program to tabulate the coordinates and velocity components of a projectile as functions of time. Have them change the initial velocity and observe changes in the coordinates of the highest point and in the range. Ask them to find the firing angle for the greatest horizontal coordinate when the landing point is above or below the firing point.
6. Laboratory
   a. Meiners Experiment 7–9: *Ballistic Pendulum — Projectile Motion* (use only the first method in connection with this chapter). Students find the initial velocity of a ball shot from a spring gun by measuring its range. Emphasizes the use of kinematic equations.
   b. *Inelastic Impact and the Velocity of a Projectile* (use only Procedure B with this chapter). In addition to using range data to find the initial velocity, students plot the range as a function of firing angle.

Chapter 5 FORCE AND MOTION — I

BASIC TOPICS

I. Overview
   A. Explain that objects may interact with each other and their velocities change as a result. State that the strength of an interaction depends on properties of the objects and their relative positions. Gravitational mass is responsible for gravitational interactions, electric charge is responsible for electric and magnetic interactions.
   B. Explain that we split the problem into two parts and say that each body exerts a force on the other and that the net force on a body changes its velocity. Remark that an equation that gives the force in terms of the properties of the objects and their positions is called a force law. Force laws are discussed throughout the course. The dominant theme of this chapter, however, is the relationship between the net force and the acceleration it produces.

II. Newton's first law.
   A. State the law: if an object does not interact with any other objects, its acceleration is zero.
   B. Point out that the acceleration depends on the reference frame used to measure it and that the first law can be true for only a select set of frames. Cover the essential parts of the relative motion section of Chapter 4, if they were not covered earlier. Define an inertial frame. Tell students that an inertial frame can be constructed, in principle, by finding an object that is not interacting with other objects and then attaching a reference frame to it. Any frame that moves with constant velocity relative to an inertial frame is also an inertial frame, but one that is accelerating relative to an inertial frame is not.
   C. Explain that we may take a reference frame attached to Earth as an inertial frame for the description of most laboratory phenomena but we cannot for the description of ocean and wind currents, space probes, and astronomical phenomena.

III. Newton's second law.
   A. Explain that the environment influences the motion of an object and that force measures the extent of the interaction. The result of the interaction is an *acceleration*. Place a cart at rest on the air track. Push it to start it moving and note that it continues at constant

velocity. After it is moving, push it to increase its speed, then push it to decrease its speed. In each case note the direction of the force and the direction of the acceleration. Also give an eraser a shove across a table and note that it stops. Point out that the table top exerts a force of friction while the eraser is moving. Push the eraser at constant velocity and explain that the force of your hand and the force of friction sum to zero.

B. Define force in terms of the acceleration imparted to the standard 1 kg mass. Explain how this definition can be used to calibrate a spring, for example. Point out that force is a vector, in the same direction as the acceleration. If two or more forces act on the standard mass, its acceleration is the same as when a force equal to the resultant acts.
Unit: newton. Explain that 1 N is $1\,\text{kg·m/s}^2$.

C. Have three students pull on a rope, knotted together as shown. Ask one to increase his or her pull and ask the others to report what they had to do to remain stationary.

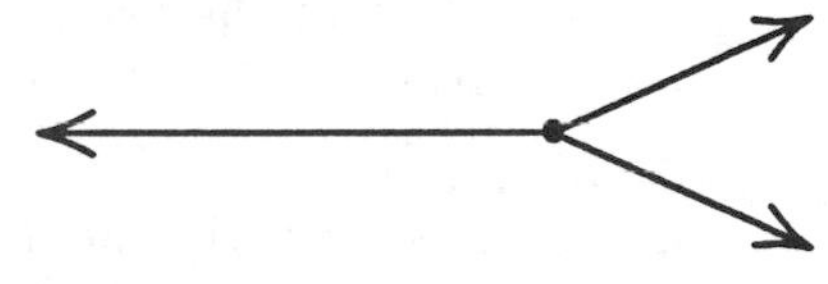

D. Define mass in terms of the ratio of the acceleration imparted to the standard mass and to the unknown mass, with the same force acting. Attach identical springs to two identical carts, one empty and the other containing a lead brick. Pull with the same force (same elongation of the springs) and observe the difference in acceleration. Unit: kilogram.

E. State the second law. Stress that the force that appears is the net or resultant force. Explain that the law holds in inertial frames. Point out that this is an experimentally established law and does not follow as an identity from the definitions of force and mass. Emphasize that $m\vec{a}$ is not a force.

F. Discuss examples: calculate the constant force required to stop an object in a given time, given the mass and initial velocity; calculate the force required to keep an object in uniform circular motion, given its speed and the radius of its orbit. Calculate the acceleration of an object being pushed by two forces in opposite directions and note that the acceleration vanishes if the forces have equal magnitudes. Emphasize that the forces continue to act but their sum vanishes. Some students believe that the forces literally cancel each other and no longer act.

IV. Newton's third law.

A. State the law. Stress that the two forces in question act on different bodies and each helps to determine the acceleration of the body on which it acts. Explain that the third law describes a characteristic of force laws. State that the two forces in an action-reaction pair are of the same type: gravitational, for example.

B. Discuss examples. Hold a book stationary in your hand, identify action-reaction pairs (hand-book, book-Earth). Now allow your hand and the book to accelerate downward with an acceleration less than g and again identify action-reaction pairs. Note that you can control the acceleration of the book by means of the force you exert but once you exert a given force you cannot control the force that the book exerts on you.

C. Attach a force probe to each of two air-track carts. Use a computer to plot the force that each exerts on the other as the carts collide. Point out that at each instant the forces have the same magnitude and are in opposite directions.

V. Applications of Newton's laws involving a single object.

A. Go over the steps used to solve a one-body problem: identify the body and all forces acting on it, draw a free-body diagram, choose a coordinate system, write the second law in component form, and finally solve for the unknown.

B. Some special forces should be explained. They are important for many of the problems but are rarely mentioned explicitly. Warn students they must take these forces into account if they act.

1. Point out that the magnitude of the gravitational force is mg, where g is the local acceleration due to gravity and m is the mass of the object. It is directed toward the center of Earth. Explain that the magnitude of this force is the weight of the object. Explain that weight varies with altitude and slightly from place to place on the surface of Earth, but mass does not vary. Emphasize that the appearance of g in the formula for the gravitational force does not imply that the acceleration of the body is g.
2. Point out that a massless rope transmits force unaltered in magnitude and that the magnitude of the force it exerts on objects at each end is called the tension force. If a person pulls an object by exerting a force on a string attached to the object, the motion is as if the person pulled directly on the object. The string serves to define the direction of the force. A frictionless, massless pulley serves to change the direction but not the magnitude of the tension force of the rope passing over it.
3. Explain that the normal force of a surface on an object originates in elastic and ultimately electric forces. It prevents the object from moving through the surface. State that it is perpendicular to the surface. If the surface is at rest, the normal force adjusts so the acceleration component perpendicular to the surface vanishes. More generally, the object and the surface have the same perpendicular acceleration component. Place a book on the table and press on it. State that the normal force is greater than when you were not pressing. Hold the book against the wall by pressing on it and mention that the normal force is horizontal.

C. Set up the situation described in Sample Problem 5–7 using an inclined air track but attach a calibrated spring scale to the support at the top of the incline and tie the other end of the scale to the block. Calculate the tension force of the string and compare the result to the reading on the scale. Cut the string, then calculate the acceleration.

D. Consider a person standing on a scale in an elevator. State that the scale measures the normal force and calculate its value for an elevator at rest, one accelerating upward, one accelerating downward with $a < g$, and one in free fall. See Sample Problem 5–8.

VI. Applications of Newton's laws involving more than one object.

A. Explain that when two or more objects are involved, a free-body diagram must be drawn for each. A Newton's second law equation, in component form, is also written for each object. Point out that differently oriented coordinate systems may be used for different bodies. Show how to invoke the third law when necessary. Explain that the same symbol should be used for the magnitude of the two forces of an action-reaction pair and that their opposing directions are taken into account when drawing the free-body diagram and in writing the second law equations.

B. Explain that in some cases both objects can be considered as a single object. Say that the objects must have the same acceleration and that the forces they exert on each other must not be requested. The mass of the single object is then the sum of the masses of the constituent objects and internal forces are not included in the analysis.

C. Use examples to show how rods, strings, and pulleys relate the motions of bodies in various cases. Explain that, in addition to the second law equations, there will be equations relating the accelerations of the objects. Show that these equations depend on the choice of coordinate systems.

D. Consider several examples, carefully explaining each step. If you have not developed an application of your own, work Sample Problem 5–9 in the text. If possible, give a demonstration.

SUGGESTIONS

1. Assignments
   a. Use questions 1 through 6 to help students think about the influence of forces on bodies. Assign one or two of problems 1, 2, and 3.
   b. Use question 8 to help students think about normal forces.
   c. Use questions 7 and 9 and problem 9 to help students with tensions in ropes.
   d. Discuss question 11 in connection with Newton's third law of motion.
   e. Discuss question 12 in connection with inclined plane problems.
   f. Assign problem 1 to emphasize the definition of force and problem 6 or 7 to demonstrate Newton's second law.
   g. Use problems 11 and 12 to discuss mass and weight.
   h. Use problems 27 and 31 to discuss Newton's third law.
   i. Assign a few applications problems from the group 13 through 56, according to the needs and interests of the class.
   j. As a prelude to Chapter 9 (where the center of mass and conservation of momentum are discussed) assign problem 25.
2. Demonstrations
   a. Inertia: Freier and Anderson Mc1 — 5, Me1.
   b. $\vec{F} = m\vec{a}$: Freier and Anderson Md2, Ml1.
   c. Third-law pairs: Freier and Anderson Md1, 3, 4.
   d. Mass and weight: Freier and Anderson Mf1, 2.
   e. Tension in a string: Freier and Anderson Ml1.
3. Books and Monographs
   Resource Letters, Book Four; American Association of Physics Teachers, One Physics Ellipse, College Park, MD 20740–3845. Contains a resource letter on mechanics.
4. Audio/Visual
   a. *Frames of Reference Videodisc*; video disk; available from Ztek Co., PO Box 11768, Lexington, KY 40577–1768.
   b. *Human Mass Measurement* from Skylab Physics Videodisc; video disk; available from Ztek Co. (see above for address).
   c. *Newton's First and Second Laws*; *Newton's Third Law*; *Inertial Forces*; *Translational Acceleration*; from AAPT collection 2 of single-concept films; video tape; available from Ztek Co., PO Box 11768, Lexington, KY 40577–1768.
   d. *Inertial Forces - Translational Acceleration*; from AAPT Miller collection of single-concept films; video tape; available from Ztek Co. (see above for address).
   e. *Forces, Newton's Laws*; Side A: Mechanics (I) of Cinema Classics; video disk; available from Ztek Co. (see above for address).
   f. *Newton's First Law*; Demonstrations of Physics: Motion; VHS video tape (5:40); Media Design Associates, Inc., Box 3189, Boulder, CO 80307–3190.
   g. *Newton's Second Law*; Demonstrations of Physics: Motion; VHS video tape (6:31); Media Design Associates, Inc. (see above for address).
   h. *Inertia*; VHS video tape (10 min); Films for the Humanities & Sciences, PO Box 2053, Princeton, NJ 08543–2053.
   i. *Mechanics in Action*; VHS video tape (25 min); Films for the Humanities & Sciences (see above for address).
   j. *Energy and Force: Part 1*; VHS video tape (24 min); Films for the Humanities & Sciences (see above for address).
   k. *The Laws of Motion*; VHS video tape (21 min); Films for the Humanities & Sciences (see above for address).

l. *The Laws of Motion Applied*; VHS video tape (21 min); Films for the Humanities & Sciences (see above for address).
m. *Newton's 1st Law*; *Newton's 2nd Law*; *Newton's 3rd Law*; Physics Demonstrations in Mechanics, Part II; VHS video tape; ≈3 min each; Physics Curriculum & Instruction, 22585 Woodhill Drive, Lakeville, MN 55044.
n. *Newton's 1st Law*; Physics Demonstrations in Mechanics, Part III; VHS video tape or video disk; ≈3 min; Physics Curriculum & Instruction (see above for address).
o. *Fundamental Forces*; *Mass and Weight*; *Newton's 2nd Law*; *Newton's 2nd and 3rd Laws*; *Force Components*; Physics Demonstrations in Mechanics, Part IV; VHS video tape or video disk; ≈3 min each; Physics Curriculum & Instruction (see above for address).

5. Computer Software
   a. *Freebody*; Graham Oberum; available from Physics Academic Software, Box 8202, North Carolina State University, Raleigh, NC 27695–8202. Students draw force vectors and can change the length and orientation of the vectors in response to questions. The screen gives the components.
   b. *Force and Motion Microworld*; Ping-Kee L. Tao and Ming-Wai Tse; available from Physics Academic Software (see above for address). Uses velocity graphs to display the effects of force on the motion of an object. Includes drag forces.
   c. *Dynamic Analyzer*; Roger Sipson; Windows; available from Physics Academic Software (see above for address). The program allows students to explore one- and two-dimensional motion, harmonic motion, and motion in electric and magnetic fields. The motions are simulated on the screen and various graphs are drawn side-by-side with the simulations. The user can change the parameters of the motion.
   d. *Newtonian Sandbox*. See Chapter 2 SUGGESTIONS.
   e. *Motion*; Macintosh, Windows; Cross Educational Software, Inc., 508 E. Kentucky Avenue, PO Box 1536, Ruston, LA 71270. Contains sections on second law problems, including projectiles, friction, and uniform circular motion.

6. Laboratory

   Meiners Experiment 8–2: *Concept of Mass: Newton's Second Law of Motion.* Students measure the accelerations of two pucks that interact via a spring on a nearly frictionless surface and compare the ratio to the ratio of their masses. This experiment may be done with dry ice pucks or on an air table or air track.

Chapter 6 FORCE AND MOTION — II

BASIC TOPICS

I. Frictional forces.
   A. Place a large massive wooden block on the lecture table. Attach a spring scale, large enough to be read easily. If necessary, tape sandpaper to the table under the block. Pull weakly on the scale and note that the reading is not zero although the block does not move. Pull slightly harder and note that the reading increases but the block still does not move. Remark that there must be a force of friction opposing the pull and that the force of friction increases as the pull increases. Now increase your pull until the block moves and note the reading just before it starts to move. Pull the block at constant speed and note the reading. Have the students repeat the experiment in a qualitative manner, using books resting on their chair arms. To show that the phenomenon depends on the nature of the surface, the demonstration can be repeated after waxing the wooden block and table top.

B. Give a brief qualitative discussion about the source of frictional forces. Stress that the force of static friction has whatever magnitude and direction are required to hold the two bodies in contact at rest relative to each other, up to a certain limit in magnitude. Define the coefficient of static friction and explain the use of $f_s < \mu_s N$. In particular, explain that if the surface is stationary the force of static friction is determined by the condition that the object on it has zero acceleration. To test if an object remains at rest, the frictional force required to produce zero relative acceleration is calculated and compared with $\mu_s N$.

C. Define the coefficient of kinetic friction and explain that $f_k = \mu_k N$ gives the frictional force as long as the object is sliding on the surface. Also explain that if the surface is stationary the force of kinetic friction is directed opposite to the velocity of the object sliding on it.

D. Work some examples:

1. Find the angle of an inclined plane for which sliding starts; find the angle at which the body slides at constant speed. These examples can be analyzed in association with a demonstration and the students can use the data to find the coefficients of friction.
2. Analyze an object resting on the floor, with a person applying a force that is directed at some angle above the horizontal. Find the minimum applied horizontal force that will start the object moving and point out that it is a function of the angle between the applied force and the horizontal.
3. Consider the same situation but with the object moving. Find its acceleration. This and the previous example demonstrate the dependence of the normal force and the force of friction on the externally applied force.
4. To give an illuminating variant, consider a book being held against the wall by a horizontal force. Calculate the minimum applied force that will keep the book from falling.

II. Drag forces and terminal speed.

A. Make or buy a small toy parachute. Drop two weights side by side and note they reach the floor at the same time. Attach the parachute to one and repeat. Explain that the force of the air reduces the acceleration.

B. State that for turbulent flow of air around an object the magnitude of the drag force is given by $D = \frac{1}{2}C\rho A v^2$, where A is the effective cross-sectional area, ρ is the density of air, and v is the speed of the object relative to the air. C is a drag coefficient, usually determined by experiment. Remark that a parachute increases the cross-sectional area. State that the drag force is directed opposite to the velocity.

C. Explain that as an object falls its speed approaches terminal speed as a limit. Write down Newton's second law for a falling object and point out that the drag and gravitational forces are in opposite directions. Suppose the object is dropped from rest and point out that the acceleration is g at first but as the object gains speed its acceleration decreases in magnitude. At terminal speed the acceleration is zero and remains zero, so the velocity no longer changes. Show that zero acceleration leads to $v_t = \sqrt{2mg/C\rho A}$. Point out Table 6–1, which gives some terminal speeds.

D. Remark that if an object is thrown downward with a speed that is greater than terminal speed it slows down until terminal speed is reached.

E. Qualitatively discuss projectile motion with drag. The horizontal component of the velocity tends to zero while the vertical component tends to the terminal speed. Contrast the trajectory with one in the absence of air resistance.

III. Uniform circular motion.

A. Point out that for uniform circular motion to occur there must be a radially inward force of constant magnitude and that something in the environment of the body supplies the force. Whirl a mass tied to a string around your head and explain that the string supplies the force. Set up a loop-the-loop with a ball or toy cart on a track and explain that the combination of the normal force of the track and the force of gravity supplies the centripetal force. Have students identify the source of the force in examples and problems as they are discussed.

B. Point out that $F = mv^2/r$ is just $F = ma$ with the expression for centripetal acceleration substituted for a.

C. Discuss problem solving strategy. After identifying the forces, find the radial component of the resultant and equate it to mv^2/r.

D. Examples:
   1. Find the speed and period of a conical pendulum.
   2. Find the speed with which a car can round an unbanked curve, given the coefficient of static friction.
   3. Find the angle of banking required to hold a car on a curve without aid of friction.
   4. Analyze the loop-the-loop and point out that the ball leaves the track when the normal force vanishes. Show that the critical speed at the top is given by $v^2/r = g$.

SUGGESTIONS

1. Assignments
   a. Discuss some or all of questions 1, 3, 4, 5, and 6 in connection with the force of static friction and the onset of sliding. Kinetic friction is the subject of questions 2, 7, and 8. Consider asking question 8 after assigning problem 25.
   b. Ways in which the coefficient of static friction is used are emphasized in problems 1, 2, 8, 9, and 11. Problem 25 is more challenging. To help students understand the role played by the normal force in the onset of sliding, assign problem 12. Problem 16 deals with the role of the normal force in kinetic friction.
   c. Problems 17, 26, and 27 provide some interesting applications of the laws of friction.
   d. Use questions 9 and 10 in your discussion of centripetal acceleration and force. Problems 37 and 39 cover the role of friction in rounding a level curve. Also consider problems 44 and 45.
2. Demonstrations
   a. Friction: Freier and Anderson Mk.
   b. Inclined plane: Freier and Anderson Mj2.
   c. Centripetal acceleration: Freier and Anderson Mb29, 31, Mm1, 2, 4 — 8, Ms5.
3. Audio/Visual
   a. *Trajectories*; from AAPT collection 2 of single-concept films; video tape; available from Ztek Co., PO Box 11768, Lexington, KY 40577–1768.
   b. *Inertial Forces — Centripetal Acceleration*; from AAPT Miller collection of single-concept films; video tape; available from Ztek Co. (see above for address).
   c. *Uniform Circular Motion*; Demonstrations of Physics: Motion; VHS video tape (5:57); Media Design Associates, Inc., Box 3189, Boulder, CO 80307–3190.
   d. *Circular Motion*; VHS video tape (10 min); Films for the Humanities & Sciences, PO Box 2053, Princeton, NJ 08543–2053.
   e. *Terminal Velocity*; Physics Demonstrations in Mechanics, Part II; VHS video tape; $\approx$3 min; Physics Curriculum & Instruction, 22585 Woodhill Drive, Lakeville, MN 55044.
4. Computer Software
   a. *Dynamic Analyzer*. See Chapter 5 SUGGESTIONS.
   b. *Motion*; Macintosh, Windows; Cross Educational Software, Inc., 508 E. Kentucky Avenue, PO Box 1536, Ruston, LA 71270. Contains sections on second law problems, including projectiles, friction, and uniform circular motion.

5. Computer Projects
   a. Have students use a computer program to investigate objects that are subjected to time dependent forces. To check the program first have them consider a constant force and compare machine generated functions with the known kinematic equations.
   b. Have students modify the program to integrate Newton's second law for velocity dependent forces, then have them investigate the motion of an object subjected to a force that is proportional to v or v^2. It is instructive to have them plot the velocity components as functions of time for a projectile fired straight up or down, subject to air resistance. Consider initial velocities that are both greater and less than the terminal velocity. Also have them study the maximum height and range of projectiles with various coefficients of air resistance. Some projects are given in the Computer Projects section of this manual.
6. Laboratory
   a. Meiners Experiment 7–6: *Coefficient of Friction — The Inclined Plane.* Students determine the coefficients of static and sliding friction for three blocks on an inclined plane. They devise their own experimental procedures.
   b. Meiners Experiment 7–7: *Radial Acceleration* (Problem I only). The centripetal force and the speed of a ball on a string, executing uniform circular motion, are measured for various orbit radii. Essentially a verification of $F = mv^2/r$.
   c. Meiners Experiment 7–8: *Investigation of Uniform Circular Motion*, or Bernard Experiment 13: *Centripetal Force.* Students measure the force acting on a body undergoing uniform circular motion, with the centripetal force provided by a spring.
   d. Meiners Experiment 8–3: *Centripetal Force.* Students measure the speed of a puck undergoing uniform circular motion on a nearly frictionless surface. The data is used to calculate the centripetal force.

Chapter 7 KINETIC ENERGY AND WORK

BASIC TOPICS

I. Kinetic energy and the work-kinetic energy theorem.
   A. Define kinetic energy for a particle. Remind students that kinetic energy is a scalar and depends on the speed but not on the direction of the velocity. Point out that $v^2 = v_x^2 + v_y^2$ for two-dimensional motion and remark that the appearance of velocity components in the expression does *not* mean K has components.
   B. Consider a ball thrown into the air. Neglect air resistance and point out that during the upward part of the motion the force of gravity slows the ball and the kinetic energy decreases. As the ball falls, the force of gravity speeds the ball and the kinetic energy increases. Remind students that for a constant force (and acceleration) $v^2 = v_0^2 + 2a\,\Delta x$ (which was derived in the study of kinematics). Multiply by $m/2$ to obtain $K = K_0 + F\,\Delta x$. Say that for a constant force acting on a particle that moves in one dimension $W = F\,\Delta x$ is the work done by the force F as the particle travels through the displacement Δx. State that $K = K_0 + W$ is an example of the work-kinetic energy theorem: the change in the kinetic energy of a particle during a given interval equals the work done on the particle by the total force during the interval.
   C. Point out that only the component of a force parallel or antiparallel to the velocity changes the speed. Other components change the direction of motion. Positive total work results in an increase in kinetic energy and speed, negative total work results in a decrease. Remind students of previous examples in which the object moved with constant speed (including uniform circular motion). The total work was zero and the kinetic energy did not change. Avoid quantitative calculations involving frictional forces.

D. Explain that the work-kinetic energy theorem can be applied only to particles and objects that can be treated as particles. To give an example in which it cannot be applied directly, consider a car crashing into a rigid barrier: the barrier does no work but the kinetic energy of the car decreases.

E. Explain that observers in different inertial frames will measure different values of the net work done and for the change in the kinetic energy but both will find $W_{\text{net}} = \Delta K$.

F. Use Newton's second law to prove the theorem for motion in one dimension. If the students are mathematically sophisticated, extend the theorem to the general case. Stress that it is the total work (done by the resultant force) that enters the theorem.

II. Work done by a constant force.

A. Write down $W = \vec{F} \cdot \vec{d} = Fd\cos\phi$ and point out ϕ on a diagram. Explain that this is the work done *on* a particle *by* the constant force $\vec{F}$ as the particle undergoes a displacement $\vec{d}$. Explain that work can be calculated for each individual force and that the total work done on the particle is the work done by the resultant force. Point out that work is a scalar quantity. Also point out that work is zero for a force that is perpendicular to the displacement and that, in general, only the component of $\vec{F}$ tangent to the path contributes to the work. The force does no work if the displacement is zero. Emphasize that work can be positive or negative, depending on the relative orientation of $\vec{F}$ and $\vec{d}$. For a constant force, the work depends only on the displacement, not on details of the path. Unit: joule.

B. Calculate the work done by the force of gravity as a mass falls a distance h and as it rises a distance h. Emphasize the sign. Calculate the work done by a non-horizontal force used to pull a box across a horizontal floor. Point out that the work done by the normal force and the work done by the force of gravity are zero. Consider both an accelerating box and one moving with constant velocity. Repeat the calculation for a crate being pulled up an incline by a force applied parallel to the incline. Show the work done by gravity is $-mgh$, where h is the change in the height of the crate.

III. Work done by a variable force.

A. For motion in one dimension, discuss the integral form for work as the limit of a sum over infinitesimal path segments. Explain that the sum can be carried out by a computer even if the integral cannot be evaluated analytically.

B. Examples: derive expressions for the work done by an ideal spring and a force of the form k/x^2. If you have not yet discussed the force of an ideal spring, do so now as a preface to the calculation of work. Explain how the spring constant can be found by hanging a mass from the spring and measuring the extension. Demonstrate changes in the spring length during which the spring does positive work and during which the spring does negative work.

C. As an example consider a stone dropped onto a vertical spring and calculate the maximum compression of the spring, given the mass of the stone, the height from which it is dropped, and the spring constant of the spring.

D. For motion in more than one dimension, write down the expression for the work in the form $\int_i^f \vec{F} \cdot d\vec{r}$ and explain its interpretation as the limit of a sum over infinitesimal path segments. Explain that this is the general definition of work. Calculate the work done by the applied force, the force of gravity, and the tension in the string as a simple pendulum is pulled along its arc until it is displaced vertically through a height h by a horizontal applied force $\vec{F}$.

IV. Power.

A. Define power as $P = dW/dt$. Unit: watt.

B. Show that $P = \vec{F} \cdot \vec{v}$. Explain that the work done over a time interval is given by $\int P\,dt$.

SUGGESTIONS

1. Assignments
   a. The idea of kinetic energy is covered in question 1. Assign one or more of problems 1, 2, 3, and 5 to exercise the concept.
   b. The idea of work is covered in questions 2, 3, and 6. Problems 6, 7, and 10 are good examples of the quantitative aspects. The work done by the net force, as opposed to individual forces, is emphasized in question 3 and problems 11 and 13.
   c. To discuss the work done by gravity, ask questions 7 and 8, then assign problems 14, 16, and 17. To discuss the force exerted by an ideal spring and the work done by it, ask questions 10 and 11, then assign problem 21.
   d. Questions 4, 5, 9, 12, and 13 and problem 10 are a good introduction to the work-kinetic energy theorem. Also assign problem 22 or 23.
   e. Assign problem 24 or 29 in connection with the work done by a variable force.
   f. Assign one or more of problems 30, 32, 34, and 36 in connection with power.
2. Demonstrations
   Work: Freier and Anderson Mv1.
3. Audio/Visual
   a. *Work and Energy*; Side E: Conservation Laws of Cinema Classics; video disk; available from Ztek Co., PO Box 11768, Lexington, KY 40577–1768.
   b. *Energy and Force: Part 2*; VHS video tape (25 min); Films for the Humanities & Sciences, PO Box 2053, Princeton, NJ 08543–2053.
   c. *Potential Energy and Kinetic Energy*; VHS video tape (10 min); Films for the Humanities & Sciences (see above for address).
   d. *Work and Energy*; Physics Demonstrations in Mechanics, Part VI; VHS video tape or video disk; ≈3 min; Physics Curriculum & Instruction, 22585 Woodhill Drive, Lakeville, MN 55044.
4. Computer Project
   Have students use a commercial math program or write a program to numerically evaluate the integral for work, then use the program to calculate the work done by various forces, given as functions of position. Include a nonconservative force and use the program to show the work done on a round trip does not vanish. Some projects can be found in the Computer Projects section of this manual.
5. Laboratory
   a. Meiners Experiment 7–16: *Elongation of an Elastomer.* Students measure the elongation of an elastomer for a succession of applied forces and use a polar planimeter to calculate the work done by the force. The experiment may also be done in connection with Chapter 13.
   b. Bernard Experiment 10: *Mechanical Advantage and Efficiency of Simple Machines.* This experiment can be used to broaden the course to include these topics. A lever, an inclined plane, a pulley system, and a wheel and axle are studied. In each case the force output is measured for a given force input and the work input is compared to the work output.

Chapter 8 POTENTIAL ENERGY AND CONSERVATION OF ENERGY

BASIC TOPICS

I. Potential energy, conservative and nonconservative forces
   A. Explain that potential energy is an energy of configuration. The potential energy of a

system of objects depends on the relative positions of the objects. A system consisting of an object and Earth has a potential energy that depends on the separation of Earth and the object, for example.

B. State that a potential energy can be associated with a force only if that force is conservative and explain that a force is conservative if the work done by the force when the system starts and ends with the same configuration is zero, no matter what the configurations and no matter what motions occur between the beginning and end of the interval. Show that this implies that the work done by the force between any given starting and ending configurations is the same no matter what intervening configurations are assumed by the system.

C. Discuss the force of gravity and the force of an ideal spring as examples. For either or both of these, show that the work done depends only on the end points and not on the path between, then argue that the work vanishes for a round trip. Point out that on some parts of the path the force does positive work while on other parts it does negative work. Demonstrate that the work done by a spring is independent of the path by considering two different motions with the same end points. For the first motion, have the mass go directly from the initial point to the final point; for the second, have it first go away from the final point before going there.

D. Use a force of friction with constant magnitude as an example of a nonconservative force. Consider a block on a horizontal table top and argue that the work done by the force cannot vanish over a round trip since it is negative for each segment. Suppose the block moves around a circular path and friction is the only force that does work. Argue that the object returns to its initial position with less kinetic energy than it had when it started. State that a potential energy cannot be associated with a frictional force.

E. Use a cart on a linear air track to demonstrate these ideas. Couple each end of the cart, via a spring, to a support at the corresponding end of the air track. Give the cart an initial velocity and tell students to observe its speed each time it returns to its initial position. Point out that the kinetic energy returns to nearly the same value and that the springs do zero work during a round trip. Reduce or eliminate air flow to show the influence of a nonconservative force. If this is done rapidly and skillfully, you can cause the cart to stop far from the starting point.

II. Potential energy.

A. Give the definition of potential energy in terms of work for motion in one dimension. See Eq. 8–6 and emphasize that the change in potential energy is the negative of the work done *by* the force responsible for the potential energy.

B. Discuss the following properties:

1. The zero is arbitrary. Only potential energy differences have physical meaning.
2. The potential energy is a *scalar* function of position.
3. The force is given by $F = -\mathrm{d}U/\mathrm{d}x$ in one dimension.
4. Unit: joule.

C. Derive expressions for the potential energy functions associated with the force of gravity (uniform gravitational field) and the force of an ideal spring. Stress that the potential energy is a property of the object-Earth or spring-mass system and depends on the *configuration* of the system.

D. Use the work-kinetic energy theorem to show that $W = \Delta U$ is the work that must be applied by an external agent to increase the potential energy by ΔU if the kinetic energy does not change. Show that ΔU is recovered as kinetic energy when the external agent is removed. Example: raising an object in a gravitational field.

III. Conservation of energy.

A. Explain that if all the forces acting between the objects of a system are conservative and the net work done by external forces on objects of the system is zero then $K + U =$ constant. This follows from the work-kinetic energy theorem with the work of the conservative forces represented by the change in potential energy. The negative sign in Eq. 8–6 is essential to obtain this result. Emphasize that U is the sum of the individual potential energies if more than one conservative force acts. Define the total mechanical energy as $E_{\text{mec}} = K + U$.

B. Discuss the conversion of kinetic to potential energy and vice versa. Drop a superball on a rigid table top and point out when the potential and kinetic energies are maximum and when they are minimum. The question of elasticity can be glossed over by saying that to a good approximation the ball rebounds with unchanged speed. Also discuss the energy in a spring-mass system. Return to the cart on the air track and discuss its motion in terms of $K + U =$ constant. To avoid later confusion in the students' minds, start the motion with neither K nor U equal to zero. Emphasize that the energy remains in the system but changes its form during the motion. The agent of the change is the work done by the forces of the springs.

C. Show how to calculate the total energy for a spring-mass system from the initial conditions. Write the conservation principle in the form $\frac{1}{2}mv^2 + U(x) = \frac{1}{2}mv_0^2 + U(x_0)$. Use conservation of energy to find expressions for the maximum speed, maximum extension, and maximum compression, given the total energy.

D. Use the example of a ball thrown upward to demonstrate that conservation of energy must be applied to a system rather than to a single particle. Remark that Earth does work on the ball, the ball does work on Earth, and the change in potential energy is the negative of the sum. Show it is mgh, where h is the change in their separation. Remark that both kinetic energies change and the total change is the negative of the change in the potential energy. Explain that because Earth is so massive the change in its kinetic energy is small and may be neglected.

E. Discuss potential energy curves. Use the curve for a spring-mass system, then a more general one, and show how to calculate the kinetic energy and speed from the coordinate and total energy. Point out the turning points on the curves and discuss their physical significance. Use $F = -\mathrm{d}U/\mathrm{d}x$ to argue that the particle turns around at a turning point. For an object on a frictionless roller coaster track, find the speed at various points and identify the turning points.

F. Define stable, unstable, and neutral equilibrium. Use a potential energy curve (a frictionless roller coaster, say) to illustrate. Emphasize that $\mathrm{d}U/\mathrm{d}x = 0$ at an equilibrium point.

IV. Potential energy in two and three dimensions.

A. Define potential energy as a line integral and explain that it is the limit of a sum over infinitesimal path segments. Remark that conservation of energy leads to $\frac{1}{2}mv^2 + U(x, y, z) =$ constant. Explain that $v^2 = v_x^2 + v_y^2 + v_z^2$ and that v^2 is a scalar.

B. Example: simple pendulum. Since the gravitational potential energy depends on height, in the absence of nonconservative forces the pendulum has the same swing on either side of the equilibrium point and always returns to the same turning points. Demonstrate with a pendulum hung near a blackboard and mark the end points of the swing on the board. For a more adventurous demonstration, suspend a bowling ball pendulum from the ceiling and release the ball from rest in contact with your nose. Stand very still while it completes its swing and returns to your nose.

V. External work and thermal energy.

A. Explain that when forces due to objects external to the system do work W on the system the energy equation becomes $\Delta K + \Delta U = W$ if the internal forces are conservative. K is the total kinetic energy of all objects in the system and U is the total potential energy of their interactions with each other.

B. Show that if the external force is conservative the system can be enlarged to include the (previously) external agent. Then, $W = 0$ and U must be augmented to include the new interactions. Give the example of a ball thrown upward in Earth's gravitational field.

C. Explain that if some or all of the internal forces are nonconservative then the total energy must include an thermal energy term to take account of energy that enters or leaves some or all of the objects and contributes to the energy (kinetic and potential) of the particles that make up the objects. Distinguish between thermal energy and the energy associated with the motion and interactions of the object as a whole. Write $\Delta K + \Delta U + \Delta E_{\text{th}} = W$.

D. Refer back to the block sliding on the horizontal table top, discussed earlier. Explain that when the block stops all the original kinetic energy has been converted to thermal energy. As an example assume the table top exerts a constant frictional force of magnitude f. Explain that $K_i = fd$, where K_i is the initial kinetic energy and d is the distance the block slides before stopping. Say that if the system consists of only the block, then $-K_i + \Delta E_{\text{th}} = W$, where W is the work done by friction on the block and ΔE_{therma} is the change in the thermal energy of the block. Note that $-fd$ is not the work done by friction. Now take the system to consist of the both the block and table top. Then $-K_i + \Delta E_{\text{th}} = 0$, where ΔE_{th} is the total change in the thermal energies of the block and table top. Argue that $\Delta E_{\text{th}} = fd$. Explain that the division of thermal energy between the block and the table top cannot be calculated without a detailed model of friction.

E. Explain that the quantity $\int \vec{f} \cdot d\vec{r}$ along the path of an object does NOT give the work done by friction but it does contribute to the change in kinetic energy of the object, along with similar contributions from other forces, if any. For a block sliding on a table top, the work done by friction is algebraically greater than the value of the integral and the difference is the increase in the thermal energies of the block and table top.

F. Explain that there may be other forms of internal energy. For example, the chemical energy stored in the fuel of a car and the kinetic energy of the moving pistons are forms of internal energy if the car is taken to be an object of a system. Write $E = K + U + E_{\text{th}} + E_{\text{int}}$, where E_{int} is the total internal energy of the system, exclusive of the thermal energy. E_{int} is not considered in detail in this chapter but is considered in Chapter 9 and other chapters.

SUGGESTIONS

1. Assignments
   a. Use question 1 to discuss the idea of a conservative force.
   b. Problems 2, 3, 4, and 7 test basic understanding of gravitational potential energy. Use question 2 and problem 1 to test basic understanding of elastic potential energy.
   c. Test for understanding of the conservation of mechanical energy by asking questions 3, 4, and 5 and assigning some of problems 9, 10, 11, 12, 16, 17, 18, 19, 21, 27, and 34. Some of these are related to previous problems. Some combine gravitational and elastic potential energies.
   d. Draw several potential energy curves and have the class analyze the particle motion for various values of the total energy. This can provide particularly useful feedback as to how well the students have mastered the idea of energy conservation. Also ask question 6 and assign problems 36 and 38.
   e. Assign a few problems dealing with applied and dissipative forces. Consider 41, 42, 43, 45, 50, 51, 56, 61, and 65.
2. Demonstrations
   a. Conservation of energy: Freier and Anderson Mn1 — 3, 6.

b. Nonconservative forces: Freier and Anderson Mw1.

3. Books and Monographs

   The Bicycle by Phillip DiLavore; available from AAPT, One Physics Ellipse, College Park MD 20740–3845. Deals chiefly with energy. Students do not need to know about rotational motion.

4. Audio/Visual
   a. *Gravitational Potential Energy*; *Conservation of Energy — Pole Vault and Aircraft Take-off*; from AAPT collection 1 of single-concept films; video tape; available from Ztek Co., PO Box 11768, Lexington, KY 40577–1768.
   b. *Energy Conservation*; Side E: Conservation Laws of Cinema Classics; video disk; available from Ztek Co. (see above for address).
   c. *Conservation of Energy*; *Work and Conservation of Energy*; Physics Demonstrations in Mechanics, Part I; VHS video tape; ≈3 min each; Physics Curriculum & Instruction, 22585 Woodhill Drive, Lakeville, MN 55044.

5. Computer Software

   Conservation Laws; Macintosh, Windows; Cross Educational Software, Inc., 508 E. Kentucky Avenue, PO Box 1536, Ruston, LA 71270. Includes a section on conservation of energy.

6. Laboratory

   Bernard Experiment 9: *Work, Energy, and Friction.* A string is attached to a car on an incline and passes over a pulley at the top of the incline. Weights on the free end of the string are adjusted so the car rolls down the incline at constant speed. The work done by gravity on the weights and on the car is calculated and used to find the change in mechanical energy due to friction. The coefficient of friction is computed. The experiment is repeated for the car rolling up the incline and for various angles of incline. It is also repeated with the car sliding on its top and the coefficients of static and kinetic friction are found.

Chapter 9 SYSTEMS OF PARTICLES

BASIC TOPICS

I. Center of mass.
   A. Spin a chalkboard eraser as you toss it. Point out that, if the influence of air can be neglected, one point (the center of mass) follows the parabolic trajectory of a projectile although the motions of other points are more complicated.
   B. Define the center of mass by giving its coordinates in terms of the coordinates of the individual particles in the system. As an example, consider a system consisting of three discrete particles and calculate the coordinates of the center of mass, given the masses and coordinates of the particles. Point out that no particle need be at the center of mass.
   C. Extend the definition to include a continuous mass distribution. Note that if the object has a point, line, or plane of symmetry, the center of mass must be at that point, on that line, or in that plane. Examples: a uniform sphere or spherical shell, a uniform cylinder, a uniform square, a rectangular plate, or a triangular plate. Show how to compute the coordinates of the center of mass of a complex object comprised of a several simple parts, a table for example. Each part is replaced by a particle with mass equal to the mass of the part, positioned at the center of mass of the part. The center of mass of the particles is then found. Explain how to find the center of mass of a simple shape, such as a rectangular or circular plate, with a hole cut in it.

D. Explain that the general motion of a rigid body may be described by giving the motion of the center of mass and the motion of the object around the center of mass.

E. Derive expressions for the velocity and acceleration of the center of mass in terms of the velocities and accelerations of the particles in the system.

F. Derive $\vec{F}_{\text{net}} = M\vec{a}_{\text{com}}$ and emphasize that $\vec{F}_{\text{net}}$ is the net *external* force on all objects of the system. As an example, consider a two-particle system with external forces acting on both particles and each particle interacting with the other. Invoke Newton's third law to show that the internal forces cancel when all forces are summed.

G. State that if $\vec{F}_{\text{net}} = 0$ and the center of mass is initially at rest, then it remains at the same point no matter how individual parts of the system move. Refer to the two carts of C above.Work a sample problem. For example, consider a person running from one end to the other of a slab that is free to slide on a horizontal frictionless surface or two skaters who pull themselves toward each other with a rope or pole. Ask how far the cart and running person or the two skaters each move.

II. Momentum.

A. Define momentum for a single particle. If you are including modern physics topics in the course, also give the relativistic definition of momentum.

B. Show that Newton's second law can be written $\vec{F}_{\text{net}} = d\vec{p}/dt$ for a particle. Emphasize that the mass of the particle is constant and that this form of the law does not imply that a new term $\vec{v}\,dm/dt$ has been added to $\vec{F}_{\text{net}} = m\vec{a}$.

C. State that the total momentum of a system of particles is the *vector* sum of the individual momenta and show that $\vec{P} = M\vec{v}_{\text{com}}$.

D. Show that Newton's second law for the center of mass can be written $\vec{F}_{\text{net}} = d\vec{P}/dt$, where $\vec{P}$ is the total momentum of the system. Stress that $\vec{F}_{\text{net}}$ is the net external force and that this equation is valid only if the mass of the system is constant.

III. Conservation of linear momentum.

A. Point out that $\vec{P} =$ constant if $\vec{F}_{\text{net}} = 0$. Stress that one examines the *external* forces to see if momentum is conserved in any particular situation. Point out that one component of $\vec{P}$ may be conserved when others are not.

B. Put two carts, connected by a spring, on a horizontal air track and set them in oscillation by pulling them apart and releasing them from rest. Explain that the center of mass does not accelerate and the total momentum of the system is constant. Use the conservation of momentum principle to derive an expression for the velocity of one cart in terms of the velocity of the other. Push one cart and explain that the center of mass is now accelerating and the total momentum is changing.

C. Consider a projectile that splits in two and find the velocity of one part, given the velocity of the other. Point out that mechanical energy is conserved for the cart-spring system but is not for the fragmenting projectile. The exploding projectile idea can be demonstrated with an air track and two carts, one more massive than the other. Attach a brass tube to one cart and a tapered rubber stopper to the other. Arrange so that the tube is horizontal and the stopper fits in its end. The tube has a small hole in its side, through which a firecracker fuse fits. Start the carts at rest and light a firecracker in the tube. The carts rapidly separate, strike the ends of the track, come back together again, and stop. Arrange the initial placement so the carts strike the ends of the track simultaneously. Explain that $\vec{P} = 0$ throughout the motion. For a less dramatic demonstration, tie two carts together with a compressed spring between them, then cut the string.

D. Explain that observers in two different inertial frames will measure different values of the momentum for a system but they will agree on the conservation of momentum. That is, if the net external force is found to vanish in one inertial frame, it vanishes in all inertial frames.

E. Illustrate the use of conservation of momentum to solve problems by considering the firing of a cannon initially resting on a frictionless surface. Assume the barrel is horizontal and calculate the recoil velocity of the cannon. Explain that muzzle velocity is measured relative to the cannon and that we must use the velocity of the cannonball relative to Earth.

IV. Variable mass systems.

A. Derive the rocket equations, Eqs. 9–42 and 9–43. Emphasize that we must consider the rocket and fuel to be a single constant mass system. Derive expressions for the momentum before and after a small amount of fuel is expelled, in terms of the mass of the rocket and fuel together, the mass of the fuel expelled, the initial velocity of the rocket, the change in its velocity, and the relative velocity of the expelled fuel. Assume no external forces act on the rocket-fuel system and equate the two expressions for the total momentum.

B. To demonstrate, screw several hook eyes into a toy CO_2 propelled rocket, run a line through the eyes, and string the line across the lecture hall. Start the rocket from rest and have the students observe its acceleration as it crosses the hall.

C. As a second example, consider the loading of sand on a conveyor belt and calculate the force required to keep the belt moving at constant velocity.

V. External forces and internal energy.

A. Show that if the system is replaced by a single particle with mass equal to the total mass of the system, located at the system center of mass, and acted on by a force equal to the total external force acting on the system, then the work done by the force equals the change in the kinetic energy associated with the motion of the center of mass (defined by $\frac{1}{2}mv_{\text{com}}^2$).

B. Remark that the integral $\int \vec{F}_{\text{net}} \cdot d\vec{s}$ over the path of the center of mass is NOT necessarily the work done by the net external force. In particular, it is not when the motions of the center of mass and the point of application of the force are different. In an extreme case the point of application of the external force does not move so the force does zero work, but the center of mass accelerates and the kinetic energy changes. Remind students of the energy equation $\Delta K + \Delta U + \Delta E_{\text{int}} = W$ and stress that W is the true work done by the external force. State that internal energy may be converted to center of mass kinetic energy or vice versa by the action of an external force.

C. Discuss the example of an accelerating car. Point out that the force of the road on the tires accelerates the car but does no work if the tires do not skid. Internal energy is converted to center of mass external energy. Discuss the example of a skater on ice pushing herself away from a wall and observe that the force of the wall on her accounts for the acceleration of her center of mass but this force does not work.

D. State that for a closed system $\Delta E_{\text{int}} + \Delta E_{\text{mec}} = 0$ and apply this equation to the skater pushing from a wall and to the accelerating automobile.

SUGGESTIONS

1. Assignments
   a. Use question 1, along with problems 3, 4, and 6 to generate discussion about the position of the center of mass. To present a challenge, assign problem 9.
   b. Questions 3, 4, 5, and 6 are good tests of understanding of the motion of the center of mass. Discuss them as an introduction to the problems. Assign some of problems 11, 14, 15, and 16.
   c. Assign some problems in which the center of mass does not move: 10, 12, 18, and 19, for example.
   d. To emphasize the vector nature of momentum, assign problems 23 and 24.

e. Use questions 7, 8, and 9 to discuss conservation of momentum, then assign problems such as 27, 29, 32, and 34. To emphasize the difference between conservation of energy and conservation of momentum assign problem 36.

f. Assign a problem or two such as 41, 44, 46, and 47, which are concerned with variable mass systems.

g. To generate a discussion of the role played by internal forces assign problems 52, 55, and 58.

2. Demonstrations
   a. Center of mass, center of gravity: Freier and Anderson Mp7, 12, 13.
   b. Motion of center of mass: Freier and Anderson Mp1, 2, 16 — 19.
   c. Conservation of momentum: Freier and Anderson Mg4, 5, Mi2.
   d. Rockets: Freier and Anderson Mh.

3. Books and Monographs

   Rockets by David Keeports; available from AAPT, One Physics Ellipse, College Park MD 20740–3845.

4. Audio/Visual
   a. *Finding the Speed of a Rifle Bullet*; from AAPT collection 1 of single-concept films; video tape; available from Ztek Co., PO Box 11768, Lexington, KY 40577–1768.
   b. *Center of Mass*, *Conservation of Linear/Angular Momentum*, *Conservation of Momentum*; from AAPT collection 2 of single-concept films; video tape; available from Ztek Co. (see above for address).
   c. *Linear Momentum*; Side E: Conservation Laws of Cinema Classics; video disk; available from Ztek Co. (see above for address).
   d. *Momentum*; VHS video tape (9 min); Films for the Humanities & Sciences, PO Box 2053, Princeton, NJ 08543–2053.
   e. *Human Momenta, No Initial Motion*, *Human Momenta, Initial Translation* from Skylab Physics Videodisc; video disk; available from Ztek Co. (see above for address).
   f. *Center of Mass Motion*; Physics Experiments; VHS video tape (15 min); Films for the Humanities & Sciences (see above for address).
   g. *Human Momenta*; from AAPT Skylab videotape; American Association of Physics Teachers, One Physics Ellipse, College Park, MD 20740–3845.
   h. *Motion of Center of Mass*; *Conservation of Momentum*; Physics Demonstrations in Mechanics, Part II; VHS video tape; ≈3 min; Physics Curriculum & Instruction, 22585 Woodhill Drive, Lakeville, MN 55044.
   i. *Impulse and Momentum*;*Conservation of Momentum*; Physics Demonstrations in Mechanics, Part V; VHS video tape or video disk; ≈3 min; Physics Curriculum & Instruction (see above for address).
   j. *Motion of Center of Mass*; Physics Demonstrations in Mechanics, Part V; VHS video tape or video disk; ≈3 min; Physics Curriculum & Instruction (see above for address).

5. Computer Software

   Conservation Laws; Macintosh, Windows; Cross Educational Software, Inc., 508 E. Kentucky Avenue, PO Box 1536, Ruston, LA 71270. Includes a section on conservation of momentum.

6. Computer Project

   Have students use a spreadsheet or write a computer program to follow individual particles in a two or three particle system, given the force law for the forces they exert on each other. The program should integrate Newton's second law for each particle. Have the students use their data to verify the conservation of momentum. See the Computer Projects section of this manual.

7. Laboratory
   a. Meiners Experiment 7–9: *Ballistic Pendulum — Projectile Motion.* A ball is shot into a trapping mechanism at the end of a pendulum. The initial speed of the ball is found by applying conservation of momentum to the collision and conservation of energy to the subsequent swing of the pendulum. Also see Bernard Experiment 11: *Inelastic Impact and the Velocity of a Projectile.*
   b. Meiners Experiment 8–6: *Center of Mass Motion.* Two pucks are connected by a rubber band or spring and move toward each other on a nearly frictionless surface. A spark timer is used to record their positions as functions of time. Students calculate and study the position of the center of mass as a function of time. They also find the center of mass velocity. Can be performed with dry ice pucks or on an air table or air track.
   c. Meiners Experiment 8–7: *Linear Momentum.* Essentially the same as 8–6 but data is analyzed to give the individual momenta and total momentum as functions of time. Kinetic energy is also analyzed.

Chapter 10 COLLISIONS

BASIC TOPICS

I. Properties of collisions
   A. Set up a collision between two carts on an air track. Point out the interaction interval and the intervals before and after the interaction.
   B. Explain that for the collisions considered two bodies interact with each other over a short period of time and that the times before and after the collision are well defined. The force of interaction is great enough that external forces can be ignored during the interaction time.

II. Impulse.
   A. Define the impulse of a force as the integral over time of the force. Note that it is a vector. Clearly distinguish between impulse (integral over time) and work (integral over path). Draw a force versus time graph for the force of one body on the other during a one-dimensional collision and point out the impulse is the area under the curve.
   B. Define the time averaged force and show that the impulse is the product $\vec{F}_{avg}\Delta t$, where Δt is the time of interaction. Remark that we can use the average force to estimate the strength of the interaction during the collision.
   C. Use Newton's second law to show that the impulse on a body equals the change in its momentum. Remark that the change in momentum depends not only on the force but also on the duration of the interaction. Refer to the air track collision and point out that it is the impulse of one body on the other that changes the momentum of the second body. Repeat the air track collision. Measure the velocity of one cart before and after the collision and calculate the change in its momentum. Equate this to the impulse the other cart exerts. Estimate the collision time and calculate the average force exerted on the cart.
   D. Use the third law to show that two bodies in a collision exert equal and opposite impulses on each other and show that, if external impulses can be ignored, then total momentum is conserved. Refer to the air track collision. Again stress that external forces are neglected during the collisions considered here.
   E. Force probes, with input to a computer, can be used to plot the forces acting during a collision as functions of time. The curves can be integrated to find the impulse.

III. Two-body collisions in one dimension.

A. Define the terms "elastic", "inelastic", and "completely inelastic". Distinguish between the transfer of kinetic energy from one colliding object to the another and the loss of kinetic energy to internal energy.

B. Two-body completely inelastic collisions.

1. Derive an expression for the velocity of the bodies after the collision in terms of their masses and initial velocities.
2. Demonstrate the collision on an air track, using carts with velcro bumpers. Point out that the kinetic energy of the bodies is not conserved and calculate the energy loss. Remark that the energy is dissipated by the mechanism that binds the objects to each other. Some goes to internal energy, some to deformation energy. Note that $\frac{1}{2}(m_1 + m_2)v_{\text{com}}^2$ is retained. If we use a reference frame attached to the center of mass to describe the collision, we would find the combined bodies at rest after the collision and all kinetic energy lost.

C. Two-body elastic collisions.

1. Derive expressions for the final velocities in terms of the masses and initial velocities.
2. Specialize the general result to the case of equal masses and one body initially at rest. Demonstrate this collision on the air track using carts with spring bumpers. Point out that the carts exchange velocities.
3. Specialize the general result to the case of a light body, initially at rest, struck by a heavy body. Demonstrate this collision on the air track. Point out that the velocity of the heavy body is reduced only slightly and that the light body shoots off at high speed. Relate to a bowling ball hitting a pin.
4. Specialize the general result to the case of a heavy body, initially at rest, struck by a light body. Demonstrate this collision on the air track. Point out the low speed acquired by the heavy body and the rebound of the light body. Relate to a ball rebounding from a wall. A nearly elastic collision can be obtained with a superball.
5. Point out that, although the total kinetic energy does not change, kinetic energy is usually transferred from one body to the other. Consider a collision in which one body is initially at rest and calculate the fraction that is transferred. Show that the fraction is small if either mass is much greater than the other and that the greatest fraction is transferred if the two masses are the same. This is important, for example, in deciding what moderator to use to thermalize neutrons from a fission reactor.

D. Point out that while the greatest energy loss occurs when the interaction is completely inelastic, there are many other inelastic collisions in which less than the maximum energy loss occurs. Note that it is possible to have a collision in which kinetic energy increases (an explosive impact, for example).

IV. Two-body collisions in two dimensions.

A. Write down the equations for the conservation of momentum, in component form, for a collision with one body initially at rest. Mention that these can be solved for two unknowns.

B. Consider an elastic collision for which one body is at rest initially and the initial velocity of the second is given. Write the conservation of kinetic energy and conservation of momentum equations. Point out that the outcome is not determined by the initial velocities but that the impulse of one body on the other must be known to determine the velocities of the two bodies after the collision. State that the impulse is usually not known and in practice physicists observe one of the outgoing particles to determine its direction of motion. Carry out the calculation: assume the final direction of motion of one body is known and calculate the final direction of motion of the other and both final speeds.

C. State, or perhaps prove, that if the particles have the same mass then their directions of motion after the collision are perpendicular to each other.

D. Consider a completely inelastic collision for which the two bodies do not move along the same line initially. Mention that the outcome of this type collision *is* determined by the initial velocities. Calculate the final velocity. Calculate the fraction of energy dissipated.

SUGGESTIONS

1. Assignments
   a. To help students with the concept of impulse and the impulse-momentum theorem ask question 1. Also consider problems 3, 5, 9, 14, and 16.
   b. Use questions 3, 4, and 5 to discuss the motion of the center of mass during a collision. Use questions 2 and 6 to discuss conservation of momentum.
   c. Assign some problems dealing with inelastic collisions, such as 23 and 28. Assign and discuss some problems for which the mechanism of kinetic energy loss is given explicitly. See problems 31 and 33.
   d. Assign a problem, such as 35 or 52, that ask students to determine if a collision is elastic.
   e. Ask question 7 in support of the discussion of elastic collisions. Assign problems 37, 39, and 44. For some fun, carry out the demonstration described in question 8 and assign problem 45.
   f. Demonstrate the ballistic pendulum and show how it can be used to measure the speed of a bullet. Assign problems 22 and 24.
   g. If you intend to cover the thermodynamics chapters, assign and discuss problem 13. Also consider problem 12. They will prove helpful for the discussion of the microscopic basis of pressure.
   h. Assign some problems that deal with two-dimensional collisions: the group 46 through 55. Include both elastic and inelastic collisions.
2. Demonstrations
   Freier and Anderson Mg1 — 3, Mi1, 3, 4, Mw3, 4.
3. Audio/Visual
   a. *One-Dimensional Collisions*, *Two-Dimensional Collisions*, *Scattering of a Cluster of Objects*, *Dynamics of a Billiard Ball*, *Inelastic One-Dimensional Collisions*, *Inelastic Two-Dimensional Collisions*, and *Colliding Freight Cars*; from AAPT collection 1 of single-concept films; video tape; available from Ztek Co., PO Box 11768, Lexington, KY 40577–1768.
   b. *Drops and Splashes*, *Collisions in Two Dimensions*, *Inelastic Collisions*; from AAPT collection 2 of single-concept films; video tape; available from Ztek Co. (see above for address).
   c. *Elastic Collisions*, *Inelastic Collisions*, *Collisions*; Side E: Conservation Laws of Cinema Classics; video disk; available from Ztek Co. (see above for address).
   d. *Air Track Collisions*; Demonstrations of Physics: Energy and Momentum; VHS video tape (7:24); Media Design Associates, Inc., Box 3189, Boulder, CO 80307–3190.
   e. *The Dynamics of a Karate Punch*; Demonstrations of Physics: Energy and Momentum; VHS video tape (4:55); Media Design Associates, Inc. (see above for address).
   f. *Collisions* from Skylab Physics Videodisc; video disk; available from Ztek Co. (see above for address).
   g. *Physics and Automobile Collisions*; interactive video disk by Dean Zollman (John Wiley, 1984). The disk shows collisions of cars with fixed barriers and two car collisions (head-on, at 90°, and at 60°). One sequence shows the influence of bumper design, others show the influence of air bags and shoulder straps on manikins. All are slow motion movies of manufacturers' tests and many show grids and clocks. Students can stop the action to

take measurements, then make calculations of momentum and energy transfers. For most exercises, a standard disk player is satisfactory; for a few, a computer-controlled player is required.

h. *Characteristics of Collisions*; *Elastic Collision*; Physics Demonstrations in Mechanics, Part V; VHS video tape or video disk; ≈3 min; Physics Curriculum & Instruction, 22585 Woodhill Drive, Lakeville, MN 55044.

4. Computer Software

   Objects in Motion. See Chapter 2 SUGGESTIONS.

5. Computer Project

   Have students use a commercial math program or write a program to graph the total final kinetic energy as a function of the final velocity of one object in a two-body, one-dimensional collision, given the initial velocities and masses of the two objects. Ask them to run the program for specific initial conditions and identify elastic, inelastic, completely inelastic, and explosive collisions on their graphs. Details of this and other projects are given in the Computer Projects section of this manual.

6. Laboratory

   a. Meiners Experiment 7–10: *Impulse and Momentum.* Students use a microprocessor to measure the force as a function of time as a toy truck hits a force transducer. They numerically integrate the force to find the impulse, then compare the result with the change in momentum, found by measuring the velocity before and after the collision.

   b. Bernard Experiment 8: *Impulse, Momentum, and Energy.* Part A deal with a mass that is hung on a string passing over a pulley and attached to an air track glider. The glider accelerates from rest for a known time and a spark timer is used to find its velocity at the end of the time. The impulse is calculated and compared with the momentum. In part B a glider is launched by a stretched rubber band and a spark record of its position as a function of time is made while it is in contact with the rubber band. A static technique is used to measure the force of the rubber band for each of the recorded glider positions and the impulse is approximated. The result is again compared with the final momentum of the glider.

   c. Meiners Experiment 7–11: *Scattering* (for advanced groups). The deflection of pellets from a stationary disk is used to investigate the scattering angle as a function of impact parameter and to find the radius of the disk.

   d. Meiners Experiment 8–5: *One-Dimensional Collisions.* A puck moving on a nearly frictionless surface collides with a stationary puck. A spark timer is used to record the positions of the pucks as functions of time. Students calculate the velocities, momenta, and energies before and after the collision. May be performed with dry ice pucks or on an air table or track.

   e. Meiners Experiment 8–8: *Two-Dimensional Collisions.* Same as Meiners 8–5 but the pucks are allowed to scatter out of the original line of motion. Students must measure angles and calculate components of the momenta. The experiment may be performed with dry ice pucks or on an air table.

   f. Bernard Experiment 12: *Elastic Collision — Momentum and Energy Relations in Two Dimensions.* A ball rolls down an incline on a table top and strikes a target ball initially at rest at the edge of the table. The landing points of the balls on the floor are used to find their velocities just after the collision. The experiment is run without a target ball to find the velocity of the incident ball just before the collision. Data is used to check for conservation of momentum and energy. Both head-on and grazing collisions are investigated. A second experiment, similar to Meiners 8–8, is also described.

Chapter 11 ROTATION

BASIC TOPICS

I. Rotation about a fixed axis.

   A. Spin an irregular object on a fixed axis. A bicycle wheel or spinning platform with the object attached can be used. Draw a rough diagram, looking along the rotation axis. Explain that each point in the body has a circular orbit and that, for any selected point, the radius of the orbit is the perpendicular distance from the point to the rotation axis. Contrast to a body that is simultaneously rotating and translating. See Figs. 11–2 and 11–3.

   B. Define angular position θ (in radians and revolutions), angular displacement $\Delta\theta$, angular velocity ω (in rad/s, deg/s, and rev/s), and angular acceleration α (in rad/s^2, deg/s^2, and rev/s^2). Treat both average and instantaneous quantities but emphasize that the instantaneous quantities are most important for us. Remind students of radian measure.

   C. Use Fig. 11–4 to show how an angular displacement is measured. Note that θ is positive for an angular position in one direction from the fixed axis and negative for an angular position in the other. By convention in this text position angles are positive in the counterclockwise direction. Remark that as the body rotates θ continues increase beyond 2π rad.

   D. Interpret the signs of ω and α. Give examples of spinning objects for which ω and α have the same sign and for which they have opposite signs.

   E. Point out the analogy to one-dimensional linear motion. θ corresponds to x, ω to v, and α to a.

   F. Point out that ω and α can be thought of as the components of vectors $\vec{\omega}$ and $\vec{\alpha}$, respectively. For fixed axis rotation, the vectors lie along the rotation axis, with the direction of $\vec{\omega}$ determined by a right hand rule: if the fingers curl in the direction of rotation, then the thumb points in the direction of $\vec{\omega}$. If $d\omega/dt > 0$, then $\vec{\alpha}$ is in the same direction; if $d\omega/dt < 0$, then it is in the opposite direction. Use Fig. 11–7 to explain that a vector cannot be associated with a finite angular displacement because displacements do not add as vectors.

II. Rotation with constant angular acceleration.

   A. Emphasize that the discussion here is restricted to rotation about a fixed axis but that the same equations can be used when the rotation axis is in linear translation. This type motion will be discussed in the next chapter.

   B. Write down the kinematic equations for $\theta(t)$ and $\omega(t)$. Make a comparison with the analogous equations for linear motion (see Table 11–1).

   C. Point out that the problems of rotational kinematics are similar to those for one-dimensional linear kinematics and that the same strategies are used for their solution.

   D. Go over examples. Calculate the time and number of revolutions for an object to go from some initial angular velocity to some final angular velocity, given the angular acceleration. If time permits, consider both a body that is speeding up and one that is slowing down. For the latter, calculate the time to stop and the number of revolutions made while stopping. Calculate the time to rotate a given number of revolutions and the final angular velocity, again given the angular acceleration.

III. Linear speed and acceleration of a point rotating about a fixed axis.

   A. Write down $s = \theta r$ for the arc length. Explain that it is a rearrangement of the defining equation for the radian and that θ must be in radians for it to be valid.

   B. Wrap a string on a large spool that is free to rotate about a fixed axis. Mark the spool so the angle of rotation can be measured. Slowly pull out the string and explain that the length of string pulled out is equal to the arc length through which a point on the rim moves. Compare the string length to θr for $\theta = \pi/2$, π, $3\pi/2$, and 2π rad. Show that $s = \theta r$ reduces to the familiar result for $\theta = 2\pi$ rad.

C. Differentiate $s = \theta r$ to obtain $v = r\omega$ and $a_t = \alpha r$. Emphasize that radian measure *must* be used. Point out that v gives the speed and a_t gives the acceleration of the string as it is pulled provided it does not slip on the spool. Point out that all points in a rotating rigid body have the same value of ω and the same value of α but points that are different distances from the rotation axis have different values of v and different values of a_t.

D. Point out that the velocity is tangent to the circular orbit but that the total acceleration is not. a_t gives the tangential component while $a_r = v^2/r = \omega^2$ gives the radial component. The tangential component is not zero only when the point on the rim speeds up or slows down in its rotational motion while the radial component is not zero as long as the object is turning. For students who have forgotten, reference the derivation of $a_r = v^2/r$, given in Chapter 4.

E. Explain how to find the magnitude and direction of the total acceleration in terms of ω, α, and r.

IV. Kinetic energy of rotation and rotational inertia.

A. By substituting $v = r\omega$ into $K = \frac{1}{2}mv^2$, show that $K = \frac{1}{2}mr^2\omega^2$ for a particle moving around a circle with angular velocity ω and, by summing over all particles in a rigid body, show that $K = \frac{1}{2}I\omega^2$, where $I = \sum m_i r_i^2$ if the body is rotating about a fixed axis. Explain that I is called the rotational inertia of the body. Mention that many texts call it the moment of inertia.

B. Point out that rotational inertia depends on the distribution of mass and on the position and orientation of the rotation axis. Explain that two bodies may have the same mass but quite different rotational inertias. State that Table 11–2 gives the rotational inertia for various objects and axes. Particularly point out the rotational inertia of a hoop rotating about the axis through its center and perpendicular to its plane. Note that all its mass is the same distance from the rotation axis. Also point out the rotational inertias of a cylinder rotating about its axis and a sphere rotating about a diameter. Note that the mass is now distributed through a range of distances from the rotation axis and the rotational inertia is less than that for a hoop with the same mass and radius.

C. Optional: show how to convert the sum for I to an integral. Use the integral to find the rotational inertia for a thin rod rotating about an axis through its center and perpendicular to its length. If the students have experience with volume integrals using spherical coordinates, derive the expression for the rotational inertia of a sphere.

D. Prove the parallel axis theorem. The proof can be carried out using a sum for I rather than an integral. Explain its usefulness for finding the rotational inertia when the rotation axis is not through the center of mass. Emphasize that the actual axis and the axis through the center of mass must be parallel for the theorem to be valid. Use the parallel axis theorem to obtain the rotational inertia for the rotation of a thin rod about one end from the rotational inertia for rotation about the center, given in Table 11–2.

V. Torque.

A. Define torque for a force acting on a single particle. Consider forces that lie in planes perpendicular to the axis of rotation and take $\tau = rF\sin\phi$, where $\vec{r}$ is a vector that is perpendicular to the rotation axis and points from the axis to the point of application of the force. ϕ is the angle between $\vec{r}$ and $\vec{F}$ when they are drawn with their tails at the same point. The definition will be generalized in the next chapter. Explain that $\tau = rF_t = r_\perp F$, where F_t is the tangential component of $\vec{F}$ and $r_\perp$ is the moment arm.

B. Explain that the torque vanishes if $\vec{F}$ is along the same line as $\vec{r}$ and that only the component of $\vec{F}$ that is perpendicular to $\vec{r}$ produces a torque. This is a mechanism for picking out the part of the force that produces angular acceleration, as opposed to the part that produces centripetal acceleration. Also explain that the same force can produce a larger torque if it is applied at a point farther from the rotation axis.

C. Use a wrench tightening a bolt as an example. The force is applied perpendicular to the wrench arm and long moment arms are used to obtain large torques.

D. Explain the sign convention for torques applied to a body rotating about a fixed axis. For example, torques tending to give the body a counterclockwise (positive) angular acceleration are positive while those tending to give the body a clockwise angular acceleration are negative. Remark that the convention is arbitrary and the opposite convention may be convenient for some problems.

VI. Newton's second law for rotation.

A. Use a single particle on a circular orbit to introduce the topic. Start with $F_t = ma_t$ and show that $\tau = I\alpha$, where $I = mr^2$. Explain that this equation also holds for extended bodies, although I is then the sum given above.

B. Remark that problems are solved similarly to linear second law problems. Tell students to identify torques, draw a force diagram, choose the direction of positive rotation, and substitute the total torque into $\tau_{\text{net}} = I\alpha$. Remark that the point of application of a force is important for rotation, so the object cannot be represented by a dot on a force diagram. Tell students to sketch the object and place the tails of force vectors at the application points.

C. Wrap a string around a cylinder, free to rotate on a fixed horizontal axis. Attach the free end of the string to a mass and allow the mass to fall from rest. Note that its acceleration is less than g, perhaps by dropping a free mass beside it. See Sample Problem 11–7.

VII. Work-kinetic energy theorem for rotation.

A. Use $dW = \vec{F} \cdot d\vec{s}$ to show that the work done by a torque is given by $W = \int \tau \, d\theta$ and that the power delivered is given by $P = \tau\omega$.

B. Use $\tau \, d\theta = I\alpha \, d\theta = \frac{1}{2} I \, d(\omega^2)$ to show that $W = \frac{1}{2} I(\omega_f^2 - \omega_i^2)$.

C. For the situation of Sample Problem 11–7 use conservation of energy to find the angular velocity of the cylinder after the mass has fallen a distance h. Use rotational kinematics and the value for the angular acceleration found in the text to check the answer.

SUGGESTIONS

1. Assignments

a. Use questions 1 and 2 to discuss graphical interpretations of angular position and velocity.

b. Use techniques of the calculus to derive the kinematic equations for constant angular acceleration. That is, integrate α = constant twice with respect to time. Assign one or two problems from the group 1, 4, and 5.

c. Assign some problems that deal with rotation with constant angular acceleration: 12, 15, 17, and 18, for example. Also consider question 4.

d. To discuss the relationship between angular and linear variables, assign some of problems 22, 23, 24, and 29. Questions 5 and 6 might help with the ideas of the radial and tangential components of the linear acceleration.

e. Use question 7 to guide students through a qualitative discussion of rotational inertia. Assign problems 37, 39, and 41. Use problem 43 to discuss the radius of gyration, if desired.

f. Use problem 47 or 48 to discuss the calculation of torque.

g. To help students think about torque and $\tau_{\text{net}} = I\alpha$, discuss questions 8, 9, 10, and 11. Assign some of problems 52, 56, and 57. To deal with a situation in which the dynamics of more than one object is important, demonstrate the Atwood machine and discuss problem 55.

h. Use question 12 to discuss the work done by a torque and changes in rotational kinetic energy. Discuss conservation of mechanical energy and assign problems 65 and 66.

2. Demonstrations
   a. Rotational dynamics: Freier and Anderson Ms7, Mt 5, 6, Mo5.
   b. Rotational work and energy: Freier and Anderson Mv2, Mr5, Ms2.
3. Audio/Visual
   a. *Newton's Second Law for Rotational Motion*; Demonstrations of Physics: Motion; VHS video tape (8:35); Media Design Associates, Inc., Box 3189, Boulder, CO 80307–3190.
4. Computer Project

   Ask students to use a commercial math program or their own root finding programs to solve rotational kinematic problems.
5. Laboratory

   Meiners Experiment 7–14: *Rotational Inertia.* The rotational inertia of a disk is measured dynamically by applying a torque (a falling mass on a string wrapped around a flange on the disk). A microprocessor is used to measure the angular acceleration. Small masses are attached to the disk and their influence on the rotational inertia is studied. The acceleration of the mass can also be found by timing its fall through a measured distance. Then, $a_t = \alpha r$ is used to find the angular acceleration of the disk. Also see Bernard Experiment 14: *Moment of Inertia.*

Chapter 12 ROLLING, TORQUE, AND ANGULAR MOMENTUM

BASIC TOPICS

I. Rolling.

A. Remark that a rolling object can be considered to be rotating about an axis through the center of mass while the center of mass moves. The text considers the special case for which the axis of rotation does not change direction. Point out that the rotational motion obeys $\tau_{\text{net}} = I_{\text{com}}\alpha$ and the translational motion of the center of mass obeys $\vec{F}_{\text{net}} = m\vec{a}_{\text{com}}$, where τ_{net} is the sum of external torques and $\vec{F}_{\text{net}}$ is the sum of external forces. Emphasize that one of the forces acting may be the force of friction produced by the surface on which the object rolls.

B. Explain that the speed of a point at the top of a rolling object is $v_{\text{com}} + \omega R$ and the speed of a point at the bottom is $v_{\text{com}} - \omega R$. Specialize to the case of rolling without slipping. Point out that the point in contact with the ground has zero velocity, so $v_{\text{com}} = \omega R$. Use Fig. 12–4 as evidence. Also point out that tire tracks in the snow are clean (not smudged) if the tires do not slip.

C. Explain that a wheel rolling without slipping can be viewed as rotating about an axis through the point of contact with the ground. Use this and the parallel axis theorem to show that the kinetic energy is $\frac{1}{2}Mv_{\text{com}}^2 + \frac{1}{2}I_{\text{com}}\omega^2$.

D. Consider objects rolling down an inclined plane and show how to calculate the speed at the bottom using energy considerations. If time permits, carry out an analysis using the equations of motion and show how to find the frictional force that prevents slipping.

E. Roll a sphere, a hoop, and a cylinder, all with the same radius and mass, down an incline. Start the objects simultaneously at the same height and ask students to pick the winner.

Point out that the speed at the bottom is determined by the dimensionless parameter $\beta = I/MR^2$ and not by I, M, and R alone. All uniform cylinders started from rest reach the bottom in the same time and have the same speed when they get there.

F. Consider a ball striking a bat. Show how to find the point at which the ball should hit so the instantaneous center of rotation is at the place where the bat is held. The striking point is called the center of percussion. When the ball hits there the batter feels no sting.

II. Torque and angular momentum.

A. Define torque as $\vec{\tau} = \vec{r} \times \vec{F}$ and explain that this is the general definition. Review the vector product, give the expression for the magnitude ($\tau = rF\sin\phi$), and give the right hand rule for finding the direction. Explain that $\vec{\tau} = 0$ if $\vec{r} = 0$, $\vec{F} = 0$, or $\vec{r}$ is parallel (or antiparallel) to $\vec{F}$.

B. Consider an object going around a circle and suppose a force is applied tangentially. Take the origin to be on the rotation axis but not at the circle center and show the general definition reduces to the expression used in Chapter 11 for the component of the torque along the rotation axis: $\tau = F_t R$, where R is the radius of the circle (not the distance from the origin to the point of application).

C. Use vector notation to define angular momentum for a single particle ($\vec{\ell} = m\vec{r} \times \vec{v}$). Give the expression for the magnitude and the right-hand rule for the direction.

D. Derive the relationship $\ell = mr^2\omega$ between the magnitude of the angular momentum and the angular velocity for a particle moving on a circle centered at the origin. Also find the angular momentum if the origin is on a line through the circle center, perpendicular to the circle, but not at the center. Explain that the component along the rotation axis is $mr^2\omega$ and is independent of the position of the origin along the line and that the component perpendicular to the axis rotates with the particle.

E. To show that a particle may have angular momentum even if it is not moving in a circle, calculate the angular momentum of a particle moving with constant velocity along a line not through the origin. Point out that the angular momentum depends on the choice of origin. In preparation for G below you might want to find the time rate of change of $\vec{\ell}$.

F. Use Newton's second law to derive $\vec{\tau} = d\vec{\ell}/dt$ for a particle. Consider a particle moving in a circle, subjected to both centripetal and tangential forces. Take the origin to be at the center of the circle and show that $\vec{\tau} = d\vec{\ell}/dt$ reduces to $F_t = ma_t$, as expected. Take the origin to be on the line through the center, perpendicular to the circle, but not at the center. Show that the torque associated with the centripetal force produces the change in $\vec{\ell}$ expected from the discussion of D.

G. Show that the magnitude of the torque about the origin exerted by gravity on a falling mass is mgd, where d is the perpendicular distance from the line of fall to the origin. Write down the velocity as a function of time and show that the angular momentum is $mgtd$. Remark that $\tau_{net} = d\ell/dt$ by inspection. See Sample Problem 12–5.

III. Systems of particles.

A. Explain that the total angular momentum for a system of particle is the vector sum of the individual momenta.

B. Show that $\vec{\tau}_{net} = d\vec{L}/dt$ for a system of particles for which internal torques cancel. Emphasize that $\vec{\tau}_{net}$ is the result of summing all torques on all particles in the system and that $\vec{L}$ is the sum of all individual angular momenta. Demonstrate in detail the cancellation of internal torques for two particles that interact via central forces. Point out that this equation is the starting point for investigations of the rotational motion of bodies.

C. Show that the component along the rotation axis of the total angular momentum of a rigid body rotating about a fixed axis is $I\omega$. Use the example of a single particle to point out that the angular momentum vector is along the rotation axis if the body is symmetric about the axis but that otherwise it is not. Emphasize that for fixed axis rotation we are chiefly interested in the components of angular momentum and torque along the rotation axis.

D. Make a connection to material of the last chapter by showing that $L = I\omega$ and $\tau_{\text{net}} = \text{d}L/\text{d}t$ lead to $\tau_{\text{net}} = I\alpha$ for a rigid body rotating about a fixed axis. Here τ_{net} is the component of the total external torque along the rotation axis.

IV. Conservation of angular momentum.

A. Point out that $\vec{L}$ =constant if $\vec{\tau}_{\text{net}} = 0$. State that different objects in a system may change each other's angular momentum but the changes sum vectorially to zero. Also explain that the rotational inertia of an object may change while it is spinning. Then, $I_i\omega_i = I_f\omega_f$ if the net external torque vanishes.

B. As examples consider a mass dropped on the rim of a freely spinning platform, a person running tangent to the rim of a merry-go-round and jumping on, and a spinning skater whose rotational inertia is changed by dropping her arms.

C. The third example can be demonstrated easily if you have a rotating platform that can hold a person. Have a student hold weights in each hand to increase the rotational inertia. Start him spinning with arms extended, then have him bring his arms in toward the sides of his body. See Fig. 12–16. Also carry out the spinning bicycle wheel demonstration described in the text. See Fig. 12–19.

SUGGESTIONS

1. Assignments

   a. In connection with rolling without sliding ask questions 1 through 3 and assign problems 6, 7, and 11. To emphasize the condition for no slipping assign problem 8 or 14.

   b. Assign problems 20 and 26 to stress the importance of the origin in calculations of torque and angular momentum. Problem 25 asks students to calculate the angular momentum if the cartesian components of the position and momentum vectors are given. Problem 24 deals with both angular momentum and torque. Be sure students can calculate the angular momentum of an object moving along a straight line and the angular momentum of a rigid body rotating about a fixed axis. See problems 23, 27, and 35. Also consider discussing the angular momentum of a projectile.

   c. Use question 9 to start the discussion of conservation of angular momentum. Problems 43, 46, and 48 include motion along a straight line. Problems 39 and 44 deal with changes in rotational inertia. Problems 41 and 42 deal with inelastic rotational collisions. Assign one or more from each of these groups. To include a situation for which angular momentum is not conserved, assign problem 37.

2. Demonstrations

   a. Rolling: Freier and Anderson Mb4, 7, 30, Mo3, Mp3, Mr1, 4, Ms1, 3, 4, 6.

   b. Conservation of angular momentum: Freier and Anderson Mt1 — 4, 7, 8, Mu1.

   c. Gyroscopes: Freier and Anderson Mu2 — 18.

3. Audio/Visual

   a. *Human Momenta, Initial Translation and Rotation* from Skylab Physics Videodisc; video disk; available from Ztek Co., PO Box 11768, Lexington, KY 40577–1768.

   b. *Conservation of Linear/Angular Momentum*; from AAPT collection 2 of single-concept films; video tape; available from Ztek Co. (see above for address).

   c. *Angular Momentum*; Side F: Angular Momentum and Modern Physics of Cinema Classics; video disk; available from Ztek Co. (see above for address).

   d. *Conservation of Angular Momentum*; Demonstrations of Physics: Energy and Momentum; VHS video tape (5:52); Media Design Associates, Inc., Box 3189, Boulder, CO 80307–3190.

e. *Why a Spinning Top Doesn't Fall*; Demonstrations of Physics: Energy and Momentum; VHS video tape (4:24); Media Design Associates, Inc. (see above for address).
f. *Conservation Laws in Zero-G*; Skylab Demonstrations; VHS video tape (18 min); Media Design Associates, Inc. (see above for address).
g. *Rotational Dynamics*; Physics Demonstrations in Mechanics, Part VI; VHS video tape or video disk; ≈3 min; Physics Curriculum & Instruction, 22585 Woodhill Drive, Lakeville, MN 55044.
h. *Conservation of Angular Momentum*; *Center of Percussion*; Physics Demonstrations in Mechanics, Part II; VHS video tape; ≈3 min; Physics Curriculum & Instruction (see above for address).

4. Computer Software
   a. *Conservation Laws*; Macintosh, Windows; Cross Educational Software, Inc., 508 E. Kentucky Avenue, PO Box 1536, Ruston, LA 71270. Includes a section on conservation of angular momentum.
   b. *Circular Motion*; Macintosh, Windows; Cross Educational Software, Inc. (see above for address).
5. Computer Project
   Given the law for the torque between two rotating rigid bodies a computer program or spreadsheet can be used to integrate Newton's second law for rotation and tabulate the angular positions and angular velocities as functions of time. The data can be used to verify the conservation of angular momentum. See the Computer Projects section of this manual.
6. Laboratory
   a. Meiners Experiment 7–12: *Rotational and Translational Motion.* Students measure the center of mass acceleration of various bodies rolling down an incline and calculate the center of mass velocities at the bottom. Results are compared to measured velocities. It is also instructive to use energy methods to find the final speeds.
   b. Meiners Experiment 7–13: *Rotational Kinematics and Dynamics.* Students find the velocity and acceleration of a ball rolling around a loop-the-loop and analyze the forces acting on it.
   c. Meiners Experiment 8–9: *Conservation of Angular Momentum.* Uses the Pasco rotational dynamics apparatus. A ball rolls down a ramp and becomes coupled to the rim of a disk that is free to rotate on a vertical axis. Students measure the velocity of the ball before impact and the angular velocity of the disk-ball system after impact, then check for conservation of angular momentum.

Chapter 13 EQUILIBRIUM AND ELASTICITY

BASIC TOPICS

I. Conditions for equilibrium.
   A. Write down the equilibrium conditions for a rigid body: $\vec{F}_{\text{net}} = 0$, $\vec{\tau}_{\text{net}} = 0$ (about any point). Remind students that only external forces and torques enter. Explain that these conditions mean that the acceleration of the center of mass and the angular acceleration about the center of mass both vanish. The body may be at rest or its center of mass may be moving with constant velocity or the body may be rotating with constant angular momentum. Point out that the equilibrium conditions form six equations that are to be solved for unknowns, usually the magnitudes of some of the forces or the angles made by some of the forces with fixed lines. Explain that we will be concerned chiefly with static

equilibrium for which $\vec{P} = 0$ and $\vec{L} = 0$. Remark that the subscript "ext" is usually omitted.

B. Show that, for a body in equilibrium, $\vec{\tau}_{\text{net}} = 0$ about *every* point.

C. Explain that the gravitational forces and torques, acting on individual particles of the body, can be replaced by a single force acting at a point called the center of gravity. If the gravitational field is uniform over the body, the center of gravity coincides with the center of mass and the magnitude of the replacement force is Mg, where M is the total mass. It points downward.

II. Solution of problems.

A. Give the problem solving steps: isolate the body, identify the forces acting on it, draw a force diagram, choose a reference frame for the resolution of the forces, choose a reference frame for the resolution of the torques, write down the equilibrium conditions in component form, and solve these simultaneously for the unknowns. Point out that the two reference frames may be different and that the reference frame for the resolution of torques can often be chosen so that one or more of the torques vanish.

B. Work examples. Consider a ladder leaning against a wall (Sample Problem 13–2) or a rock climber wedged in a crevice (Sample Problem 13–4). In each case show how the situation can be analyzed qualitatively to find the directions of the forces, then solve quantitatively.

C. Use four or more bricks to set up the situation shown in Fig. 13–36. Point out how large the overhang can be, then assign problem 24 or discuss the solution in the lecture.

III. Elasticity.

A. Point out that you have been considering mainly rigid bodies until now. Real objects deform when external forces are applied. Explain that deformations are often important for determining the equilibrium configuration of a system.

B. Consider a rod of unstrained length L subjected to equal and opposite forces F applied uniformly at each end, perpendicular to the end. Define the stress as F/A, where A is the area of an end. Define strain as the fractional change in length $\Delta L/L$ caused by the stress. Explain that stress and strain are proportional if the stress is sufficiently small. Define Young's modulus E as the ratio of stress to strain and show that $\Delta L = FL/EA$. Explain that Young's modulus is a property of the object and point out Table 13–1.

C. Explain that if the stress is small, the object returns to its original shape when the stress is removed and it is said to be *elastic*. Explain what happens if the stress is large and define *yield strength* and *ultimate strength*.

D. Calculate the fractional change in length for compressional forces acting on rods made of various materials. Use data from Table 13–1.

E. Explain that shearing occurs when the forces are parallel to the ends. Define the stress as F/A and the strain as $\Delta x/L$ where Δx is the displacement of one end relative to the other. Define the shear modulus G as the ratio of stress to strain and show that $\Delta x = FL/GA$.

F. Explain hydraulic compression. Define pressure as the force per unit area exerted by the fluid on the object. Explain that the pressure is now the stress and the fractional volume change $\Delta V/V$ is the strain. Define the bulk modulus B by $p = B\Delta V/V$.

G. To show how elastic properties are instrumental in determining equilibrium go over Sample Problem 13–6 or a similar problem.

SUGGESTIONS

1. Assignments

a. Use questions 1, 2, 3, and 4 to help students gain understanding of the equilibrium conditions in specific situations. Assign a few problems, such as 5 and 10, for which only the total force is important. Assign others, such as 9, 13, and 14, for which torque is also important. To provide a greater challenge assign a few of problems 23, 24, 28, and 29.

b. The fundamentals of elasticity are covered in problems 36 (Young's modulus), and 37 (shear). Also assign one or both of problems 38 and 39, in which the laws of elasticity are used in conjunction with the equilibrium conditions to solve for forces and their points of application.

2. Demonstrations

   Freier and Anderson Mo1, 2, 4, 6 — 9, Mp4 — 6, 9, 11, 14, 15, Mq1, 2.

3. Audio/Video

   a. *Equilibrium of Forces*; VHS video tape (30 min); Films for the Humanities & Sciences, PO Box 2053, Princeton, NJ 08543–2053.

4. Computer Software

   Statics; Macintosh, Windows; Cross Educational Software, Inc., 508 E. Kentucky Avenue, PO Box 1536, Ruston, LA 71270.

5. Laboratory

   a. Bernard Experiment 5: *Balanced Torques and Center of Gravity.* A non-uniform rod is pivoted on a fulcrum. A single weight is hung from one end and the pivot point moved until equilibrium is obtained. The data is used to find the center of gravity and mass of the rod. Additional weights are hung and equilibrium is again attained. The data is used to check that the net force and net torque vanish.

   b. Bernard Experiment 6: *Equilibrium of a Crane.* Students study a model crane: a rod attached to a wall pivot at one end and held in place by a string from the other end to the wall. Weights are attached to the crane and the equilibrium conditions are used to calculate the tension in the rod and in the string. The latter is measured with a spring balance.

   c. Meiners Experiment 7–16: *Elongation of an Elastomer* (see Chapter 7 notes).

   d. Meiners Experiment 7–17: *Investigation of the Elongation of an Elastomer with a Microcomputer.* Same as Meiners 7–16 but a microprocessor is used to plot the elongation as a function of applied force. A polar planimeter is used to calculate the work done.

Chapter 14 GRAVITATION

BASIC TOPICS

I. Newton's law of gravity.

   A. This is an important chapter. It is the first chapter devoted to a force law and its ramifications. Students get a glimpse of how a force law and the laws of motion are used together. It reviews the concepts of potential energy, angular momentum, and centripetal acceleration in the context of some important applications. In addition, the discussion of the gravitational fields of continuous mass distributions is a precursor to Gauss' law.

   B. Write down the equation for the magnitude of the force of one point mass on another. Explain that the force is one of mutual attraction and is along the line joining the masses. Give the value of G ($6.67 \times 10^{-11}\ \mathrm{N{\cdot}m^2/kg^2}$) and explain that it is a universal constant determined by experiment. If you have a Cavendish balance, show it but do not take the time to demonstrate it. As a thought experiment dealing with the magnitude of G, consider a pair of 100-kg spheres falling from a height of 100 m, initially separated by a bit more than their radii. As they fall, their mutual attraction pulls them only slightly closer together. Air resistance has more influence.

C. Explain that the same mathematical form holds for bodies with spherically symmetric mass distributions (this was tacitly assumed in B) if r is now the separation of their centers. Explain that the force on a point mass anywhere inside a uniform spherical shell is zero. (Optional: use integration to prove that this follows from Newton's law for point masses.) Use this to derive an expression for the force on a point mass inside a spherically symmetric mass distribution. See Sample Problem 14–4.

D. Point out the assumed equivalence of gravitational and inertial mass.

E. Use Newton's law of gravity to calculate the acceleration a_g due to gravity for objects near the surface of Earth and justify the use of a constant acceleration due to gravity in previous chapters. Remark that the acceleration due to gravity is independent of the mass of the body.

F. Optional: Discuss factors that influence a_g and apparent weight.

II. Gravitational potential energy.

A. Use integration to show that the gravitational potential energy of two point masses is given by $U = -GMm/r$ if the zero is chosen at $r \to \infty$. Demonstrate that this result obeys $F = -\mathrm{d}U/\mathrm{d}r$.

B. Argue that the work needed to bring two masses to positions r apart is independent of the path. Divide an arbitrary path into segments, some along lines of gravitational force and others perpendicular to the gravitational force.

C. Consider a body initially at rest far from Earth and calculate its speed when it gets to Earth's surface. Calculate the escape velocity for Earth and for the Moon.

D. Show how to calculate the gravitational potential energy of a collection of discrete masses. Warn the students about double counting the interactions — a term of the sum is associated with each *pair* of masses. Relate this energy to the binding energy of the system.

III. Planetary motion and Kepler's laws.

A. Consider a single planet in orbit about a massive sun. The center of mass for the system is essentially at the sun and it remains stationary.

B. Explain that the orbit is elliptical with the sun at one focus. This is so because the force is proportional to $1/r^2$ and the planet is bound. Draw a planetary orbit and point out the semimajor axis, the perihelion point, and the aphelion point. Define eccentricity. Show that $R_p = a(1-e)$ and $R_a = a(1+e)$, where a is the semimajor axis, R_p is the perihelion distance, and R_a is the aphelion distance.

C. Explain that the displacement vector from the sun to the planet sweeps out equal areas in equal time intervals. Sketch an orbit to illustrate. Show that the torque acting on the planet is zero because the force is along the displacement vector; then show that conservation of angular momentum leads to the equal area law. Note that the result is true for any central force.

D. For circular orbits, show that the square of the period is proportional to the cube of the orbit radius and that the constant of proportionality is independent of the planet's mass. State that the result is also true for elliptical orbits if the radius is replaced by the semimajor axis. Verify the result for planets in nearly circular orbits. The data can be found in Table 14–3.

E. For a body held by gravitational force in circular orbit about another, much more massive body, show that the kinetic energy is proportional to $1/r$ and that the total mechanical energy is $-GMm/2r$. Explain that the energy is zero for infinite separation with the bodies at rest, that a negative energy indicates a bound system, and that a positive energy indicates an unbound system. Describe the orbits of recurring and non-recurring comets. Explain that the expression for the energy is valid for elliptical orbits if r is replaced by the semimajor axis. Remark that the energy of a satellite cannot be altered without changing the semimajor axis of its orbit.

F. Remark that the laws of planetary motion hold for moons (including artificial satellites) traveling around planets, binary star systems, stars traveling around the center of a galaxy, and for galaxies in clusters. Explain that when the masses of the two objects are comparable, both objects travel around the center of mass and it is the relative displacement that obeys Kepler's laws. When discussing stars in galaxies you might show how the law of periods has been used to argue for the existence of dark matter.

SUPPLEMENTARY TOPICS

1. Detailed calculations of the gravitational force of a spherical distribution of mass on a point mass.
2. The general theory of relativity. Include a discussion of the distinction between gravitational and inertial mass.

SUGGESTIONS

1. Assignments
   a. To stress Newton's force law, ask question 1 and assign problem 1. Also assign problem 15 to test if students know the source of the value for a_g. To discuss symmetry, ask questions 2 and 4.
   b. Use one or two of problems 8 through 12 to test for understanding of the superposition principle.
   c. Discuss problems 2, 3, and 23 in connection with calculations of the gravitational force of a spherically symmetric mass distribution on a point mass. Problem 22 is fundamental to the shell theorem.
   d. The essentials of gravitational potential energy are covered in problems 26 and 31. Conservation of mechanical energy is important for the solution to problems 32 and 39. Some of these can be used later as models for electrostatic potential energy. Questions 7 and 8 cover some important qualitative aspects of gravitational work and potential energy. Escape velocity and energy are covered in many problems. Consider problems 29, 30, 33, and 35.
   e. To discuss planetary orbits assign some of problems 41, 46, 47, 54, 58, 59, and 60.
2. Books and Monographs
   Measurements of Newtonian Gravitation, edited by George T. Gillies. A collection of journal articles covering the measurement of G and the measurement of gravitational effects.
3. Audio/Visual
   a. *Retrograde Motion — Heliocentric Model and Geocentric Model*; *Kepler's Laws*; *Jupiter Satellite Orbits*; from AAPT collection 1 of single-concept films; video tape; available from Ztek Co., PO Box 11768, Lexington, KY 40577–1768.
   b. *Measurement of "G" — The Cavendish Experiment*; from AAPT collection 2 of single-concept films; video tape; available from Ztek Co., PO Box 11768, Lexington, KY 40577–1768.
   c. *Planetary Motion*; Side B: Mechanics (II) and Heat of Cinema Classics; video disk; available from Ztek Co. (see above for address).
   d. *Zero-G*; Skylab Demonstrations; VHS video tape (15 min); Media Design Associates, Inc., Box 3189, Boulder, CO 80307–3190.
   e. *The Determination of the Newtonian Constant of Gravitation*; VHS video tape (15 min); Films for the Humanities & Sciences, PO Box 2053, Princeton, NJ 08543–2053.

f. *Newton's Law of Universal Gravitation*; Physics Demonstrations in Mechanics, Part IV; VHS video tape or video disk; ≈3 min; Physics Curriculum & Instruction, 22585 Woodhill Drive, Lakeville, MN 55044.

4. Computer Software

   a. *Orbits*; James B. Harold, Kenneth Hennacy, and Edward Redish; Windows; available from Physics Academic Software, North Carolina State University, PO Box 8202, Raleigh, NC 27690–0739. Calculates and plots the trajectories of up to seven bodies. Two can be massive and influence the motions of the others. The rest are light. The user can change the value of the gravitational constant and can shift the view to various reference frames.

   b. *Orbital Maneuvers*; Windows; Cross Educational Software, Inc., 508 E. Kentucky Avenue, PO Box 1536, Ruston, LA 71270.

   c. *Orbits*; Vernier Software, 8565 S.W. Beaverton-Hillside Hwy., Portland OR, 97225–2429. The student can place a satellite in orbit and change the orbit with thruster rockets. The motion is plotted and the position and speed of the satellite are continuously displayed.

   d. *Objects in Motion*. See Chapter 2 SUGGESTIONS.

5. Computer Project

   Have students use a spreadsheet or write a computer program to integrate Newton's second law for a $1/r^2$ central force and use it to investigate satellite motion. Some projects are described in the Computer Projects section of this manual.

6. Laboratory

   Meiners Experiment 7–21: *Analysis of Gravitation.* Students use the Leybold-Heraeus Cavendish torsional balance to determine G. Requires extremely careful work and a solid vibration free wall to mount the apparatus.

Chapter 15 FLUIDS

BASIC TOPICS

I. Pressure and density.

   A. Introduce the subject by giving a few examples of fluids, including both liquids and gases. Remark that fluids cannot support shear.

   B. Define density as the mass per unit volume in a region of the fluid. Point out that the limit is a macroscopic limit: the limiting volume still contains many atoms. The density is a scalar and is a function of position in the fluid.

   C. Explain that fluid in any selected volume exerts a force on the material across the boundary of the volume. The boundary may be a mathematical construct and the material on the other side may be more of the same fluid. The boundary may also be a container wall or an interface with another fluid. Explain that, for a small segment of surface area, the force exerted by the fluid is normal to the surface and is proportional to the area. The pressure is the force per unit area and $\vec{F} = p\vec{A}$, where the magnitude of $\vec{A}$ is the area and the direction of $\vec{A}$ is outward, normal to the surface. Units: Pa (= N/m^2), atmosphere, bar, torr, mm of Hg. Give the conversions or point out Appendix D in the text.

   D. Show that in equilibrium with y measured positive above some reference height $\mathrm{d}p/\mathrm{d}y = -\rho g$, where ρ is the fluid density. Then, note that $p_2 - p_1 = -\int \rho g\,\mathrm{d}y$, where the integral limits are y_1 and y_2. Point out that the difference in pressure arises because a fluid surface is supporting the fluid above it. Finally, point out that if the fluid is incompressible and homogeneous, then ρ is a constant. If $y_2 - y_1$ is sufficiently small that g is also constant, then $p_2 - p_1 = -\rho g(y_2 - y_1)$. Point out that if p_0 is the surface pressure, then the pressure a distance h below the surface is $p = p_0 + \rho gh$. Note that the pressure is the same at all points at the same depth in the fluid. Explain that p_0 is atmospheric pressure if the surface is open to the air and is zero if the fluid is in a tube with the region above the surface evacuated.

E. Connect a length of rubber tubing to one arm of a U-tube partially filled with colored water. Blow into the tube, then suck on it. In each case note the change in water level. Insert the U-tube into a deep beaker of water, with the free end of the tubing out of the water. As the open end is lowered, the change in the level of the colored water will indicate the increase in pressure. Go over Sample Problem 15–3 to show the equilibrium positions of two immiscible liquids of different densities. Show how to obtain the pressure at the top of one arm in terms of the pressure at the top of the other arm, the densities, and the quantities of fluids. Point out that the pressures are the same and are the atmospheric pressure if the U-tube is open. Explain that the pressure is always the same at two points that are at the same height *and* can be joined by a line along which neither ρ nor g vary. Use the diagram associated with the problem to point out two places at the same height where the pressure is the same and two places at the same height where the pressures are different.

II. Measurement of pressure.

A. This section not only describes some pressure measuring instruments but also provides some applications of previous material, especially the variation of pressure with depth in a fluid.

B. Show a mercury barometer. A lens system or an overhead projector suitably propped on its side can be used to project an image of the mercury column on a screen for viewing by a large class. Use $p = p_0 + \rho g h$ to show why the height of the column is proportional to the pressure at the mercury pool. Emphasize that the pressure at the top of the column is nearly zero and that this is important for the operation of the barometer.

C. Show a commercial open-tube manometer or explain that such an instrument is similar to the U-tube demonstration done earlier. Explain gauge pressure and emphasize that the instrument measures gauge pressure.

III. Pascal's and Archimedes' principles.

A. State Pascal's principle. Start with $p = p_0 + \rho g h$, consider a change in p_0, and show $\Delta p = \Delta p_0$ if the fluid is incompressible. You can demonstrate the transmission of pressure with a soda bottle full of water, fitted with a tight rubber stopper. Wrap a towel around the neck of the bottle and hit the stopper sharply. With some practice you can blow the bottom out of the bottle cleanly.

B. Apply the principle to a hydraulic jack. Show that $F_1/A_1 = F_2/A_2$. Also explain that if the fluid is incompressible, F_1 and F_2 do the same work. The point of application of the smaller force moves the greater distance. A hydraulic jack can be made from a hot water bottle, fitted with a narrow rubber tube. Put the bottle on the floor and fasten the tube to a tall ringstand so it is vertical. Place a thin wooden board on the bottle to distribute the weight and have a student stand on it. To change the pressure, use a plunger or rubber squeeze ball from an atomizer or blow into the tube.

C. State Archimedes' principle. Stress that the force is due to the surrounding fluid. Contrast the case of an immersed body surrounded by fluid with one placed on the bottom of the container. Consider a flat board floating on the surface of a liquid, compute the net upward force in terms of the difference in pressure and use $p = p_0 + \rho g h$ to show that this is the weight of the displaced liquid.

D. Explain why some objects sink while others float.

E. Fill a large mouthed plastic vessel with water precisely up to an overflow pipe. Immerse a dense object tied by string to a spring balance. Weigh the object while it is immersed and weigh the displaced water. Observe that the buoyant force is the same as the weight of the displaced water.

F. Explain that for purposes of calculating torque the buoyant force can be taken to act through the center of gravity of the fluid that will be displaced by the object. The force of gravity acts through the center of gravity of the object. Show that these points may not be the same and that the two forces may produce a net torque. Show the relative positions of the center of buoyancy and the center of gravity for stable and unstable equilibrium.

IV. Fluids in motion.

A. Describe:

1. Steady and non-steady flow. Emphasize that the velocity and density fields are independent of time if the flow is steady. They may depend on position, however.
2. Compressible and incompressible flow. Emphasize that the density is independent of both position and time if the flow is incompressible.
3. Rotational and irrotational flow.
4. Viscous and nonviscous flow.

B. Describe streamlines for steady flow and point out that streamlines are tangent to the fluid velocity and that no two streamlines cross. Remark that the velocity is not necessarily constant along a streamline. Describe a tube of flow as a bundle of streamlines. Sketch a tube of flow with streamlines far apart at one end and close together at the other. Explain that since streamlines do not cross the boundaries of a tube of flow they are close together where the tube is narrow and far apart where the tube is wide. Remark that particles do not cross the boundaries of a tube of flow.

V. Equation of continuity

A. Define volume flow rate (volume flux) and mass flow rate (mass flux). Consider a tube of flow with cross-sectional area A at one point and give the physical significance of $A\rho v$ and Av. Remark that the first can be measured in kg/s and the second in m^3/s. Show how to convert m^3/s to gal/s and li/s.

B. State the equation of continuity: $A\rho v =$ constant along a streamline if there are no sources or sinks of fluid and if the flow is steady. Argue that if the equation were not true there would be a build up or depletion of fluid in some regions and the flow would not be steady.

C. Discuss the special case of an incompressible fluid and explain that the fluid speed is great where the tube of flow is narrow and vice versa. Point out that the fluid velocity is great where the streamlines are close together and small where they are far apart. Use the diagram of section IVB above as an example.

VI. Bernoulli's equation.

A. Apply the work-kinetic energy theorem to a tube of flow to show that for steady, nonviscous, incompressible flow $p + \frac{1}{2}\rho v^2 + \rho g y =$ constant along a streamline. Point out that this equation also gives the pressure variation in a static fluid ($v = 0$ everywhere).

B. Remark that a typical fluid dynamics problem gives the conditions v, p, y at one point on a streamline and asks for conditions at another. The equation of continuity and Bernoulli's equation can be solved simultaneously for two quantities.

C. Work a sample problem. Consider horizontal flow ($y =$ const) through a pipe that narrows. Give the fluid velocity where the pipe is wide and use the equation of continuity to calculate the velocity where it is narrow. Then, use Bernoulli's equation to calculate the pressure difference. Emphasize that the pressure must decrease to provide the force that accelerates the fluid as it passes into the narrow region.

D. Now work the same problem but suppose the height of the pipe increases along the direction of flow. Point out the difference in the answers for the pressure.

SUGGESTIONS

1. Assignments
   a. Use question 1 to discuss pressure. Problems 2, 3, and 6 cover the definition of pressure. Problems 9, 11 and 14 deal with the variation of pressure with depth. Problem 19 includes torque.
   b. Use question 4 and problems 22 and 23 in connection with Pascal's principle.
   c. Questions 5 through 9 all provide good examples of Archimedes' principle. Pick several to illustrate applications of the principle. Also assign problems 24 and 25 and some of problems 29 through 38.
   d. Use question 10 and problems 40 and 42 as part of the discussion of the equation of continuity.
   e. The fundamentals of Bernoulli's equation are covered in problems 43, 44, and 45. Also consider problems 52, 54, and 55. Some of these require students to combine the equation of continuity and Bernoulli's equation. Work one or two of these as examples in lecture and assign others.
2. Demonstrations
   a. Force and pressure: Freier and Anderson Fa, Fb, Fc, Fd, Fe, Ff, Fh.
   b. Archimedes' principle: Freier and Anderson Fg.
   c. Bernoulli's principle: Freier and Anderson Fj, Fl1.
3. Books and Monographs
   Hydraulic Devices; by Malcolm Goldber, John P. Ouderkirk, and Bruce B. Marsh ; available from AAPT, One Physics Ellipse, College Park MD 20740–3845.
4. Audio/Visual
   a. *Why Divers Exhale While Surfacing*; Demonstrations of Physics: Liquids and Gases; VHS video tape (2:25); Media Design Associates, Inc., Box 3189, Boulder, CO 80307–3190.
   b. *Archimedes' Principle for Gases*; Demonstrations of Physics: Liquids and Gases; VHS video tape (3:18); Media Design Associates, Inc. (see above for address).
   c. *The Cartesian Diver*; Demonstrations of Physics: Liquids and Gases; VHS video tape (6:27); Media Design Associates, Inc. (see above for address).
   d. *Bernoulli's Equation and Streamlines*; Demonstrations of Physics: Liquids and Gases; VHS video tape (5:33); Media Design Associates, Inc. (see above for address).
   e. *Fluids in Weightlessness*; Skylab Demonstrations; VHS video tape (15 min); Media Design Associates, Inc. (see above for address).
   f. *Pressure*; VHS video tape (30 min); Films for the Humanities & Sciences, PO Box 2053, Princeton, NJ 08543–2053.
   g. *Atmospheric Pressure*; VHS video tape (9 min); Films for the Humanities & Sciences (see above for address).
   h. *Fluid Pressure*; VHS video tape (9 min); Films for the Humanities & Sciences (see above for address).
   i. *Liquid Drops*, *Water Bridges*, *Soap and Water* from AAPT Skylab videotape; American Association of Physics Teachers, One Physics Ellipse, College Park, MD 20740–3845.
5. Computer Software
   Fluids; Macintosh, Windows; Cross Educational Software, Inc.; 508 E. Kentucky Avenue, PO Box 1536, Ruston, LA 71270.
6. Laboratory
   a. Meiners Experiment 7-7: *Radial Acceleration* (Problem II only). Students measure the orbit radii of various samples floating on the surface of water in a spinning globe and

analyze the forces on the samples. This experiment is an application of buoyancy forces to rotational motion.

b. Bernard Experiment 16: *Buoyancy of Liquids and Specific Gravity.* Archimedes' principle is checked by weighing the water displaced by various cylinders. Buoyant forces are measured by weighing the cylinders in and out of water. The same cylinder is immersed in various liquids and the results are used to find the specific gravities of the liquids.

Chapter 16 OSCILLATIONS

BASIC TOPICS

I. Oscillatory motion.

A. Set up an air track and a cart with two springs, one attached to each end. Mark the equilibrium point, then pull the cart aside and release it. Point out the regularity of the motion and show where the speed is the greatest and where it is the least. By reference to the cart define the terms periodic motion, equilibrium point, period, frequency, cycle, and amplitude.

B. Explain that $x(t) = x_m \cos(\omega t + \phi)$ describes the coordinate of the cart as a function of time if $x = 0$ is taken to be the equilibrium point, where the force of the springs on the cart vanishes. State that this type motion is called simple harmonic. Show where $x = 0$ is on the air track, then show what is meant by positive and negative x. Sketch a mass on the end of a single spring and explain that the mass also moves in simple harmonic motion if dissipative forces are negligible.

C. Discuss the equation for $x(t)$.

1. Explain that x_m is the maximum excursion of the mass from the equilibrium point and that the spring is compressed by x_m at one point in a cycle. x_m is called the *amplitude* of the oscillation. Explain that the amplitude depends on initial conditions. Draw several $x(t)$ curves, identical except for amplitude. Illustrate with the air track apparatus.

2. Note that ω is called the angular frequency of the oscillation and is given in radians/s. Define the *frequency* by $f = \omega/2\pi$ and the *period* by $T = 1/f$. Show that $T = 2\pi/\omega$ is in fact the period by direct substitution into $x(t)$; that is, show $x(t) = x(t+T)$. Explain that the angular frequency does not depend on the initial conditions. For the cart on the track, use a timer to show that the period, and hence ω, is independent of initial conditions. Draw several $x(t)$ curves, for oscillations with different periods. Replace the original springs with stiffer springs and note the change in period. Also replace the cart with a more massive cart and note the change in period.

3. Define the phase of the motion and explain that the phase constant ϕ is determined by initial conditions. Draw several $x(t)$ curves, identical except for ϕ, and point out the different conditions at $t = 0$. Remark that the curves are shifted copies of each other. Illustrate various initial conditions with the air track apparatus.

D. Derive expressions for the velocity and acceleration as functions of time for simple harmonic motion. Show that the speed is a maximum at the equilibrium point and is zero when $x = \pm x_m$. Also show that the magnitude of the acceleration is a maximum when $x = \pm x_m$ and is zero at the equilibrium point. Relate these results to $F(x)$.

E. Show that the initial conditions are given by $x_0 = x_m \cos\phi$ and $v_0 = -x_m\omega \sin\phi$. Solve for x_m and ϕ: $x_m^2 = x_0^2 + v_0^2/\omega^2$ and $\tan\phi = -v_0/\omega x_0$. Calculate x_m and ϕ for a few special cases: $x_0 = 0$ and v_0 positive, $x_0 = 0$ and v_0 negative, x_0 positive and $v_0 = 0$, x_0 negative and $v_0 = 0$. Tell students how to test the result given by a calculator for ϕ to see if π must be added to it.

II. The force law.

A. State the force law for an ideal spring: $F = -kx$. Point out that the negative sign is necessary for the force to be a restoring force. Hang identical masses on springs with different spring constants, measure the elongations, and calculate the spring constants. Remark that stiff springs have larger spring constants than weak springs. Remark that the expression for the force is an idealization. It is somewhat different for real springs.

B. Start with Newton's second law and derive the differential equation for $x(t)$. Show that $x = x_m \cos(\omega t + \phi)$ satisfies the equation if $\omega = \sqrt{k/m}$ and explain that this is the most general solution for a given spring constant and mass.

C. Show a vertical spring-mass system. Point out that the equilibrium point is determined by the mass, force of gravity, and the spring constant. Show, both analytically and with the apparatus, that the force of gravity does not influence the period, phase, or amplitude of the oscillation.

III. Energy considerations.

A. Derive expressions for the kinetic and potential energies as functions of time. Show that the total mechanical energy is constant by adding the two expressions and using the trigonometric identity $\sin^2\alpha + \cos^2\alpha = 1$. Remark that the energy is wholly kinetic at the equilibrium point and wholly potential at a turning point. It changes from one form to the other as the mass moves between these points.

B. Show how to use the conservation of energy to find the amplitude, given the initial position and velocity, to find the maximum speed, and to find the speed as a function of position.

IV. Applications.

A. Demonstrate a torsional pendulum and discuss it analytically. Derive the differential equation for the angle as a function of time and compare with the differential equation for a spring to obtain the angular frequency and period in terms of the spring constant and the rotational inertia.

B. Demonstrate a simple pendulum and discuss it analytically in the small amplitude approximation. Derive the differential equation for the angle as a function of time and obtain expressions for the angular frequency and period from the equation. Emphasize that the angular displacement must be measured in radians for the small amplitude approximation to be valid. Have students use their calculators to find the sines of some angles, in radians, starting with large angles and progressing to small angles.

C. Demonstrate a physical pendulum. Use Newton's second law for rotation to obtain the differential equation for the angular displacement. Obtain expressions for its angular frequency and period in the small amplitude approximation. Remind students that the rotational inertia depends on the position of the pivot and show them how to use the parallel axis theorem to find its value.

V. Simple harmonic motion and uniform circular motion.

A. This section is particularly important if you intend to include any of Chapters 33, 36, and 37 in the course.

B. Mount a bicycle wheel vertically and arrange for it to be driven slowly with uniform angular speed. Attach a tennis ball to the rim and project the shadow of the ball on the wall. Note that the shadow moves up and down in simple harmonic motion. Point out that the period of the wheel and the period of the shadow are the same. It is possible to suspend a mass on a spring near the wall and adjust the angular speed and initial conditions so the mass and shadow move together for several cycles. A period of about 1 s works well.

C. Analytically show that the projection of the position vector of a particle in uniform circular motion undergoes simple harmonic motion. Mention the converse: if an object simultaneously undergoes simple harmonic motion in two orthogonal directions, with the same amplitude and frequency, but a $\pi/2$ phase difference, the result is a circular orbit.

VI. Damped and forced harmonic motion.

A. Write the differential equation for a spring-mass oscillator with a damping term proportional to the velocity. Treat the case $(b/2m)^2 < k/m$ and write the solution, including the expression for the angular frequency in terms of k, m, and b. If there is time, prove it is the solution by direct substitution into the differential equation or leave the proof as an exercise for the students. Remark that the natural angular frequency is nearly $\sqrt{k/m}$ if damping is small.

B. Show a graph of the displacement as a function of time. See Fig. 16–16. Point out the exponential decay of the amplitude. Mention that the oscillator loses mechanical energy to dissipative forces.

C. Explain that if $(b/2m)^2 > k/m$ then the mass does not oscillate but rather moves directly back to the equilibrium point. The displacement is a decreasing exponential function of time. To demonstrate under- and over-damping, attach a vane to a pendulum. Experiment with the size so the pendulum oscillates in air but does not when the vane is in water.

D. Write the differential equation for a forced spring-mass system, including a damping term. Assume an applied force of the form $F_m \cos(\omega_f t)$ and point out that ω_f is not necessarily the same as the natural angular frequency of the oscillator.

E. Mention that when the system is first started transients are present and the motion is somewhat complicated. However, it settles down to a sinusoidal motion with an angular frequency that is the same as that of the impressed force.

F. Also point out that in steady state the amplitude is constant in time but that it depends on the frequency of the applied force. Illustrate with Fig. 16–20, which shows the amplitude as a function of the forcing frequency for various values of the damping coefficient. Mention that the amplitude is the greatest when the forcing frequency nearly matches the natural frequency and say this is the resonance condition. Also mention that at resonance the amplitude is greater for smaller damping and that small damping produces a sharper resonance than large damping.

G. Resonance can be demonstrated with three identical springs and two equal masses, as shown. Fasten the bottom spring to a heavy weight on the floor and drive the upper spring by hand (perhaps standing on a table). Obtain resonance at each of the normal modes (masses moving in the same and opposite directions). After showing the two resonances, drive the system at a low frequency to show a small response, then drive it at a high frequency to again show a small response. Repeat at a resonance frequency to show the larger response. To show pronounced damping effects, attach a large stiff piece of aluminum plate to each mass.

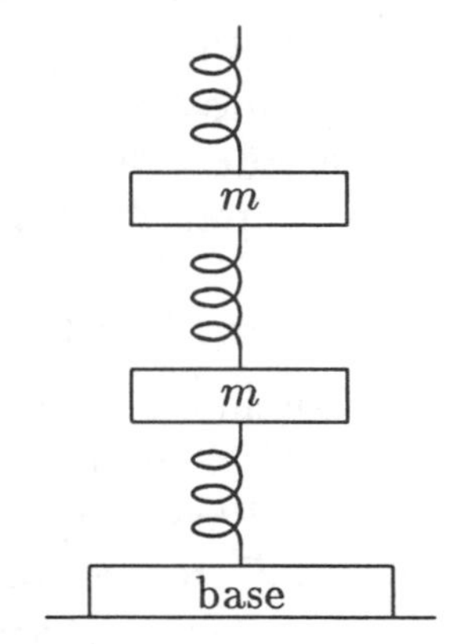

SUGGESTIONS

1. Assignments

a. Ask some of questions 1 through 8 as part of the discussion of the conditions for simple harmonic motion and of the parameters of that motion. Also assign problems 16, 17, and 32.

b. Assign question 10 and problems 3 and 14 in support of the spring-mass demonstration and discussion. Ask question 9 and assign problem 16. Problems 18 and 21 deal with vertical oscillators. Assign one of these for variety.

c. Springs in parallel and series test understanding of the spring force law. Consider problems 24, 25, and 27. Also consider problems 28 and 30.
d. Assign problems 33 and 37 in support of the discussion of energy. When assigning problem 33 also ask for the maximum speed of the mass.
e. If oscillators other than a spring-mass system are considered, assign problems 40 and 41 (angular simple harmonic motion), 43 and 44 (simple pendulum), and 45 and 52 (physical pendulum). Problem 58 is more challenging.
f. Use problems 61 and 62 to test for understanding of damped harmonic motion and problems 63 and 64 to test for understanding of forced harmonic motion.

2. Computer Project

Have students use the numerical integration program developed in connection with Chapter 6 to investigate forced and damped harmonic motion. Have them plot the coordinate as a function of time to see transients. To study resonance, have them plot the amplitude as a function of forcing frequency. They can also investigate the influence of damping on the amplitude and resonance width. Details are given in the Computer Projects section of this manual.

3. Demonstrations
   a. Simple harmonic motion: Freier and Anderson Mx1, 2, 3, 4, 7.
   b. Pendulums: Freier and Anderson Mx6, 9, 10, 11, 12, My1, 2, 3, 8, Mz1, 2, 3, 6, 7, 9.

4. Audio/Visual
   a. *Simple Harmonic Motion*; *The Stringless Pendulum*; *Sand Pendulum*; from AAPT collection 2 of single-concept films; video tape; available from Ztek Co., PO Box 11768, Lexington, KY 40577–1768.
   b. *Oscillations* from Skylab Physics Videodisc; video disk; available from Ztek Co. (see above for address).
   c. *Tacoma Narrows Bridge Collapse*; from AAPT Miller collection of single-concept films; video tape; available from Ztek Co. (see above for address).
   d. *Coupled Oscillators — Equal Masses*; *Coupled Oscillators — Unequal Masses*; from AAPT Miller collection of single-concept films; video tape; available from Ztek Co. (see above for address).
   e. *Periodic Motion*; Side B: Mechanics (II) and Heat of Cinema Classics; video disk; available from Ztek Co. (see above for address).
   f. *The Tacoma Narrows Bridge Collapse*; video disk; American Association of Physics Teachers, One Physics Ellipse, College Park, MD 20740–3845.

5. Computer Software
   a. *Dynamic Analyzer*. See Chapter 5 SUGGESTIONS.
   b. *Physics of Oscillation*; Eugene L. Butikov; Windows; Physics Academic Software, North Carolina State University, PO Box 8202, Raleigh, NC 27690–0739.
   c. *Simple Harmonic Motion*; Socrates Software, 6187 Rosecommon Drive, Norcross, GA 30092. User controls the parameters of mass-spring oscillator. The motion is shown and graphed.

6. Laboratory
   a. Meiners Experiment 7–2: *The Vibrating Spring*. Students time a vertical vibrating spring with various masses attached, then use the data and a logarithmic plot to determine the relationship between the period and mass.
   b. Bernard Experiment 15: *Elasticity and Vibratory Motion*. The experiment is much the same as Meiners 7–2, in that a graph is used to determine the relationship between the

mass on a spring and the period of oscillation. This measurement is preceded by a static determination of the spring constant.

c. Meiners Experiment 7–4: *The Vibrating Ring.* Students time the oscillations of various diameter rings, hung on a knife edge, then use the data and a logarithmic plot to determine the relationship between the period and ring diameter. A good example of a physical pendulum.

d. Meiners Experiment 7–15: *Investigation of Variable Acceleration.* A pendulum swings above a track and a spark timer is used to record its position as a function of time. Its velocity and acceleration are investigated.

e. Meiners Experiment 7–19: *Harmonic Motion Analyzer.* This apparatus allows students to vary the spring constant, mass, driving frequency, driving amplitude, and damping coefficient of a spring-mass system. They can measure the amplitude, period, and relative phase of the oscillating mass. A variety of experiments can be performed.

f. Meiners Experiment 8–4: *Linear Oscillator.* A spark timer is used to record the position of an oscillating mass on a spring, moving horizontally on a nearly frictionless surface. The period as a function of mass can be investigated and the conservation of energy can be checked.

g. Meiners Experiment 7–18: *Damped Driven Linear Oscillator.* The amplitude and relative phase of a driven damped spring-mass system are measured as functions of the driving frequency and are used to plot a resonance curve.

h. Meiners Experiment 7–20: *Analysis of Resonance with a Driven Torsional Pendulum.* The driving frequency and driving amplitude of a driven damped torsional pendulum are varied and the frequency, amplitude, and relative phase are measured. Damping is electromagnetic and can be varied or turned off. A variety of experiments can be performed.

Chapter 17 WAVES — I

BASIC TOPICS

I. General properties of waves.

A. Explain that wave motion is the mechanism by which a disturbance created at one place travels to another. Use the example of a pulse on a taut string and point out that the displaced string causes neighboring portions of the string to be displaced. Stress that the individual particles have limited motion (perhaps perpendicular to the direction of wave travel), whereas the pulse travels the length of the string. Demonstrate by striking a taut string stretched across the room. Point out that energy is transported by the wave from one place to another. Ask the students to read the introductory section of the chapter for other examples of waves.

B. Point out that a wave on a string travels in one dimension, water waves produced by dropping a pebble travel in two, and sound waves emitted by a point source travel in three.

C. Explain the terms longitudinal and transverse. Demonstrate longitudinal waves with a slinky.

D. State that waves on a taut string of uniform density travel with constant speed and that this course deals chiefly with idealized waves that do not change shape. Take the string to lie along the x axis and draw a distortion in the shape of a pulse, perhaps a sketch of $\exp[-\alpha(x-x_0)^2]$. Remark that the initial displacement of the string can be described by giving a function $f(x)$. Now suppose the pulse moves in the positive x direction and draw the string at a later time. Point out that the maximum has moved from x_0 to x_0+vt, where v is the wave speed. Remark that the displacement can be calculated by substituting $x-vt$ for x in the function $f(x)$. Substantiate the remark by showing that $x-vt=x_0$ if

x is the coordinate of the pulse maximum at time t. Explain that $x + vt$ is substituted if the pulse travels in the negative x direction. Emphasize the relative signs of kx and ωt.

II. Sinusoidal traveling waves.

A. Write $f(x) = y_m \sin(kx)$ for the initial displacement of the string and sketch the function. Identify the amplitude as giving the limits of the displacement and point it out on the sketch. Also point out the periodicity of the function and identify the wavelength on the sketch. Show that k must be $2\pi/\lambda$ for $f(x)$ to equal $f(x + n\lambda)$ for all integers n. Remark that k is called the angular wave number of the wave.

B. Substitute $x - vt$ for x in $f(x)$ and explain you will assume the wave travels in the positive x direction. Show that the result is $y(x,t) = y_m \sin(kx - \omega t)$, where $\omega = kv$.

C. State that the motion of the string at any point is simple harmonic and that ω is the angular frequency. Show that at a given place on the string the motion repeats in a time equal to $2\pi/\omega$. This is the period T. Remind students that the frequency is $f = 1/T = \omega/2\pi$.

D. Remark that any given point on the string reaches its maximum displacement whenever a maximum on the wave passes that point. Since the time interval is one period a sinusoidal wave travels one wavelength in one period and $v = \lambda/T = \lambda f = \omega/k$, in agreement with the derivation of $y(x,t)$.

E. Explain that $y(x,t) = y_m \sin(kx + \omega t)$ represents a sinusoidal wave traveling in the negative x direction.

F. Show that the string velocity is $u(x,t) = \partial y/\partial t = -\omega y_m \cos(kx - \omega t)$. Point out that x is held constant in taking the derivative since the string velocity is proportional to the difference in the displacement of the *same* piece of string at two slightly different times. Remark that different points on the string may have different velocities at the same time and the same point may have a different velocity at different times. Contrast this behavior with that of the wave velocity. Point out that for a transverse wave u is transverse.

G. Explain that the wave speed for an elastic medium depends on the inertia and elasticity of the medium. State that, for a taut string, $v = \sqrt{\tau/\mu}$, where τ is the tension in the string and μ is the linear density of the string. Show how to measure μ for a homogeneous, constant radius string. The expression for v may be derived as in Section 17–6 of the text.

H. Point out that the frequency is usually determined by the source and that doubling the frequency for the same string with the same tension halves the wavelength. The product λf remains the same. Remark that if a wave goes from one medium to another the speed and wavelength change but the frequency remains the same. Work an example: given the two densities and the frequency, calculate the wave speed and wavelength in each segment. Draw a diagram of the wave.

III. Energy considerations.

A. Point out that the energy in the wave is the sum of the kinetic energy of the moving string and the potential energy the string has because it is stretched in the region of the disturbance. Energy moves with the disturbance.

B. Show that the kinetic energy of an infinitesimal segment of string is given by $\mathrm{d}K = \frac{1}{2}\,\mathrm{d}m\, v^2 = \frac{1}{2}(\mu\, \mathrm{d}x)(\omega^2 y_m^2)\cos^2(kx - \omega t)$. State that this energy is transported to a neighboring portion in time $\mathrm{d}t = v\,\mathrm{d}x$, so $\mathrm{d}K/\mathrm{d}t = \frac{1}{2}\mu v \omega^2 y_m^2 \cos^2(kx - \omega t)$ gives the rate at which kinetic energy is transported past the point with coordinate x, at time t. Explain that when this averaged over a cycle the result is $(\mathrm{d}K/\mathrm{d}t)_{\mathrm{avg}} = \frac{1}{4}\mu v \omega^2 y_m^2$. Remark that this is not zero. Although kinetic energy moves back and forth as the string oscillates, there is a net flow.

C. State without proof that the average rate at which potential energy is transported is exactly the same as the rate for kinetic energy, so the average rate of energy flow is

$P_{avg} = \frac{1}{2}\mu v\omega^2 y_m^2$. Note that this depends on the square of the amplitude and on the square of the frequency.

IV. Superposition and interference.

A. Stress that displacements, not intensities, add. State that if y_1 and y_2 are waves that are simultaneously present, then $y = y_1 + y_2$ is the resultant wave. Using diagrams of two similar sinusoidal waves, show that the resultant amplitude can be twice the amplitude of one of them, can vanish, or can have any value in between. Mention that the medium must be linear.

B. Start with the waves $y_1 = y_m \sin(kx - \omega t + \phi)$ and $y_2 = y_m \sin(kx - \omega t)$ and show that $y = 2y_m \cos(\phi/2) \sin(kx - \omega t + \phi/2)$. Show that maximum constructive interference occurs if $\phi = 2n\pi$, where n is an integer and maximum destructive interference occurs if $\phi = (2n+1)\pi$, where n is again an integer. Remark that the maximum amplitude is $2y_m$ and the minimum is zero. The derivation depends heavily on the trigonometric identity given as Eq. 17–37. You may wish to verify this identity for the class. Use the expressions for the sine and cosine of the sum of two angles to expand the right side of Eq. 17–37.

C. Explain that a phase difference can arise if waves start in phase but travel different distances to get to the same point. Show that $\phi = k\Delta x$ and find expressions for the path differences that result in fully constructive and fully destructive interference. In the first case Δx is a multiple of λ, while in the second it is an odd multiple of $\lambda/2$.

D. Interference can easily be demonstrated with a monaural amplifier, a signal generator, a microphone, an oscilloscope, and a pair of speakers. Fix the position of speaker S_1 and, with S_2 disconnected, show the wave form on the oscilloscope. Then, connect S_2 and show the wave form as S_2 is moved. Because both speakers are driven by the same amplifier, the only phase difference is due to the path difference.

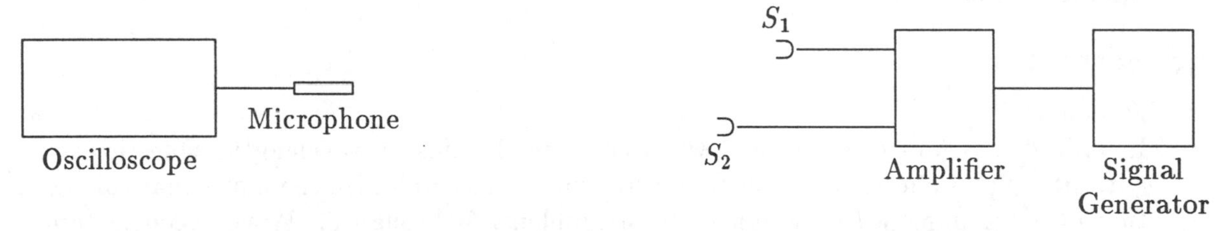

E. Explain that a phasor is an arrow that rotates around the location of its tail. Its length, to some scale, is taken to be the amplitude of a sinusoidal traveling wave and its angular velocity is taken to the angular frequency of the wave. Show that its projection on an axis through the tail behaves like the displacement in a wave. Point out the significance of the phase constant ϕ for the phasor rotation. Show how to use phasors to add two sinusoidal waves with the same frequency and wavelength but with different amplitudes and phase constants. Develop the expression $y_m^2 = y_{1m}^2 + y_{2m}^2 + 2y_{1m}y_{2m}\cos\phi$ for the amplitude of the resultant wave. Show how to use the law of sines to obtain the phase constant of the resultant wave.

V. Standing waves.

A. Use a mechanical oscillator to set up a standing wave pattern on a string. Otherwise, draw the pattern. Point out nodes and antinodes. Explain that all parts of the string vibrate either in phase or 180° out of phase and that the amplitude depends on position along the string. The disturbance does not travel. If possible, use a stroboscope to show the standing wave pattern. CAUTION: students with epilepsy should not watch this demonstration.

B. Explain that a standing wave can be constructed from two sinusoidal traveling waves of the same frequency and amplitude, traveling with the same speed in opposite directions.

Use the trigonometric identity of Eq. 17–37 to show that $y_1 + y_2 = 2y_m \sin(kx)\cos(\omega t)$ if the phase constant for each wave is zero. Find the coordinates of the nodes and show they are half a wavelength apart. Also find the coordinates of the antinodes and show they lie halfway between nodes.

C. Point out that standing waves can be created by a wave and its reflection from a boundary. By means of a diagram show how the incident and reflected waves cancel at the fixed end of a string.

D. Remark that for a string fixed at both ends, each end must be a node. Derive the expression for the standing wave frequencies of such a string. Draw diagrams showing the string at maximum displacement for the lowest three or four frequencies. Remark that for a string with one end fixed and the other free, the fixed end is a node and the free end is an antinode. Derive the expression for the standing wave frequencies. Draw diagrams for the lowest three or four frequencies.

E. Place two speakers, driven by the same signal generator and amplifier, well apart on the lecture table, facing the class. Standing waves are created throughout the room. Have each student place a finger in one ear and move his head slowly from side to side in an attempt to find the nodes and antinodes. Use a frequency of about 1 kHz.

F. Consider a driven string and describe resonance. Explain that the amplitude becomes large when the driving frequency matches a standing wave frequency. Explain that at resonance the energy supplied by the driving force is dissipated and that off resonance the string does work on the driving mechanism.

G. You may wish to explain that when the string is driven at a non-resonant frequency, each traveling wave and its reflection from an end produce a standing wave, just as at resonance. The standing waves produced by successive reflections, however, do not coincide and a jumble results.

SUGGESTIONS

1. Assignments
   a. Include discussions of question 1 when covering the idea of wavelength. Question 3 deals with phase and question 4 deals with wave speed. To emphasize the mathematical description of a traveling wave, assign some of problems 5 through 8. Wave speed in terms of tension and linear mass density is covered in problems 11 through 16 and 18 through 23. Assign a few of these.
   b. Assign problem 25 or 27 when discussing energy transport.
   c. Question 5 deals with the superposition of waves and questions 6 and 7 deal with wave interference. The fundamentals of interference are covered in problems 26 and 27. Include problem 30 if you discuss phasors.
   d. Assign questions 8 through 11 and problems 32, 33, and 38 in connection with standing waves. The superposition of traveling waves to form a standing wave is covered in problems 41, 42, 47, and 48. For a challenge assign problem 51.

2. Demonstrations
   a. Traveling waves: Freier and Anderson Sa3, 4, 5, 6, 12, 13.
   b. Reflection: Freier and Anderson Sa7, 12, 14.
   c. Standing waves: Freier and Anderson Sa8, 9.

3. Audio/Visual
   a. *Superposition*; *Vibrations of a Wire*; *Vibrations of a Drum*; from AAPT collection 1 of single-concept films; video tape; available from Ztek Co., PO Box 11768, Lexington, KY 40577–1768.

b. *Nonrecurrent Wavefronts*; from AAPT Miller collection of single-concept films; video tape; available from Ztek Co. (see above for address).
c. *Wave Propagation*, *Periodic Waves*, *Superposition*, *Standing Waves*; Side C: Chapter 10 of Cinema Classics; video disk; available from Ztek Co. (see above for address).
d. *Wave Characteristics*; Demonstrations of Physics: Waves; VHS video tape (7:50); Media Design Associates, Inc., Box 3189, Boulder, CO 80307–3190.
e. *Chladni Plates*; Demonstrations of Physics: Waves; VHS video tape (4:46); Media Design Associates, Inc. (see above for address).
f. *Resonance*; Demonstrations of Physics: Waves; VHS video tape (4:32); Media Design Associates, Inc. (see above for address).
g. *The Nature of Waves*; VHS video tape (11 min); Films for the Humanities & Sciences, PO Box 2053, Princeton, NJ 08543–2053.
h. *Standing Waves*; VHS video S video tape (15 min); Films for the Humanities & Sciences (see above for address).
i. *The Propagation of Waves*; VHS video tape (9 min); Films for the Humanities & Sciences (see above for address).
j. *The Puzzle of the Tacoma Narrows Bridge Collapse Videodisc*; video disk; available from Ztek Co. (see above for address).
k. *Mechanical Resonance*; *Velocity/Wavelength & Frequency*; *Standing Waves*; *Change in Medium/Interference*; Physics Demonstrations in Sound & Waves, Part I; VHS video tape; ≈3 min each; Physics Curriculum & Instruction, 22585 Woodhill Drive, Lakeville, MN 55044.

4. Computer Software
   a. *Physics Simulation Programs*; Robert H. Good; Windows, Apple II; requires GW-BASIC; available from Physics Academic Software, North Carolina State University, PO Box 8202, Raleigh, NC 27690–0739. Contains simulations of traveling and standing waves.
   b. *WaveMaker*; Freeman Deutsch, Philip Sadler, Charles Whitney, Stephen Engquist, and Linda Shore; Macintosh; available from Physics Academic Software (see above for address). Beads are attached to elastic, massless strings and oscillate transversely. The user can control the masses and the spring constants. The program sill plot the position, velocity, and acceleration of any bead. Demonstrates beats, reflection at fixed and free ends, normal oscillations, wave superposition, and transmission through a boundary between two different media.
   c. *Interference*; Socrates Software, 6187 Rosecommon Drive, Norcross, GA 30092. Shows waves for constructive and destructive interference.

5. Computer Project

   Have students use a commercial math program, a spreadsheet, or their own programs to investigate energy in a string carrying a wave. The program should calculate the kinetic, potential, and total energies at a given point and time, given the string displacement as a function of position and time. Use the program to plot the energies as functions of time for a given position. Consider a pulse, a sinusoidal wave, and a standing wave. Demonstrate that energy passes the point in the first two cases but not in the third. For sinusoidal and standing waves, the program should also calculate averages over a cycle. Other projects are described in the Computer Projects section of this manual.

6. Laboratory
   a. Meiners Experiment 12–1: *Transverse Standing Waves* (Part A). Several harmonics are generated in a string by varying the driving frequency. Frequency ratios are computed and compared with theoretical values. Values of the wave speed found using λf and using

$\sqrt{\tau/\mu}$ are compared. The experiment can be repeated for various tensions and various linear densities.

b. Bernard Experiment 22: *A Study of Vibrating Strings.* A horizontal string is attached to a driven tuning fork vibrator. It passes over a pulley and weights are hung on the end. The weights are adjusted so standing wave patterns are obtained and the wavelength of each is found from the measured distance between nodes. Graphical analysis is used to find the relationship between the wave velocity and the tension in the string and to find the frequency. Several strings are used to show the relationship between the wave velocity and the linear density.

Chapter 18 WAVES — II

BASIC TOPICS

I. Qualitative description of sound waves.

A. Explain that the disturbance that is propagated is a deviation from the ambient density and pressure of the material in which the wave exists. This comes about through the motion of particles. If Chapter 15 was not covered, you should digress to discuss density and pressure briefly. Point out that sound waves in solids can be longitudinal or transverse but sound waves in fluids are longitudinal: the particles move along the line of wave propagation. Waves in crystalline solids moving in low symmetry directions are examples that are neither transverse nor longitudinal. Use a slinky to show a longitudinal wave and point out the direction of motion of the particles. State that sound can be propagated in all materials.

B. Draw a diagram, similar to Fig. 18–3, to show a compressional pulse. Point out regions of high, low, and ambient density. Also show the pulse at a later time.

C. Similarly, diagram a sinusoidal sound wave in one dimension and draw a rough graph of the pressure as a function of position for a given time. Give the rough frequency limits of audible sound and mention ultrasonic and infrasonic waves.

D. Discuss the idea that the wave velocity depends on an elastic property of the medium (bulk modulus) and on an inertia property (ambient density). Recall the definition of bulk modulus (or introduce it) and show by dimensional analysis that v is proportional to $\sqrt{B/\rho}$. Assert that the constant of proportionality is 1. Point out the wide range of speeds reported in Table 18–1.

II. Interference.

A. Remind students of the conditions for interference. Consider two sinusoidal sound waves with the same amplitude and frequency, traveling in the same direction. Explain that constructive interference occurs if they are in phase and complete destructive interference occurs if they are π rad out of phase.

B. Explain that a phase difference can occur at a detector if two waves from the same source travel different distances. Show that the phase difference is given by $k\Delta x$ $(= 2\pi\Delta x/\lambda)$.

C. Interference of sound waves can be demonstrated by wiring two speakers to an audio oscillator and putting the apparatus on a slowly rotating platform. Students will hear the changes in intensity.

III. Mathematical description of one-dimensional sound waves.

A. If desired, derive $v = \sqrt{B/\rho}$ as it is done in the text.

B. Write $s = s_m \cos(kx - \omega t)$ for the displacement of the material at x. Show how to calculate the pressure as a function of position and time. Relate the pressure amplitude to the displacement amplitude. Explain that a sinusoidal pressure wave traveling in the positive

x direction is written $\Delta p(x,t) = \Delta p_m \sin(kx - \omega t)$, where $\Delta p_m = v\rho\omega s_m$. State that Δp is the deviation of the pressure from its ambient value. Remind students that $k = 2\pi/\lambda$, $f = \omega/2\pi$, and $\lambda f = v$.

C. Remark that power is transmitted by a sound wave because each element of fluid does work on neighboring elements. Show that the kinetic energy in an infinitesimal length dx of a sinusoidal sound wave traveling along the x axis is $dK = \frac{1}{2}A\rho\omega^2 s_m^2 \sin^2(kx - \omega t)\, dx$, where A is the cross-sectional area. Show that its average over a cycle is $(dK)_{avg} = \frac{1}{4}A\rho\omega^2 s_m^2$. Argue that this energy moves to a neighboring segment in time $dt = dx/v$ and show that the rate of kinetic energy flow is, on average, $(dK/dt)_{avg} = \frac{1}{4}A\rho v\omega^2 s_m^2$, where v is the speed of sound. Tell students that the rate of flow of potential energy is exactly the same, so the rate of energy flow is $P_{avg} = \frac{1}{2}A\rho v\omega^2 s_m^2$.

D. Define intensity as the average rate of energy flow per unit area and show that it is given by $I = \frac{1}{2}\rho v\omega^2 s_m^2$. Show that conservation of energy implies that the intensity decreases as the reciprocal of the square of the distance as a spherical wave moves outward from an isotropic point source.

E. Show a scale of the range of human hearing in terms of intensity. Introduce the idea of sound level and define the bel and decibel. Discuss both absolute (relative to $10^{-12}\,\mathrm{W/m^2}$) and relative intensities. Remark that an increase in intensity by a factor of 10 means an increase in sound level by 10 db. If you have a sound level meter, use an oscillator, amplifier, and speaker to demonstrate the change of a few db in sound level.

IV. Standing longitudinal waves and sources of sound.

A. Use a stringed instrument or a simple taut string to demonstrate a source of sound. Point out that the wave pattern on the string is very nearly a standing wave, produced by a combination of waves reflected from the ends. If the string is vibrating in a single standing wave pattern, then sound waves of the same frequency are produced in the surrounding medium. Demonstrate the same idea by striking a partially filled bottle, then blowing across its mouth. Also blow across the open end of a ball point pen case. If you have them, demonstrate Chladni plates.

B. Derive expressions for the natural frequencies and wavelengths of air pipes open at both ends and closed at one end. Stress that pressure nodes occur near open ends and that pressure antinodes occur at closed ends. Define the terms fundamental and harmonic.

C. Optional: Discuss the quality of sound for various instruments in terms of harmonic content. If possible, demonstrate the instruments.

D. Demonstrate voice patterns by connecting a microphone to an oscilloscope and keeping the setup running through part or all of the lecture. This is particularly instructive in connection with part C.

V. Beats.

A. Demonstrate beats using two separate oscillators, amplifiers, and speakers, operating at nearly, but not exactly, the same frequency. If possible, show the time dependence of the wave on an oscilloscope. Remark that the sound is like that of a pure note but the intensity varies periodically. Explain that this technique is used to tune instruments in an orchestra.

B. Remark that you will consider displacement oscillations at a point in space when two sound waves of the same amplitude and nearly the same frequency are present. Write the expression for the sum of $s_1 = s_m \cos(\omega_1 t)$ and $s_2 = s_m \cos(\omega_2 t)$, where $\omega_1 \approx \omega_2$, but the two frequencies are not exactly equal. Show that $s_1 + s_2 = 2s_m \cos(\omega' t)\cos(\omega t)$, where $\omega' t = (\omega_1 - \omega_2)/2$ and $\omega = (\omega_1 + \omega_2)/2$. Remark that because the difference in frequencies is much smaller than either constituent frequency we can think of the oscillation as having an angular frequency of $\omega = (\omega_1 + \omega_2)/2$ and a time dependent amplitude. Note that the angular frequency of the amplitude is $\omega' = |\omega_1 - \omega_2|/2$ but the angular frequency of the intensity is $\omega_{\text{beat}} = |\omega_1 - \omega_2|$. The latter is the beat angular frequency.

VI. Doppler effect.

A. Explain that the frequency increases when the source is moving toward the listener, decreases when the source is moving away, and that similar effects occur when the listener is moving toward or away from the source. Use Fig. 18–19 to illustrate the physical basis of the phenomenon.

B. Derive expressions for the frequency when the source is moving and for the frequency when the listener is moving. Point out that the velocities are measured relative to the medium carrying the sound.

C. The effect can be demonstrated by placing an auto speaker and small audio oscillator (or sonalert type oscillator) on a rotating table. The sonalert can also be secured to a cable and swung in a circle. Show the effect of a passive reflector by moving a hand-held sonalert rapidly toward and away from the blackboard.

SUGGESTIONS

1. Assignments
   a. The speed of sound is emphasized in problems 1, 2, and 3 while its dependence on the ambient density is covered in question 1.
   b. Ask questions 2 through 6 and assign problems 13 and 15 in connection with interference.
   c. Use problems 18, 19, and 23 to discuss sound intensity and problems 20 and 21 to discuss sound level. They will help students with the concepts of bel and decibel. Also consider problem 20.
   d. Ask questions 7 through 12 and assign problems 31, 38, and 40 when discussing standing waves.
   e. Tuning stringed instruments is covered in problem 36.
   f. Assign problems 42 and 44 in connection with beats.
   g. Use questions 13 and 14 and problem 47 in a discussion of the Doppler effect. Assign problems 54 and 57. Assign problem 59 in connection with sonic booms.
2. Demonstrations
   a. Wavelength and speed of sound in air: Freier and Anderson Sa16, 17, 18, Sh1.
   b. Sound not transmitted in a vacuum: Freier and Anderson Sh2.
   c. Sources of sound, acoustical resonators: Freier and Anderson Sd3, Se, Sf, Sj6.
   d. Harmonics: Freier and Anderson Sj2 — 5
   e. Beats: Freier and Anderson Si4 — 6.
   f. Doppler shift: Freier and Anderson Si1 — 3.
3. Books and Monographs
   a. *Resource Letters, Book Four* and *Resource Letters, Book Five*; American Association of Physics Teachers, One Physics Ellipse, College Park, MD 20740–3845. Contain resource letters on sound and acoustics.
   b. *Musical Acoustics*; edited by Thomas D. Rossing; available from AAPT (see above for address). Reprints.
4. Audio/Visual
   a. *Measuring the Speed of Sound*; Demonstrations of Physics: Waves; VHS video tape (7:48); Media Design Associates, Inc., Box 3189, Boulder, CO 80307–3190.
   b. *Waves and Sound*; VHS video tape (30 min); Films for the Humanities & Sciences, PO Box 2053, Princeton, NJ 08543–2053.
   c. *Experiments on the Doppler Effect*; Physics Experiments; VHS video tape (15 min); Films for the Humanities & Sciences (see above for address).

d. *Longitudinal Waves*; *Longitudinal Standing Waves*; Physics Demonstrations in Sound & Waves, Part I; VHS video tape; ≈3 min each; Physics Curriculum & Instruction, 22585 Woodhill Drive, Lakeville, MN 55044.
e. *Nature of Sound Waves*; *Propagation of Sound*; *Transmission of Sound*; *Refraction of Sound*; *Interference of Sound*; *diffraction of Sound*; *Doppler Effect*; Physics Demonstrations in Sound & Waves, Part II; VHS video tape; ≈3 min each; Physics Curriculum & Instruction (see above for address).
f. *Standing Sound Waves*; *Standing Sound Waves in Two Dimensions*; *Resonance/Real Time*; *Superposition Principle*; Physics Demonstrations in Sound & Waves, Part III; VHS video tape; ≈3 min each; Physics Curriculum & Instruction (see above for address).

5. Computer Software
   a. *Sound Waves*; Macintosh, Windows; Cross Educational Software, Inc., 508 E. Kentucky Avenue, PO Box 1536, Ruston, LA 71270. Contains sections on beats, Doppler shift, resonances in open and closed tubes.
   b. *Physics Demonstrations*. See Chapter 2 SUGGESTIONS.
   c. *Doppler Effect*; Socrates Software, 6187 Rosecommon Drive, Norcross, GA 30092. Animated display of the sound waves for both subsonic and supersonic flight of an aircraft.
6. Computer Projects
   A spreadsheet or computer program can be used to add waves. Have students use it to investigate beats. See the Computer Projects section of this manual for details.
7. Laboratory
   a. Meiners Experiment 12–2: *Velocity of Sound in Air* and Bernard Experiment 23: *Velocity of Sound in Air — Resonance-Tube Method.* Resonance of an air column is obtained by holding a tuning fork of known frequency at the open end of a tube with one closed end. The length of the column is changed by adjusting the amount of water in the tube. The wavelength and speed of sound are found.
   b. Meiners Experiment 12–3: *Velocity of Sound in Metals* and Bernard Experiment 24: *Velocity of Sound in a Metal — Kundt's-Tube Method.* A Kundt's tube is used to find the frequency of sound excited in a rod with its midpoint clamped and its ends free. The wavelength is known to be twice the rod length and λf is used to find the speed of sound. In another experiment, a transducer and oscilloscope are used to time a sound pulse as it travels the length of a rod and returns.
   c. Meiners Experiment 12–4: *Investigation of Longitudinal Waves.* The amplitude and phase of a sound wave are investigated as functions of distance from a speaker source. To do this, Lissajous figures are generated on an oscilloscope screen by the source signal and the signal picked up by a microphone. To eliminate noise, the speaker and microphone should be in a large sound-proof enclosure with absorbing walls. Use Meiners Experiment 10–10 to familiarize students with the oscilloscope and Lissajous figures.

Chapter 19 TEMPERATURE, HEAT, AND THE FIRST LAW OF THERMODYNAMICS

BASIC TOPICS

I. The zeroth law of thermodynamics.
   A. Explain that if two bodies, not in thermal equilibrium, are allowed to exchange energy then they will do so and one or more of their macroscopic properties will change. When no further changes take place, the bodies are in thermal equilibrium. Explain that two bodies in thermal equilibrium are said to have the same temperature.

B. For gases, the properties of interest include pressure, volume, internal energy, and the quantity of matter. Other properties may be included for other materials. The quantity of matter may be given as the number of particles or as the number of moles.

C. Explain what is meant by diathermal and adiabatic walls and remark that diathermal walls are used to obtain thermal contact without an exchange of particles. Adiabatic walls are used to thermally isolate a system.

D. State the zeroth law: if body A and body B are each in thermal equilibrium with body C, then A is in thermal equilibrium with B. Discuss the significance of the zeroth law. State that it is the basis for considering the temperature to be a property of an object. If it were not true, then, at best, an object might have a large number of temperatures, depending on what other objects were in thermal equilibrium with it.

E. Explain that the temperature of a body is measured by measuring some property of a thermometer in thermal equilibrium with it. Illustrate by reminding students that the length of the mercury column in an ordinary household thermometer is a measure of the temperature. Explain that the zeroth law guarantees that the same temperature, as measured by the same thermometer, will be obtained for two substances in thermal equilibrium with each other.

II. Temperature measurements.

A. Mention that the value of the temperature obtained depends on the substance used for the thermometer and on the property measured but that several techniques exist that allow us to define temperature independently of the thermometric substance and property.

B. Describe a constant-volume gas thermometer. If one is available, demonstrate its use. If not, show Fig. 19–5. The gas is placed in thermal contact with the substance whose temperature is to be measured and the pressure is adjusted so that the volume has some standard value (for that thermometer). After corrections are made, the temperature is taken to be proportional to the pressure: $T = ap$, where a is the constant of proportionality.

C. Describe the triple point of water and explain that water at the triple point is assigned the temperature $T = 273.16\,\text{K}$. Solve for a and show that $T = 273.16(p/p_3)$.

D. Point out that thermometers using different gases give different values for the temperature when used as described. Explain the limit used to obtain the Kelvin temperature. See Fig. 19–6.

E. Define the Celsius and Fahrenheit scales. Give the relationships between the degree sizes and the zero points. Give equations for conversion from one scale to another and give the temperature value for the ice and steam points in each system. Use Fig. 19–7 and Table 19–1.

F. Define the Kelvin scale and explain the kelvin as a unit of temperature. Give the relationship between the Celsius and Kelvin scales. Give the ice and steam points on the Kelvin scale.

III. Thermal expansion.

A. Describe linear expansion and define the coefficient of linear expansion: $\alpha = \Delta L/L\Delta T$. Point out Table 19–2. Obtain a bimetallic strip and use both a bunsen burner and liquid nitrogen (or dry ice) to show bending. After the students see the strip bend ask which of the metals has the greater coefficient of linear expansion. Explain that these devices are often used in thermostats.

B. Discuss area and volume expansion. Consider a plate and show that the coefficient of area expansion is 2α. Consider a rectangular solid and show that the coefficient of volume expansion is 3α. In each case apply the equation for linear expansion to each dimension of the object and find ΔA or ΔV to first order in ΔT.

C. Explain that the length of a scratch on the flat face of an object increases as the temperature increases. The area of a hole also increases. Carefully drill a 1/2 inch hole in a piece of aluminum, roughly $1\frac{1}{4}$ inch thick. Obtain a 13-mm diameter steel ball bearing and place it in the hole. It will not pass through. Heat the plate on a bunsen burner and the ball passes through easily.

D. Demonstrate volume expansion of a gas using a flat bottomed flask, a bulbed tube, a two hole stopper, and some colored water. Partially evacuate the bulb so the colored water stands in the tube somewhat above the stopper. Place your hand on the bulb to warm the air inside and the water in the tube drops in response.

IV. Heat.

A. Explain that when thermal contact is made between two bodies at different temperatures, a net flow of energy takes place from the higher temperature body to the lower temperature body. The temperature of the hotter body decreases, the temperature of the cooler body increases, and the net flow continues until the temperatures are the same. Energy also flows from warmer to cooler regions of the same body. State that heat is energy that is transferred because of a temperature difference. Distinguish between heat and internal energy. Emphasize that the idea of a body having heat content is not meaningful. Also emphasize that heat is not a new form of energy. The energy transferred may be the kinetic energy of molecules or the energy in an electromagnetic wave. Examples: a bunsen burner flame, radiation across a vacuum. State that heat is usually measured in Joules but calories and British thermal units are also used. 1 kcal = 3.969 Btu = 4187 J. Remark that the unit used in nutrition, a Calorie (capitalized) is 1 kcal.

B. Remind students of the energy equation studied in Chapter 8. Tell them that for the systems considered here the center of mass remains at rest (or has a constant velocity) and changes in potential energy are ignored. Processes considered change only the internal energy. A new term, however, must be added since the environment can exchange energy as heat with the system, as well as do macroscopic work on the system. Write $\Delta E_{\text{int}} = Q - W$, where Q is the energy absorbed as heat *by* the system and W is the work done *by* the system.

C. Stress the sign convention for heat and work: Q is positive if the system takes in energy, W is positive if the system does positive work.

D. Stress that heat and work are alternate means of transferring energy and explain that, for example, temperature changes can be brought about by both heat and mechanical work. To demonstrate this, connect a brass tube, fitted with a rubber stopper, to a motor as shown. Make a wooden brake or clamp that fits tightly around the tube. Put a few drops of water into the tube, start the motor, and exert pressure on the tube with the clamp. Soon the stopper will fly off. Note that mechanical work was done and steam was produced.

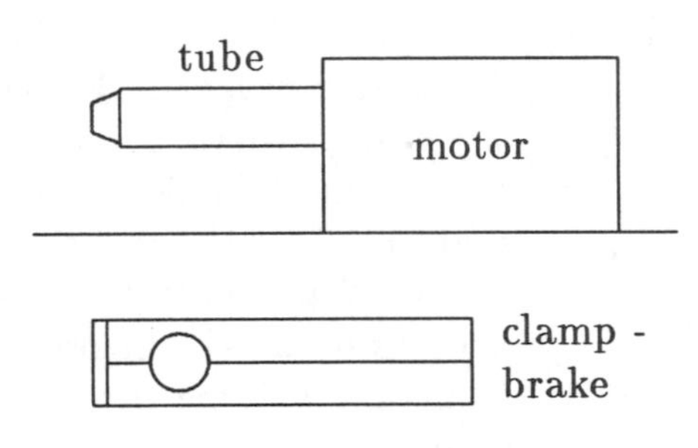

V. Heat capacity.

A. Define the heat capacity of a body as the amount of energy absorbed as heat per degree of temperature change: for a small temperature change $C = Q/\Delta T$. Point out that it depends on the temperature and on the constraints imposed during the transfer. The heat capacity at constant volume is different from the heat capacity at constant pressure because positive work is done by the system when the temperature is increased at constant pressure. More energy is therefore required as heat to obtain the same increase in internal energy and temperature.

B. Point out that the heat capacity depends on the amount of material. Define the specific heat c and the molar specific heat. Explain they are independent of the amount of material. Point out Table 19–3. You might use C' to denote a molar specific heat.

C. Do a simple calorimetric calculation (see Sample Problem 19–4). Stress the fundamental idea: the energy that leaves one body enters another, so the sum of the energies absorbed by all objects in a closed system vanishes.

D. Explain that energy must be transferred to or from a body when it changes phase (liquid to gas, etc.). The energy per unit mass is called the heat of transformation or latent heat. Point out Table 19–4. If time permits, work a calorimetric problem that involves a change in phase. Consider, for example, dropping ice into warm water and calculate the final temperature. Work a problem for which all the ice is melted and a problem for which only part of the ice is melted.

VI. Heat, work, and the first law of thermodynamics.

A. Describe a gas in a cylinder fitted with a piston. Remind students that as the piston moves the gas volume changes and the gas does work $W = \int p\,dV$ on the piston. Explain that the gas might exchange energy with its environment through both work and heat.

B. Draw a p-V diagram (such as Fig. 19–14) and mark initial and final states, with $V_f > V_i$. Explain that p and V are thermodynamic state variables and have definite, well defined values for a given thermodynamic state. They can be used to specify the state. Point out there are many paths from the initial to the final state. Define the term "quasi-static process" and explain that the various paths on the diagram represent quasi-static processes, for which the system is always infinitesimally close to equilibrium states. Point out that for different paths p is a different function of V and different amounts of work are done by the gas. Also explain that different amounts of heat are transferred for different paths. Work and heat are not thermodynamic state variables.

C. Explain that $Q - W$ is independent of the process. Define the internal energy by $\Delta E_{\text{int}} = Q - W$ and point out that ΔE_{int} is the same for any two selected states regardless of the path used to get from one to the other. State that ΔE_{int} is the change in mechanical energy (kinetic and potential energy) of all the particles that make up the system. Stress that the first law $\Delta E_{\text{int}} = Q - W$ is an expression of the conservation of energy.

VII. Applications of the first law.

A. Adiabatic process. Explain that $Q = 0$ and $\Delta E_{\text{int}} = -W$. As an example, consider a gas in a thermally insulated cylinder and allow the volume to change by moving the piston. Explain that when the internal energy increases the temperature goes up for most materials. This can be achieved by compressing the gas. The opposite occurs when the piston is pulled out. Stress that no heat has been exchanged. Illustrate an adiabatic process on a p-V diagram.

B. Constant volume process. Explain that $W = 0$ and $\Delta E_{\text{int}} = Q$. Illustrate on a p-V diagram.

C. Isobaric process. Explain that $W = p(V_f - V_i)$ for a quasi-static isobaric process. For a change in phase, show that $\Delta E_{\text{int}} = mL - p\Delta V$. Illustrate on a p-V diagram.

D. Describe adiabatic free expansion and note that $\Delta E_{\text{int}} = 0$. Explain that this process is not quasi-static and cannot be shown on a p-V diagram. The end points, however, are well defined thermodynamic states and are points on a p-V diagram.

E. Cyclical process. Explain that all state variables return to their original values at the end of each cycle and, in particular, $\Delta E_{\text{int}} = 0$. Thus, $Q = W$. Illustrate on a p-V diagram.

For later reference, stress that heat may be absorbed (or rejected) and work done during a cyclic process.

VIII. Transfer of heat.

A. Explain that steady state heat flow can be obtained if both ends of a slab are held at different temperatures. Define the thermal conductivity k of the material using $P_{\text{cond}} = -kA\,dT/dx$ for a slab of uniform cross section A. Here P_{cond} is the rate of heat flow. Emphasize that the negative sign appears because heat flows from hot to cold. Stress that P_{cond} and T are constant in time in the steady state. Explain that $P_{\text{cond}} = kA(T_H - T_C)/L$ for a uniform bar of length L, with the cold end held at temperature T_C and the hot end held at temperature T_H.

B. A demonstration that shows both thermal conductivity and heat capacity can be constructed from three rods of the same size, one made of aluminum, one made of iron, and one made of glass. Use red wax to attach small ball bearings at regular intervals along each rod. Clamp the rods so that each has one end just over a bunsen burner. The rate at which the wax melts and the ball bearings drop off is mostly dictated by the thermal conductivity of the rods, but it is influenced a bit by the specific heats.

C. For a practical discussion, introduce the idea of R value and discuss home insulation.

D. Qualitatively discuss radiation as a means of energy transfer. Place a heating element at the focal point of one spherical reflector and some matches, stuck in a cork, at the focal point of the another. Place the reflectors several meters apart and adjust the positions so that the heater is imaged at the matches. Use a 1 kW or so heater. The matches will ignite in about a minute.

E. Give $P_{\text{rad}} = \sigma\epsilon AT^4$ for the rate with which a surface with area A at Kelvin temperature T emits radiative energy. Here σ $(= 5.603 \times 10^{-8}\,\text{W/m}^2 \cdot \text{K}^4)$ is the Stefan-Boltzmann constant and ϵ is the emissivity of the surface. Explain that ϵ has a value between 0 and 1 and depends on the composition of the surface.

F. Qualitatively discuss convection as a means of heat transfer.

SUGGESTIONS

1. Assignments
   a. After discussing gas thermometers assign problems 1 and 3. Temperature scales are covered in question 1 and problems 4, 5, and 9.
   b. Use question 2 and several of problems 12, 14, 16, and 17 to discuss thermal expansion. Use problem 19 in connection with the ball and hole demonstration.
   c. Introduce Newton's law of cooling and assign problems 7 and 8.
   d. To test the fundamental concepts of heat capacity and heat of transformation assign questions 3 and 4 and some of problems 26 through 47. Problems 26, 28, 29, 34, 36, 39, 43, 44, 45, 46 involve changes of phase. Include one or two of these.
   e. Problems 48 and 50 are good tests of understanding of the first law. Also assign questions 5 and 6 and problems such as 49, 50, 51, 52, and 53, which involve the interpretation of p-V diagrams. Tell students to pay attention to signs.
   f. Following the discussion of thermal conductivity, ask question 7. Assign problems 57, 59, and 63 in connection with heat conduction.

2. Demonstrations
   a. Thermometers: Freier and Anderson Ha1 — 4.
   b. Thermal expansion: Freier and Anderson Ha5 — 12.
   c. Heat capacity and calorimetry: Freier and Anderson Hb1, 2.
   d. Work and heat: Freier and Anderson He1 — 6.
   e. Heat transfer: Freier and Anderson Hc, Hd1 — 7, Hf.

f. p-V relations: Freier and Anderson Hg1 — 3.

3. Books and Monographs

   Resource Letters, Book Five; American Association of Physics Teachers, One Physics Ellipse, College Park, MD 20740–3845. Contains a resource letter on heat and thermodynamics.

4. Audio/Visual

   a. *Heat and Temperature*; Side B: Mechanics (II) and Heat of Cinema Classics; video disk; available from Ztek Co., PO Box 11768, Lexington, KY 40577–1768.
   b. *Expansion Due to Heating*; Demonstrations of Physics: Thermal Effects; VHS video tape (7:26); Media Design Associates, Inc., Box 3189, Boulder, CO 80307–3190.
   c. *Heat of Fusion*; Demonstrations of Physics: Thermal Effects; VHS video tape (6:14); Media Design Associates, Inc. (see above for address).
   d. *Temperature*; VHS video tape (30 min); Films for the Humanities & Sciences, PO Box 2053, Princeton, NJ 08543–2053.
   e. *The Thermal Expansion of Metals*; VHS video tape (20 min); Films for the Humanities & Sciences (see above for address).
   f. *Heat*; VHS video tape (30 min); Films for the Humanities & Sciences (see above for address).
   g. *Heat and Energy Systems*; VHS video tape (15 min); Films for the Humanities & Sciences (see above for address).
   h. *The Conduction of Heat*; VHS video tape (15 min); Films for the Humanities & Sciences (see above for address).
   i. *The Convection of Heat*; VHS video tape (30 min); Films for the Humanities & Sciences (see above for address).

   Entropy; Physics Demonstrations in Heat, Part III; VHS video tape; ≈3 min; Physics Curriculum & Instruction, 22585 Woodhill Drive, Lakeville,, MN 55044.

5. Computer Software

   a. *Thermodynamics*; Macintosh, Windows; Cross Educational Software, Inc., 508 E. Kentucky Avenue, PO Box 1536, Ruston, LA 71270.
   b. *Thermodynamics Lecture Demonstrations*; Kurt Wick and Philip Johnson; Windows; available from Physics Academic Software, North Carolina State University, PO Box 8202, Raleigh, NC 27690–0739. Simulations of ten thermodynamic processes, including isochoric, adiabatic, and isothermal processes, the Carnot cycle, and the Otto and diesel engines. The processes are also diagramed on p-V and T-S diagrams.

6. Laboratory

   a. Meiners Experiment 9–3: *Linear Expansion* and Bernard Experiment 18: *Linear Coefficient of Expansion of Metals.* The length of a metal rod is measured at room temperature and at 100°C (in a steam jacket), then the data is used to compute the coefficient of thermal expansion. The experiment can be repeated for several different metals and the results compared.
   b. Meiners Experiment 9–1: *Calorimetry — Specific Heat and Latent Heat of Fusion.* Students use a calorimeter to find the specific heat of water and a metal sample. They also measure the latent heat of fusion of ice. Since the specific heat of the stirring rod and the calorimeter must be taken into account, this is a good exercise in experimental design.
   c. Meiners Experiment 9–2: *Calorimetry — Mechanical Equivalent of Heat* and Bernard Experiment 30: *The Heating Effect of an Electric Current.* A calorimeter is used to find the relationship between the energy dissipated by a resistive heating element and the temperature rise of the water in which it is immersed. Students must accept $P = i^2R$ for

the power output of the heating element. With slight revision these experiments can also be used in conjunction with Chapter 27.

d. Bernard Experiment 19: *Specific Heat and Temperature of a Hot Body.* A calorimeter is used to obtain the specific heat of metal pellets. In a second part, a calorimeter and a metal sample with a known specific heat are used to find the temperature of a Bunsen burner flame.
e. Bernard Experiment 20: *Change of Phase — Heat of Fusion and Heat of Vaporization.* A calorimeter is used to measure the heat of fusion and heat of vaporization of water. If the lab period is long or writeups are done outside of lab, experiments 19 and 20 may be combined nicely.
f. Meiners Experiment 9–6: *Calorimetry Experiments* (with a microprocessor).
g. Meiners Experiment 9–4: *Thermal Conductivity.* The sample is sandwiched between a thermal reservoir and a copper block. The rate at which energy passes through the sample is found by measuring the rate at which the temperature of the copper increases. Temperature is monitored by means of a thermocouple.
h. Meiners Experiment 9–5: *Thermal Conductivity with Microprocessor.*

Chapter 20 THE KINETIC THEORY OF GASES

BASIC TOPICS

I. Macroscopic description of an ideal gas.
   A. Explain that kinetic theory treats the same type problems as thermodynamics but from a microscopic viewpoint. It uses averages over the motions of individual particles to find macroscopic properties. Here it is used to clarify the microscopic basis of pressure and temperature.
   B. Define the mole. Define Avogadro's number N_A and give its value, $6.02 \times 10^{23}\,\text{mol}^{-1}$. Explain the relationships between the mass of a molecule, the mass of the sample, the molar mass, the number of moles, the number of molecules, and Avogadro's number. These often confuse students.
   C. Write down the ideal gas equation of state in the form $pV = nRT$ and in the form $pV = NkT$. Here N is the number of molecules and n is the number of moles. Give the values of R and k and state that $k = R/N_A$. Explain that for real gases at low density pV/T is nearly constant. Point out that the equation of state connects the thermodynamic variables n (or N), p, V, and T. Draw some ideal gas isotherms on a p-V diagram.
   D. To show how the equation of state is used in thermodynamic calculations, go over Sample Problem 20–1. Also consider a problem in which the pressure and volume of an ideal gas are changed. Calculate the change in temperature.
   E. Derive expressions for the work done by an ideal gas during an isothermal process and during an isobaric process.

II. Kinetic theory calculations of pressure and temperature.
   A. Go over the assumptions of kinetic theory for an ideal gas. Consider a gas of molecules with only translational degrees of freedom. Assume the molecules are small and are free except for collisions of negligible duration. Also assume collisions with other molecules and with walls of the container are elastic. At the walls the molecules are specularly reflected.
   B. Discuss a gas in a cubic container and explain that the pressure at the walls is due to the force of molecules as they bounce off. By considering the change in momentum at the wall per unit time, show that the pressure is given by $p = nMv_{\text{rms}}^2/3V$, where M is the molar mass. Define the rms speed. Use Table 20–1 to give some numerical examples of v_{rms}^2 and

calculate the corresponding pressure. For many students, the rms value of a quantity needs clarification. Consider a system of five or so molecules and select numerical values for their speeds, then calculate v_{rms}^2 numerically.

C. Substitute $p = nMv_{rms}^2/3V$ into the ideal gas equation of state and show that $v_{rms} = \sqrt{3RT/M}$. Remark that this equation can be used to calculate the rms speed for a particular (ideal) gas at a given temperature.

D. Rearrange the equation for the rms speed to obtain $\frac{1}{2}Mv_{rms}^2 = \frac{3}{2}RT$ and use $M/m = N_A$ to show this can be written $\frac{1}{2}mv_{rms}^2 = \frac{3}{2}kT$, where m is the mass of a molecule. Remark that the left side is the mean kinetic energy of the molecules and point out that the temperature is proportional to the mean kinetic energy.

III. Internal energy and equipartition of energy.

A. Explain that the internal energy of a monatomic ideal gas is the sum of the kinetic energies of the molecules and write $E_{int} = \frac{1}{2}Nmv_{rms}^2 = \frac{3}{2}NkT = \frac{3}{2}nRT$, where N is the number of molecules and n is the number of moles. Stress that for an ideal gas the internal energy is a function of temperature alone, not the pressure and volume individually. This is an approximation for a real gas. Emphasize that the velocities used in computing the internal kinetic energy are measured relative to the center of mass and that the internal energy does not include the kinetic energy associated with motion of the system as a whole.

B. Point out that if adiabatic work W is done on the gas the internal energy increases by W and the temperature increases by $\Delta T = 2W/3nR$.

C. Point out that the expression obtained above for ΔT agrees closely with experimental values for monatomic gases but gives values that are too high for gases of diatomic and polyatomic molecules. Draw diagrams of these types of molecules and explain that they have two and three degrees of rotational freedom, respectively. Some of the energy goes into motions other than the translational motion of the molecules. Define the term degree of freedom and show how to count the number for monatomic, diatomic, and polyatomic molecules.

D. State the equipartition theorem: in thermal equilibrium the energy is distributed equally among all degrees of freedom, with each receiving $\frac{1}{2}kT$ for each molecule. Point out that this agrees with the previous result for monatomic gases: there are three degrees of freedom per molecule and an energy of $\frac{1}{2}kT$ is associated with each.

E. Discuss diatomic molecules and explain there are two new degrees of freedom, both rotational in nature. Show that $E_{int} = \frac{5}{2}nRT = \frac{5}{2}NkT$. Explain that $\frac{3}{2}nRT$ is in the form of translational kinetic energy and nRT is in the form of rotational kinetic energy.

F. Discuss polyatomic molecules. State that there are now three rotational degrees of freedom and show that $E_{int} = 3nRT = 3NkT$. Explain that $\frac{3}{2}nRT$ is in the form of translational kinetic energy and $\frac{3}{2}nRT$ is in the form of rotational kinetic energy.

G. Explain that vibrational motions may also contribute to the internal energy and that, since a vibration has both kinetic and potential energy, there are two degrees of freedom and energy kT associated with each vibrational mode. Explain that, in fact, for most materials vibrational modes generally do not contribute to the internal energy except at extremely high temperatures. Quantum mechanics is required to explain why vibrational modes are frozen out.

IV. Heat capacities of ideal gases.

A. Use equations previously derived for ΔE_{int} to obtain expressions for the molar specific heat at constant volume C_V. Point out the different results for monatomic, diatomic, and polyatomic molecules. Remark that C_V is used to denote *molar* specific heats in this and the next chapter. The symbol is not used for heat capacity as it was in the last chapter.

B. Show that the molar specific heat at constant pressure is related to the molar specific heat at constant volume by $C_p = C_v + R$ and derive the formulas for C_p for monatomic, diatomic, and polyatomic ideal gases.
C. For each type ideal gas, obtain the value for the ratio of molar specific heats $\gamma = C_p/C_v$. Point out these values are independent of T.
D. Derive $pV^\gamma =$ constant for an ideal gas undergoing an adiabatic quasi-static process. Also derive the expression $W = -(p_fV_f - p_iV_i)/(\gamma - 1)$ for the work done by the gas during an adiabatic change of state. Draw an ideal gas adiabat on a p-V diagram. Suppose the initial pressure and volume and the final volume are given. Show how to calculate the final pressure and temperature.

SUPPLEMENTARY TOPICS

1. Mean free path. This topic emphasizes the collisions of molecules and adds depth to the kinetic theory discussion but it is not crucial to subsequent chapters. Discuss as much as time allows.
2. Distribution of molecular speeds. This section deals with the Maxwell distribution and provides a deeper understanding of average speed and root-mean-square speed. Include it if you intend to cover thermonuclear fusion later in the course.

SUGGESTIONS

1. Assignments
   a. Ask questions 1, 2, and 5 in connection with the ideal gas law. Assign a problem, such as 5 or 7, that is a straightforward application of the law. Then, assign problems that show how the law is used to compute changes in various quantities when the gas changes state: 7 and 8, for example.
   b. Problem 10 provides an illustration of the work done by an ideal gas and problem 14 provides an example of heat exchange. Also consider question 7.
   c. Problems 15, 16, and 17 deal with real-life applications. If possible, assign one or two. You may wish to discuss mixtures of gases and partial pressures; if so, consider problem 12.
   d. Use problem 23 in a discussion of the kinetic basis of pressure. Also assign problem 22.
   e. Assign problem 25 when you deal with the kinetic basis of temperature and the relationship between kinetic energy and temperature. Also consider question 6.
   f. After discussing the various specific heats, ask questions 8, 9, and 10 and assign problem 46. Assign problem 48 to emphasize the dependence of the heat capacity on the process. Problem 45 illustrates an isothermal process.
   g. Consider using problem 54 to discuss adiabatic processes.
2. Demonstrations
   Kinetic theory models: Freier and Anderson Hh1, 2, 4, 5.
3. Audio/Visual
   a. *Boyle's Law*, *Equipartition of Energy*, *Maxwellian Speed Distribution*, *Random Walk and Brownian Motion*, *Diffusion*, *Gas Diffusion Rates*; from AAPT collection 2 of single-concept films; video tape; available from Ztek Co., PO Box 11768, Lexington, KY 40577–1768.
   b. *Gas Laws*; Side B: Mechanics (II) and Heat of Cinema Classics; video disk; available from Ztek Co. (see above for address).
   c. *Temperature and Kinetic Theory*; Physics Experiments; VHS video tape (15 min); Films for the Humanities & Sciences, PO Box 2053, Princeton, NJ 08543–2053.
4. Computer Software
   a. *Thermodynamics Lecture Demonstrations*. See Chapter 19 SUGGESTIONS.

b. *Thermodynamics*; Macintosh, Windows; Cross Educational Software, Inc., 508 E. Kentucky Avenue, PO Box 1536, Ruston, LA 71270.

5. Laboratory
   a. Bernard Experiment 17: *Pressure and Volume Relations for a Gas.* The volume of gas in a tube is adjusted by changing the amount of mercury in the tube and a U-tube manometer is used to measure pressure. A logarithmic plot is used to determine the relationship between pressure and volume.
   b. Meiners Experiment 9–8: *Kinetic Theory Model.* The Fisher kinetic theory apparatus, consisting of a large piston-fitted tube of small plastic balls, is used to investigate relationships between pressure, temperature, and volume for a gas. A variable-speed impeller at the base allows changes in the average kinetic energy of the balls; the piston can be loaded to change the pressure. A variety of experiments can be performed.

Chapter 21 ENTROPY AND THE SECOND LAW OF THERMODYNAMICS

BASIC TOPICS

I. Entropy.
   A. Distinguish between reversible and irreversible processes. Remark that reversible processes are quasi-static but not all quasi-static processes are reversible (*i.e.* quasi-static processes involving friction). Also mention that for a gas the path of a reversible process can be plotted on a p-V diagram. As examples consider reversible and irreversible compressions of an ideal gas.
   B. Define the entropy difference between two infinitesimally close equilibrium states as $\mathrm{d}S = \mathrm{d}Q/T$ and between any two equilibrium states as $\Delta S = \int \mathrm{d}Q/T$. Explain that the integral is independent of path and that S is therefore a thermodynamic state function. Stress that a reversible path must be used to evaluate the integral but that entropy differences are defined regardless of whether the actual process is reversible or irreversible. The end points must be equilibrium states, however.
   C. Derive expressions for the change in entropy for an ideal gas undergoing processes at constant volume ($nC_V \ln(T_f/T_i)$), constant pressure ($nC_p \ln(T_f/T_i)$), and constant temperature ($nR\ln(V_f/V_i)$).
   D. Consider the adiabatic free expansion of an ideal gas. Point out that the process is irreversible, $Q = 0$, and $\Delta E_{\text{int}} = 0$. Since the gas is ideal, $T_f = T_i$. Find the change in entropy by evaluating $\int \mathrm{d}Q/T$ over a reversible isotherm through the initial and final states. Point out that the isothermal path does not represent the actual process. Show that $\Delta S = nR\ln(V_f/V_i)$ and state this is positive.
   E. Consider two identical rigid containers of ideal gas, at different temperatures, T_H and T_L. Place them in contact in an adiabatic enclosure. Show they reach equilibrium at temperature $T_m = (T_H + T_L)/2$. Then, consider a reversible, constant volume process that connects the initial and final states and show that $\Delta S = C_V \ln(T_m^2/T_H T_L)$. Remark that this is positive.

II. The second law of thermodynamics.
   A. State the second law: for processes that proceed from an initial equilibrium state to a final equilibrium state the total entropy of a closed system (or a system and its environment) does not decrease. State that if the process is reversible the total entropy does not change and if the process is irreversible it increases. Point out that the previous two examples are consistent with this statement.

B. Stress that the entropy change of the environment must be included. The entropy of a system can decrease but if it does the entropy of its environment increases by at least as much.

C. Remark that for reversible processes the total entropy of the system and its environment does not change because, for the combination of system and environment, the process is adiabatic and $dQ = 0$ for each segment of the reversible path. On the other hand, entropy increases for an adiabatic *irreversible* process.

III. Engines and refrigerators.

A. Discuss heat engines and refrigerators in general, from the point of view of the first law only. Explain that they run in cycles and that an engine absorbs energy as heat at a high temperature, rejects energy as heat at a low temperature, and does work. Describe a refrigerator in similar terms. Define the efficiency of an engine and the coefficient of performance of a refrigerator. Remark that heat engines and refrigerators may be reversible or irreversible.

B. Remind students that a cycle is a process for which the system starts and ends in the same equilibrium state and that $\Delta E_{\text{int}} = 0$, $\Delta p = 0$, $\Delta T = 0$, $\Delta V = 0$, and $\Delta S = 0$ for a cycle.

C. As an example, consider a gas undergoing a reversible cycle consisting of two isothermal processes at different temperatures and linked by two adiabatic processes (a Carnot cycle). Illustrate with a p-V diagram. Mention that, when run as a heat engine, energy enters the gas as heat during the isothermal expansion that energy leaves the gas as heat during the isothermal compression.

D. Over a cycle the change in the entropy of the working substance is zero, the change in the entropy of the high-temperature reservoir is $-Q_H/T_H$, and the change in entropy of the low temperature reservoir is $-Q_L/T_L$. Thus, the change in the total entropy of the system and its environment is $\Delta S = -(Q_L/T_L) - (Q_H/T_H)$. Since the process is reversible this must be zero. So $Q_L/Q_H = -T_L/T_H$ and the efficiency of the engine is $\varepsilon = W/Q_H = (Q_H + Q_L)/Q_H = 1 - (T_L/T_H)$. You may prefer to write these equations in terms of the absolute magnitudes of the quantities involved.

E. Remark that the efficiency is independent of the working substance. Also say that the second law of thermodynamics leads to the same expression for the efficiency of any reversible engine operating between those temperatures. The efficiencies of real engines, which are of necessity irreversible, are less.

F. Show that the second law forbids an engine with zero heat output. In particular, observe that the total entropy of the engine and reservoirs decreases if $|Q_L| < |Q_H| T_L/T_H$ and that this violates the second law.

G. Say that if the process is irreversible the total entropy must increase, so $|Q_L|/T_L > |Q_H|/T_H$ and the efficiency must be less than the ideal efficiency. Remark that no engine operating between two given temperatures can be more efficient than a reversible engine and that all reversible engines operating between the given temperatures have the same efficiency.

H. Carry out a similar analysis for an ideal refrigerator. Show that the coefficient of performance is given by $K = T_L/(T_H - T_L)$, independently of the working substance. State that all reversible refrigerators have the same coefficient of performance and that irreversible refrigerators have lower coefficients for the same reservoirs. Use an entropy argument to show that the second law forbids a refrigerator that operates with no work input.

IV. The statistical basis of entropy.

A. Explain that for any system composed of many molecules, there are many possible arrangements of the molecules. Illustrate by considering a small collection of molecules in a box. Say that each possible arrangement is called a microstate and that microstates can be grouped into configurations such that all the microstates in a given configuration are macroscopically equivalent. That is, the system has the same macroscopic properties. Use the molecules in the box to illustrate two equivalent and two non-equivalent microstates. State that the number of microstates associated with a configuration is called the multiplicity of the configuration.

B. Say that the fundamental assumption of statistical mechanics is that the system has the same probability of being in any microstate. Thus, the most likely configuration is the one with the largest multiplicity.

C. Show that if there are N molecules in the box, with n_R in the right half and n_L in the left half, the multiplicity is $W = N!/(n_R!)(n_L!)$.

D. Say that the entropy of a system when it has a given configuration is given by $S = k \ln W$, where W is the multiplicity of the configuration and k is the Boltzmann constant.

E. Use the statistical definition of entropy to show that the entropy changes by $\Delta S = nR \ln 2$ when the volume available to the molecules in the box is suddenly doubled. You will need to use Stirling's approximation ($\ln N! \approx N \ln N - N$).

SUGGESTIONS

1. Assignments
   a. To start students thinking about entropy changes as they occur in common processes, ask a few of the questions in the group 1 through 5. Assign problems 3, 5, 7, and 8. To include entropy changes in calorimetry experiments, ask problems 9, 10, and 16. Also consider problem 17.
   b. Use questions 6 and 7 to discuss real and ideal engines. Problems 27, 29, and 32 cover the fundamentals of cycles.
   c. Consider practical engines and their efficiencies by approximating their operation by reversible cycles. For a gasoline engine, $T_H \approx 1000°F$ and $T_L \approx 400°F$. Compare actual efficiencies with the ideal efficiency. Actual efficiencies can be obtained by considering the fuel energy available and the work actually obtained.
   d. Consider practical refrigerators. Look in a catalog for typical values of the coefficient of performance and compare with the ideal coefficient of performance. Also consider questions 8 and 9. Assign problems 36, 38, 40, and 41.
   e. Ask questions 10 and 11 in connection with the statistical interpretation of entropy.
2. Demonstrations
   Engines: Freier and Anderson Hm5, Hn.
3. Computer Software
   a. *Physics Simulation Programs*. See Chapter 17 SUGGESTIONS.
   b. *Thermodynamics Lecture Demonstrations*. See Chapter 19 SUGGESTIONS.
   c. *Thermodynamics*; Macintosh, Windows; Cross Educational Software, Inc., 508 E. Kentucky Avenue, PO Box 1536, Ruston, LA 71270.

Chapter 22 ELECTRIC CHARGE

BASIC TOPICS

I. Charge.

A. Explain that there are two kinds of charge, called positive and negative, and that like charges repel each other, unlike charges attract each other. Give the SI unit (coulomb) and explain that it is defined in terms of current, to be discussed later. Optional: explain that current is the flow of charge and is measured in amperes. One coulomb of charge passes a cross section each second in a wire carrying a steady current of 1 A.

B. Carry out the following sequence of demonstrations. They work best in dry weather.

1. Suspend a pith ball by a string. Charge a rubber rod by rubbing it with fur, then hold the rod near the pith ball. The ball is attracted, touches the rod, then flies away after a short time. Use the rod to push the ball around without touching it. Explain that the rod and ball carry the same type charge. Hold the fur near the pith ball and explain that they are oppositely charged.
2. Repeat using a second pith ball and a wooden rod charged by rubbing it on a plastic sheet (this replaces the traditional glass rod – silk combination and works much better). Place the two pith balls near each other and explain that they are oppositely charged.
3. Suspend a charged rubber rod by a string. Use another charged rubber rod to push it around without touching it. Similarly, pull it with the charged wooden rod. Also show that only the rubbed end of the rubber rod is charged.

II. Conductors and insulators.

A. Explain the difference between a conductor and an insulator as far as the conduction of charge is concerned. Explain that excess charge on a conductor is free to move and generally does so when influenced by the electric force of other charges. Excess charge on a conductor is distributed so the net force on any of it is zero. Any excess charge on an insulator does not move far from the place where it is deposited. Remind students of the demonstration that showed that only the rubbed end of the rubber rod remains charged. Metals are conductors. The rubber rod is an insulator. Mention semiconductors and superconductors.

B. Use an electroscope to demonstrate the conducting properties of conductors. Charge the electroscope by contact with a charged rubber rod and explain why the leaves diverge. Discharge it by touching the top with your hand. Explain why the leaves converge. Recharge the electroscope with a charged wooden rod, then bring the charged rubber rod near the electroscope, but do not let it touch. Note the decrease in deflection and explain this by pointing out the attraction of the charge on the rod for the charge on the leaves. Throughout, emphasize the motion of the charge through the metal leaves and stem of the electroscope.

C. Demonstrate charging by induction. Bring a charged rubber rod near to but not touching an uncharged electroscope. Touch your finger to the electroscope, then remove it. Remove the rubber rod and note the deflection of the leaves. Bring the rubber rod near again and note the decrease in deflection. Observe that the electroscope and rod are oppositely charged. Confirm this with the wooden rod. Explain the process.

III. Coulomb's law.

A. Assert that experimental evidence convinces us that there are only two kinds of charge and that the force between a pair of charges is along the line joining them, has magnitude proportional to the product of the magnitudes of the charges, and is inversely proportional to the square of the distance between them. Further, the force is attractive for unlike charges and repulsive for like charges.

B. Write down Coulomb's law for the magnitude of the electric force exerted by one point charge on another. Give the SI value for ϵ_0 and for $1/4\pi\epsilon_0$. Stress that the law holds for point charges. Note in detail that the mathematical form of the law contains all the qualitative features discussed previously. If Chapter 14 was covered, point out the similarity with Newton's law of gravity and mention that, unlike charge, there is no negative mass.

C. Explain that a superposition law holds for electric forces and illustrate by finding the resultant force on a charge due to two other charges. Use the analogy with the law of gravity to show that the force of one spherical distribution of charge on another obeys the same law as two point charges and that the force on a charge inside a uniformly charged spherical shell is zero. If Chapter 14 was not covered, simply state the shell theorems.

IV. Quantization and conservation of charge.

A. State that all measured charge is an integer multiple of the charge on a proton: $q = ne$. Give the value of e: 1.60×10^{-19} C. State that the charge on the proton is $+e$, the charge on the electron is $-e$, and the neutron is neutral.

B. Remark that macroscopic objects are normally neutral; they have the same number of protons as electrons. Stress that the word "neutral" describes the algebraic sum of the charges and does not indicate the absence of charged particles. Remark that when an object is charged, the charge imbalance is usually slight but significant.

C. State that charge is conserved in the sense that for a closed system the sum of all charges before an event or process is the same as the sum after the event or process. Stress that the charges in the sum must have appropriate signs. Example: rubbing a rubber rod with fur. The rod and fur are oppositely charged afterwards and the magnitude of the charge is the same on both. Also discuss the conservation of charge in the annihilation and creation of fundamental particles and note that the identity of the particles may change in an event but charge is still conserved. Examples: beta decay, electron-positron annihilation.

SUGGESTIONS

1. Assignments
   a. Discuss questions 8 and 9, perhaps in connection with demonstrations or lab experiments. Also see problems 4 and 7.
   b. Use questions 3 and 4 to test for understanding of the direction of an electrical force and the superposition of forces. Problems 6 and 10 deal with the addition of electric forces in one dimension and problems 12 and 14 deal with the addition of electric forces in two dimensions. Also see problem 11.
2. Demonstrations
   a. Charging, electroscopes: Freier and Anderson Ea1, 2, 11.
   b. Electric force: Freier and Anderson Ea5, 6, 8, 12, 15, 17, Eb3, 4, 9, 10, 12, Ec4 — 6.
   c. Induction: Freier and Anderson Ea12, 13, 14.
   d. Touch a grounded wire to several places within a small area of a wall. Rub a balloon with fur and place it in contact with that area. Ask students to explain why the balloon sticks.
3. Books and Monographs
   Teaching about Electrostatics; by Robert A. Morse; available from AAPT, One Physics Ellipse, College Park MD 20740–3845. Describes reliable and inexpensive apparatus for demonstrations and student activities.
4. Audio/Visual
   a. *Electrostatics*; Side D: Waves (II) & Electricity and Magnetism of Cinema Classics; video disk; available from Ztek Co., PO Box 11768, Lexington, KY 40577–1768.
   b. *Static Electricity*; Demonstrations of Physics: Electricity and Magnetism; VHS video tape (7:12); Media Design Associates, Inc., Box 3189, Boulder, CO 80307–3190.
   c. *Electrostatic Generators*; Demonstrations of Physics: Electricity and Magnetism; VHS video tape (10:54); Media Design Associates, Inc. (see above for address).
   d. *Electrostatics 1*; VHS video tape (21 min); Films for the Humanities & Sciences, PO Box 2053, Princeton, NJ 08543–2053.
   e. *Electrostatics 2*; VHS video tape (21 min); Films for the Humanities & Sciences (see above for address).

f. *Conductors and Insulators*; Electricity; VHS video tape (10 min); Films for the Humanities & Sciences (see above for address).
g. *Charging and Discharging*; Electricity; VHS video tape (10 min); Films for the Humanities & Sciences (see above for address).
h. *Charging by Induction*; VHS video tape (10 min); Films for the Humanities & Sciences (see above for address).
i. *Electrostatics*; *Isolation of Charges*; Physics Demonstrations in Electricity and Magnetism, Part I; VHS video tape; ≈3 min each; Physics Curriculum & Instruction, 22585 Woodhill Drive, Lakeville, MN 55044.

5. Computer Software
   Electric Field Hockey; Ruth W. Chabay; Windows, Macintosh; available from Physics Academic Software, North Carolina State University, PO Box 8202, Raleigh, NC 27690–0739. The user tries to score a goal by placing stationary charges so they guide a charged puck around obstacles and into the net. The force on the puck can be shown as the puck moves.
6. Laboratory
   Meiners Experiment 10–2: *The Electrostatic Balance.* A coulomb torsional balance is used to find the functional relationship between the electrostatic force of one small charged ball on another and the separation of balls. An electrostatic generator is used to charge the balls.

Chapter 23 ELECTRIC FIELDS

BASIC TOPICS

I. The electric field.
   A. Use a fluid to introduce the idea of a field. The temperature of the fluid $T(x, y, z, t)$ is an example of a scalar field and the velocity $\vec{v}(x, y, z, t)$ is an example of a vector field. Point out that these functions give the temperature and velocity at the place and time specified by the dependent variables.
   B. Explain that charges may be thought to create an electric field at all points in space and that the field exerts a force on another charge, if present. The important questions to be answered are: Given the charge distribution, what is the field? Given the field, what is the force on a charge?
   C. Consider two point charges and remark that each creates a field and that the field of either one exerts a force on the other. Explain that the two together produce a field that is the superposition of the individual fields and that this field exerts a force on a third charge, if present.
   D. Define the field at any point as the force per unit charge on a positive test charge at the point, in the limit of a vanishingly small test charge. Mention that the limiting process eliminates the influence of the test charge on the charge creating the field. SI units: N/C.
   E. Use Coulomb's law to obtain the expression for the field of a point charge. Explain that the field of a collection of charges is the vector sum of the individual fields.

II. Electric field lines.
   A. Explain that field lines are useful for visualizing the field. Draw field lines for a point charge and explain that, in general, the field at any point is tangent to the line through that point and that the magnitude of the field is proportional to the number of lines per unit area that pass through a surface perpendicular to the lines.

B. By considering a sphere around a point charge and calculating the number of lines per unit area through the sphere, show that the $1/r^2$ law allows us to associate lines with a charge and to take the number of lines to be proportional to the charge. Explain that lines can be thought of as directed and that they originate at positive charge and terminate at negative charge. Emphasize that they are not vectors.

C. Show Figs. 23–2, 23–3, 23–4, and 23–5 or similar diagrams that illustrate the field lines of some charge distributions.

D. Field lines can be illustrated by floating some long seeds in transformer oil in a shallow, flat-bottomed dish. Place two metal plates in the dish and connect them to an electrostatic generator. The seeds line up along the field lines. You can place the apparatus on an overhead projector and shadow project the seeds.

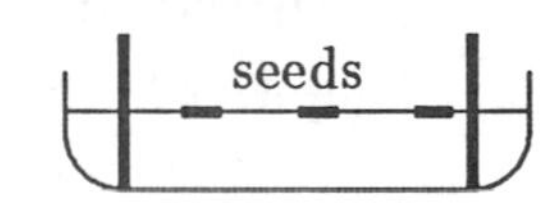

III. Calculation of the electric field.

A. Remind the students of the expression for the field of a point charge. State that the field is radially outward for a positive charge and radially inward for a negative charge. Also remind them that the total field is the vector sum of the individual fields of the charge being considered.

B. Derive an expression for the field of an electric dipole by considering the field of two charges with equal magnitudes and opposite signs. Consider a field point on a line perpendicular to the dipole moment, on a line along the dipole moment, or a general point. Evaluate the expression in the limit of vanishingly small separation and finite dipole moment. Define the dipole moment and stress that it points from the negative toward the positive charge. Point out that the field is proportional to $1/r^3$ for points far from the dipole.

C. Consider a small set of discrete charges and calculate the electric field by evaluating the vector sum of the individual fields. Example: the field at the center of a square with various charges on its corners.

D. As an introduction to the fields of continuous charge distributions, go over the ideas of linear and area charge densities. Graphically show how a line of charge is divided into infinitesimal segments and point out that a segment of length ds contains charge $dq = \lambda\, ds$. Explain that for purposes of calculating the field each segment can be treated as a point charge and that the fields of all segments are summed vectorially to find the total field.

E. Derive an expression for the field on the axis of a continuous ring of charge. Carefully explain how the integral is set up and how the vector nature of the field is taken into account by dealing with components. Explain in detail the symmetry argument used to show that the field is along the axis.

F. Extend the calculation to find an expression for the field on the axis of a charged disk and for an infinite sheet of charge. Remark that the field of a sheet is perpendicular to the sheet and is independent of distance from the sheet. This will be useful later when parallel plate capacitors are studied.

IV. Motion of a charge in an electric field.

A. Point out that the electric force on a charge is $q\vec{E}$ and explain that the electric field used is that due to all *other* charges (except q). Substitute the force into Newton's second law and remind the students that once the acceleration and initial conditions are known, kinematics can be used to find the subsequent motion of the charge.

B. Find the trajectory of a charge moving into a region of uniform field, perpendicular to its initial velocity. Compare to projectile motion problems studied in Chapter 4. See Sample Problem 23–4.

C. Show that the force on a dipole in a uniform field is zero and that the torque is $\vec{p} \times \vec{E}$. Also show that the potential energy of a dipole is $-\vec{p}\cdot\vec{E}$. Emphasize that the potential energy minimum occurs when the dipole moment is aligned with the field. To review oscillatory rotational motion calculate the angular frequency of small angle oscillations for a dipole with rotational inertia I in a uniform electric field. Assume no other forces act.

SUGGESTIONS

1. Assignments
   a. Assign problem 28 to illustrate the measurement of an electric field.
   b. Center a qualitative discussion of electric field lines on question 1. Have students sketch field lines for various charge distributions. See problems 2, 3, and 10.
   c. Ask questions 2, 3, 4, and 5 and have students work some of problems 6, 8, 9, 12, and 13. These deal with the superposition of fields.
   d. Problems 14 and 16 are a good tests of understanding of the derivation of the dipole field.
   e. Ask questions 6 and 7 and assign problems 20 and 21 in support of the calculation of the field of a ring of charge. Assign problems 23 and 24 to give students practice in deriving expressions for the field of a continuous charge distribution.
   f. Assign problem 27 to support the discussion of the field of a uniformly charged disk.
   g. Assign problems 36, 37, and 43 to help students with the motion of point charges in fields. Assign questions 10 and 11 and problems 45 (torque) and 46 (energy) in connection with the discussion of a dipole in a field.
   h. To include the Millikan oil drop experiment, assign problems 37 and 38.
2. Demonstrations
   Electric field lines: Freier and Anderson Eb1, Ec2 — 4.
3. Audio/Visual
   a. *Electrostatic Induction*; *The Van de Graaff Generator*; *Field as a Vector*; from AAPT collection 2 of single-concept films; video tape; available from Ztek Co., PO Box 11768, Lexington, KY 40577–1768.
   b. *Millikan's Oil-Drop Experiment*; Physics Experiments; VHS video tape (15 min); Films for the Humanities & Sciences, PO Box 2053, Princeton, NJ 08543–2053.
   c. *Electric Fields*; Physics Demonstrations in Electricity and Magnetism, Part II; VHS video tape; ≈3 min; Physics Curriculum & Instruction, 22585 Woodhill Drive, Lakeville, MN 55044.
4. Computer Software
   a. *Electric Field*; Socrates Software, 6187 Rosecommon Drive, Norcross, GA 30092. Graphically displays field lines, force vectors, and equipotential lines for a variety of charge distributions.
   b. *Electric Field Plotter*; Windows; Bob Nelson; Physics Academic Software, North Carolina State University, PO Box 8202, Raleigh, NC 27690–0739. Draws electric field lines and equipotential lines. Students can place up to nine charges anywhere on the screen. The program also searches for points where the electric field vanishes.
   c. *EM Field*; David Trowbridge; Windows, Macintosh; available from Physics Academic Software (see above for address). Plots electric field lines and equipotential surfaces of point and line charges, set up by the user. The electric field vectors can be shown. Plots magnetic field lines of user-selected current distributions. Users can draw gaussian surfaces or amperian paths. The program gives the flux through the surface or the contribution to $\int \vec{B}\cdot d\vec{s}$. When the surface or line is closed the program gives the charge or current enclosed. A game asks the user to find hidden charge or current.
   d. *Electric Field Plotter*: Bob Nelson; Windows; available from Physics Academic Software (see above for address). The user can place up to nine charges in the plane of the screen.

The program plots electric field lines and the intersections of equipotential surfaces with the plane of the screen. It can also search for points where the field is zero.

e. *Charged particles*; Vernier Software, 8565 S.W. Beaverton-Hillside Hwy., Portland, OR 97225–2429. A simulation program that shows the motion of charges in an electric field.

f. *Dynamic Analyzer*. See Chapter 5 SUGGESTIONS.

5. Computer Projects

a. Have students use a commercial math program or write their own programs to calculate the electric fields of discrete charge distributions. Have them use the programs to plot the magnitude of the field at various distances from a dipole, along lines that are perpendicular and parallel to the dipole moment.

b. Have students write programs to trace field lines for discrete charge distributions.

These and other projects are described in the Computer Projects Section of this manual.

Chapter 24 GAUSS' LAW

BASIC TOPICS

I. Electric flux.

A. Start by discussing some of the important concepts in a general way. Define a vector surface element. Define the flux of a vector field through a surface. Distinguish between open and closed surfaces and explain that for the latter the surface normal is taken to be *outward*. Interpret the surface integral for the flux as a sum over surface elements. If you covered Chapter 15, use the velocity field of a fluid as an example.

B. Define electric flux. Point out that it is the normal component of the field that enters. Also point out that the sign of the contribution of any surface element depends on the choice for the direction of $d\vec{A}$.

C. Interpret electric flux as a quantity that is proportional to the net number of field lines penetrating the surface. Remind students that the number of lines through a small area perpendicular to the field is taken to be proportional to the magnitude of the field. By considering surfaces with the same area but different orientations, show that the net number of penetrating lines is proportional to the cosine of the angle between the field and the normal to the surface. Conclude that $\vec{E}\cdot d\vec{A}$ is proportional to the number of lines through $d\vec{A}$.

D. Stress that lines roughly in the same direction as the normal contribute positively to the flux, lines roughly in the opposite direction contribute negatively, and lines that pass completely through a volume do not contribute to the flux through its boundary. Point out that zero flux through a surface does not imply zero field at points on the surface.

E. As an example, calculate the flux through each side of a cube in a uniform electric field. Also consider Sample Problem 24–2, which deals with a nonuniform field.

II. Gauss' law.

A. Write down the law. Stress that the surface is closed and that the charge appearing in the law is the net charge enclosed. Interpret the law as a statement that the number of (signed) lines crossing the surface is proportional to the net charge inside, and make the statement plausible by reminding students that the field of each charge is proportional to the charge and its direction depends on the sign of the charge.

B. Illustrate by considering the surface of a sphere with positive charge inside, with negative charge inside, with both positive and negative charge inside, and with charge outside. In each case draw representative field lines with the number of lines proportional to the net charge. Stress that the position of the charge inside is irrelevant for the flux through the surface. Also use Gauss' law to calculate the flux.

C. Use Gauss' law and symmetry arguments to obtain an expression for the electric field of a point charge.

III. Gauss' law and conductors.

A. Argue that the electrostatic field vanishes inside a conductor and use Gauss' law to show that there can be no net charge at interior points under static conditions. Point out that exterior charge and charges on the surface separately produce fields in the interior but that the resultant field vanishes. For contrast, point out that an insulator may have charge distributed throughout.

B. Demonstrate that any excess charge on a conductor resides on the exterior surface. Use a hollow metal sphere with a small hole cut in it. As an alternative, solder shut the top of an empty metal can and drill a small hole in it. This will not work as well because of the sharp edges. Charge a rubber rod by rubbing it with fur and touch it to the inside of the sphere, being careful not to touch the edge of the hole. Repeat several times to build up charge. Now scrape at the interior with a metal transfer rod, again being careful not to touch the edge of the hole. Touch the transfer rod to an uncharged electroscope and note the lack of deflection. Scrape the exterior of the sphere with the transfer rod and touch the electroscope. Note the deflection.

C. Show how to calculate the charge on the inner and outer surfaces of neutral and charged conducting spherical shells when charge is placed in the cavities. See Sample Problem 24–4.

D. Use Gauss' law to show that the magnitude of the field just outside a charged conductor is given by $E = \sigma/\epsilon_0$, where σ is the surface charge density.

IV. Applications of Gauss' law.

A. Derive expressions for the electric field at various points for a uniformly charged sphere and for a uniformly charged thick spherical shell. Remark that such distributions are possible if the sphere or shell is not conducting. Carefully give the symmetry argument to show the field is radial and has the same magnitude at all points on a concentric sphere.

B. Derive an expression for the electric field at a point outside an infinite sheet with a uniform charge distribution. Contrast with the field outside an infinite conducting sheet with the same area charge density on one surface. Point out that for the conductor the field is not due only to the charge on the surface being considered. Another field must be present to produce a net field of zero in the interior and this doubles the field in the exterior.

C. Consider a point charge at the center of a neutral spherical conducting shell and derive expressions for the electric field in the various regions. Repeat for a charged shell.

D. Work one problem with cylindrical symmetry. For example, consider charge distributed uniformly throughout a cylinder and find the field in all regions.

E. Note that Gauss' law can be used to find $\vec{E}$ only if there is adequate symmetry.

SUGGESTIONS

1. Assignments

a. Use questions 1 through 6 to help students understand the flux integral and charge that appear in Gauss' law. Problem 1 is a good example of a fluid flux calculation. Assign it if you covered Chapter 15. Use problems 2 and 3 to introduce electric flux. The latter problem also demonstrates the vanishing of the total flux for a closed surface in a uniform field.

b. Problem 5 illustrates the fundamental idea of Gauss' law. Problems 6 and 7 are also instructive.

c. Use questions 8 and 9 and problem 15 to discuss the electrostatic properties of conductors.
d. Assign a variety of problems dealing with applications: 20 (cylinder of charge); 26, 32, and 33 (plane of charge); 36, 40, and 43 (sphere of charge). Assign problem 41, 45, or 46 to challenge good students.

2. Demonstrations
   Charges on conductors: Freier and Anderson Ea7, 18, 23, Eb7.
3. Audio/Visual
   Charge Distribution — Faraday Ice Pail Experiment; from AAPT collection 2 of single-concept films; video tape; available from Ztek Co., PO Box 11768, Lexington, KY 40577–1768.
4. Computer Software
   a. *EM Field.* See Chapter 23 SUGGESTIONS.
   b. *Electricity & Magnetism*; Macintosh, Windows; Cross Educational Software, Inc., 508 E. Kentucky Avenue, PO Box 1536, Ruston, LA 71270. Contains a section on Gauss' law.
5. Computer Project
   Have students use a commercial math program or their own programs to evaluate the flux integral in Gauss' law. Have them separately calculate the flux through each face of a cube containing a point charge. Consider various positions of the charge within the cube to show that the flux through individual faces may change as the charge changes position but the total flux remains the same and obeys Gauss' law. Repeat for a point charge outside the cube. Details are given in the Computer Projects Section of this manual.

Chapter 25 ELECTRIC POTENTIAL

BASIC TOPICS

I. Electric potential.
   A. Define the potential difference of two points as the negative of the work per unit charge done by the electric field when a positive test charge moves from one point to the other. Stress the sign of the potential: the potential of the end point is higher than that of the initial point if the work is negative. The electric field points from regions of high potential toward regions of low potential and positive charge tends to be repelled from regions of high potential. The region near an isolated positive charge has a higher potential than regions far away. The opposite is true for a negative charge. Unit: volt. Define electron volt as a unit of energy.
   B. If you covered Chapter 14, use the similarity of Coulomb's law and Newton's law of gravity to argue that the electrostatic force is conservative and that the work is independent of path. If you did not cover Chapter 14, either derive or state these results.
   C. Show that the definition is equivalent to $V_b - V_a = -\int \vec{E} \cdot d\vec{s}$, where the integral is along a path from a to b. Point out that the potential is constant in regions of zero field. Note that the unit N/C is the same as V/m and the latter is a more common unit for $\vec{E}$.
   D. Point out that the potential is a scalar and that only potential differences are physically meaningful. One point can be chosen arbitrarily to have zero potential and the potential at other points is measured relative to the potential there. Often the potential is chosen to be zero where the field (or force) is zero. For a finite distribution of charge, the potential is usually chosen to be zero at a point far away (infinity). Show a voltmeter and remark that the meter reads the potential difference between the leads.
   E. Show that the potential a distance r from an isolated point charge is given by $V = q/4\pi\epsilon_0 r$. Remark that this is the potential energy per unit test charge of a system consisting of the

point charge q and the test charge. Explain that the equation is valid for both positive and negative charge. Show how to calculate the potential due to a collection of point charges. Derive the expression for the potential of an electric dipole.

F. Give some examples of calculations of the potential from the electric field. Start with a uniform electric field, like that outside a uniform plane distribution of charge, and show that potential is given by $-Ex + C$, where C is a constant. Since the distribution is infinite the point at infinity cannot be picked as the zero of potential.

F. As a more complicated example, consider one of the configurations discussed in the last chapter, a point charge at the center of a spherical conducting shell, say. Take the potential to be zero at infinity and compute its value at points outside the outer surface, within the shell, and inside the inner surface. As an alternative you might find expressions for the potential in various regions around and inside a sphere with a uniform charge distribution.

G. Write down the integral expressions for the potential due to a line of charge and for a surface of charge, in terms of the linear and area charge densities. Work an example, such as the potential of a uniform finite line of charge or a uniform disk of charge.

II. Equipotential surfaces.

A. Define the term equipotential surface. Show diagrams of equipotential surfaces for an isolated point charge and for the region between two uniformly charged plates. Equipotential surfaces of a dipole are shown in Fig. 25–3(c).

B. Point out that the field does zero work if a test charge is carried between two points on the same surface and note that this means that the force, and hence $\vec{E}$, is perpendicular to the equipotential surfaces. Note further that the work done by the field when a charge is carried from any point on one surface to any point on another is the product of the charge and the negative of the potential difference.

III. Calculation of $\vec{E}$ from V.

A. Remind students that $\Delta V = -E\Delta x$ for a uniform field in the positive x direction. Note that E has the form $-\Delta V/\Delta x$ and $\vec{E}$ is directed from high to low potential. Use this result to reenforce the idea of an equipotential surface and the fact that $\vec{E}$ is perpendicular to equipotential surfaces.

B. Generalize the result to $E = -\mathrm{d}V/\mathrm{d}s$, where s is the distance along a normal to an equipotential surface. Then specialize this to $E_x = -\partial V/\partial x$, $E_y = -\partial V/\partial y$, and $E_z = -\partial V/\partial z$. Verify that the prescription works for a point charge and for a dipole.

IV. Electrostatic potential energy.

A. Remark that when a charge Q moves from point a to point b the potential energy of the system changes by $Q(V_b - V_a)$, where V is the potential due to the other charges. When charge Q is brought into position from infinity (where the potential is zero), the potential energy changes by QV, where V is the potential at the final position of Q due to charge already in place.

B. Show that the potential energy of two point charges is given by $q_1 q_2/4\pi\epsilon_0 r$, where r is their separation and the zero of potential energy is taken to be infinite separation. Point out that the potential energy is positive if the charges have like signs and negative if they have opposite signs. Explain that the potential energy decreases if two charges of the same sign move apart or two charges of opposite sign move closer together.

C. Remind students that potential energy can be converted to kinetic energy. Explain what happens if the charges used in the last example are released from their positions. Consider a proton fired directly at a heavy nucleus with charge Ze and find the distance of closest approach in terms of the initial speed.

D. Calculate the potential energy of a simple system: charges at the corners of a triangle or square, for example. Assume the charges are brought in from infinity one at a time and sum the potential energies. Explain that the total is the sum over charge pairs. Show how to calculate the potential energy of any collection of point charges.

E. Explain that the potential energy of a system of charges is the work an agent must do to assemble the system from rest at infinite separation. This is the negative of the work done by the field.

V. An isolated conductor.

A. Recall that the electric field vanishes at points in the interior of a conductor. Argue that the surface must be an equipotential surface and that V at all points inside must have the same value as on the surface. State that this is true if the conductor is charged or not and if an external field exists or not.

B. Consider two spherical shells of different radii, far apart and connected by a very fine wire. Explain that $V_1 = V_2$ and show that $q_1/R_1 = q_2/R_2$. Then show that the surface charge density varies inversely with the radius: $\sigma_1/\sigma_2 = R_2/R_1$. Recall that E is proportional to σ just outside a conductor and argue that σ and E are large near places of small radius of curvature and small near places of large radius of curvature. Use an electrostatic generator to show discharge from a sharp point and from a rounded (larger radius) ball. Discuss the function of lightning rods and explain their shape.

SUGGESTIONS

1. Assignments
   a. Questions 1 through 5 can be used to help students think about some qualitative aspects of electric potential.
   b. Use question 6 and problems 5 and 14 to test for understanding of equipotential surfaces.
   c. Ask students to calculate potential differences for various situations: see problems 6, 9, 10, 11, 21, 22, and 28.
   d. Use questions 8, 9, and 10 and problems 37, 39, and 41 in connection with the discussion of electrostatic potential energy and the work done by an electric field or an external agent. Also assign some conservation of energy problems, such as 43 and 47.
   e. Assign problems 51, 52, and 56 to aid in a discussion of the field and potential of a conductor.
2. Demonstrations
   Electrostatic generators: Freier and Anderson Ea22, Ec1.
3. Audio/Visual
   a. *Potential Difference*; Electricity; VHS video tape (10 min); Films for the Humanities & Sciences, PO Box 2053, Princeton,, NJ 08543–2053.
4. Computer Software
   a. *Electric Field Plotter*; Windows; Bob Nelson; Physics Academic Software, North Carolina State University, PO Box 8202, Raleigh, NC 27690–0739. Draws electric field lines and equipotential lines. Students can place up to nine charges anywhere on the screen. The program also searches for points where the electric field vanishes.
   b. *EM Field*. See Chapter 23 SUGGESTIONS.
5. Computer Project
   Have students use a commercial math program or their own root finding programs to plot equipotential surfaces for a discrete charge distribution. It is instructive to consider two unequal charges (any combination of signs). Details are given in the Computer Projects portion of this manual.

6. Laboratory

Meiners Experiment 10–1: *Electric Fields* and Bernard Experiment 25: *Mapping of Electric Fields.* Students map equipotential lines on sheets of high resistance paper with metallic electrodes at two sides. In the Meiners experiment an audio oscillator generates the field and an oscilloscope or null detecting probe is used to find points of equal potential. If students are not familiar with oscilloscopes, you might want to preface this experiment with Part A of Meiners Experiment 10–10. In the Bernard experiment the field is generated by a battery and a galvanometer is used as a probe.

Chapter 26 CAPACITANCE

BASIC TOPICS

I. Capacitance.

A. Describe a generalized capacitor. Draw a diagram showing two separated, isolated conductors. Assume they carry charge q and $-q$, respectively, draw representative field lines, and point out that all field lines start on one conductor and terminate on the other. Explain that there is a potential difference V between the conductors and that the positively charged conductor is at the higher potential. Define capacitance as $C = q/V$. Explain that V is proportional to q and that C is independent of q and V. C does depend on the shapes, relative positions, and orientations of the conductors and on the medium between them. Unit: $1\,\text{farad} = 1\,\text{C/V}$.

B. Show a radio tuning capacitor and some commercial fixed capacitors. Mention that one usually encounters μF and pF capacitors. Capacitors on the order of 1 F have been developed for the electronics industry.

C. Remark that in circuit drawings a capacitor is denoted by ⊣⊢.

D. State that a battery can be used to charge a capacitor. The battery transfers charge from one plate to the other until the potential difference of the plates is the same as the terminal potential difference of the battery. Calculate the charge, given the battery potential difference and the capacitance.

E. In a general way, give the steps required to calculate capacitance: put charge q on one conductor, $-q$ on the other, and calculate the electric field due to the charge, then calculate the potential difference V between the conductors, and finally use $q = CV$ to find the capacitance. Except for highly symmetric situations, the charge is not uniformly distributed over the surfaces of the conductors and fairly sophisticated means must be used to calculate V. The text deals with symmetric situations for which Gauss' law can be used to calculate the electric field.

F. Examples: derive expressions for the capacitance of two parallel plates (neglect fringing) and two coaxial cylinders or two concentric spherical shells. Use Gauss' law to find the electric field, then evaluate the integral for the potential difference. Emphasize that the field is due to the charge on the plates.

G. Large demonstration parallel plate capacitors with variable plate separations are available commercially. You can also make one using two $\approx$ 1-ft diameter circular plates of 1/8 inch aluminum sheeting. Attach an aluminum disk to the center of each with a hole drilled for a support rod. Use an insulating rod on one and a metal rod on the other. By sliding the two conductors closer together, you can show the effect of changing d while holding q constant. An electroscope serves as a voltmeter.

II. Explain how the equivalent capacitance of a device can be measured. Consider a black box with two terminals. State that a potential difference V is applied and the total charge q deposited is measured. The capacitance is q/V.

I. Derive $1/C_{eq} = 1/C_1 + 1/C_2$ for the equivalent capacitance of two capacitors in series and $C_{eq} = C_1 + C_2$ for the equivalent capacitance of two capacitors in parallel. Emphasize that two capacitors in parallel have the same potential difference and that two in series have the same charge. Explain the usefulness of these equations for circuit analysis.

II. Energy storage.

A. Derive the expression $W = \frac{1}{2}q^2/C$ for the work required to charge a capacitor. Explain that, as an increment of charge is transferred, work is done by an external agent (a battery, for example) against the electric field of the charge already on the plates. Show that this expression is equivalent to $W = \frac{1}{2}CV^2$. Interpret the result as the potential energy stored in the charge system and explain that it can be recovered when the capacitor is discharged.

B. Remark that if two capacitors are in parallel the larger stores the greater energy. If two capacitors are in series, the smaller stores the greater energy.

C. Show that the energy density in a parallel plate capacitor is $\frac{1}{2}\epsilon_0 E^2$. State that this result is quite general and that its volume integral gives the work required to assemble charge to create the electric field E. Explain that the energy may be thought to reside in the field or it may be considered to be the potential energy of the charges.

D. Integrate the energy density to find an expression for the energy stored in the electric field of a charged spherical capacitor or a charged cylindrical capacitor. Compare the result with $\frac{1}{2}q^2/C$.

III. Dielectrics.

A. Explain that when the region between the conductors of a capacitor is occupied by insulating material the capacitance is greater by a factor $\kappa > 1$, called the dielectric constant of the material. Remark that $\kappa = 1$ for a vacuum.

B. Use a large commercial or homemade capacitor to show the effect of a dielectric. Charge the capacitor, then isolate it and insert a glass plate between the plates. The electroscope shows that V decreases and, since q is fixed, the capacitance increases.

C. Calculate the change in stored energy that occurs when a dielectric slab is inserted between the plates of an isolated parallel plate capacitor (see Sample Problem 26–5). Also calculate the change in stored energy when the slab is inserted while the potential difference is maintained by a battery. Explain that the battery now does work in moving charge from one plate to the other.

D. Explain that dielectric material between the plates becomes polarized, with the positively charged ends of the dipoles attracted toward the negative conductor. The field of the dipoles opposes the external field, so the electric field is weaker between the plates than it would be if the material were not there. This reduces the potential difference between the conductors for a given charge on them. Since the potential difference is less for the same charge on the plates, the capacitance is greater.

E. Explain that if the polarization is uniform, the material behaves like neutral material with charge on its surfaces.

F. Optional: Show how Gauss' law can be written in terms of $\kappa\vec{E}$ and the free charge. Show how to compute the polarization charge for a parallel plate capacitor with dielectric material between its plates.

SUGGESTIONS

1. Assignments

a. Use question 1 to emphasize the dependence of capacitance on geometry.

b. The fundamental idea of capacitance is illustrated by problem 2. Assign problem 8 to have students compare spherical and plane capacitors. Problem 6 covers the dependence of the capacitance of a parallel plane capacitor on area and separation.

c. Include some of questions 4 through 11 in the discussion of series and parallel connections of capacitors. Problems 10 and 12 cover equivalent capacitance, charge, and potential difference for series and parallel combinations. Also consider assigning some problems in which students must find the equivalent capacitance of more complicated combinations. See problems 11 and 13, for example. Problem 20 is more challenging.
d. Problem 24 covers most of the important points discussed in connection with energy storage. Also assign problem 32, which deals with the energy density around a charged metal sphere and problem 29, which deals with the energy needed to separate the plates of a parallel plate capacitor. Question 10 can be asked in connection with this topic.
e. Include question 11 in the discussion of the influence of a dielectric on capacitance. Assign problems 36 and 38. Also consider problem 47.
f. To test understanding of induced polarization charge, assign problem 43 or 45.

2. Demonstrations
   a. Charge storage: Freier and Anderson Eb8, Ed3, 7.
   b. Capacitance and voltage: Freier and Anderson Ed1.
   c. Energy storage: Freier and Anderson Ed8
   d. Dielectrics: Freier and Anderson Ed2, 4.
3. Audio/Video
   a. *Capacity of a Condenser*; VHS video tape (14 min); Films for the Humanities & Sciences, PO Box 2053, Princeton, NJ 08543–2053.
4. Computer Software
   Electricity & Magnetism; Macintosh, Windows; Cross Educational Software, Inc., 508 E. Kentucky Avenue, PO Box 1536, Ruston, LA 71270. Contains a section on capacitors.
5. Laboratory
   a. Meiners Experiment 10–7 (Part B): *Measuring Capacitance with a Ballistic Galvanometer.* A ballistic galvanometer is used to measure the capacitance of individual capacitors and capacitors in series and parallel. Students must temporarily accept on faith that the deflection of the galvanometer is proportional to the total charge that passes through it.
   b. Meiners Experiment 11–2 (Part C): *Coulomb Balance Attachment (to the current balance).* Students use gravitational force to balance the force of one capacitor plate on the other. The voltage and plate separation are used to find the charge on the plates, then ϵ_0 is calculated.

Chapter 27 CURRENT AND RESISTANCE

BASIC TOPICS

I. Current and current density.
   A. Explain that an electric current is moving charge. Draw a diagram of a long straight wire with positive charge moving in it. Consider a cross section and state that the current is dq/dt if charge dq passes the cross section in time dt. Give the sign convention: both positive charge moving to the right and negative charge moving to the left constitute currents to the right. Early on, use the words "conventional current" quite often. Later "conventional" can be dropped. Many high school courses now take the current to be in the direction of electron flow and it is worthwhile making the effort to reduce confusion in students' minds. Unit: $1\,\text{ampere} = 1\,\text{C/s}$.
   B. Explain that under steady state conditions, in which no charge is building up or being depleted anywhere in the wire, the current is the same for every cross section. Remark that current is a scalar, but arrows are used to show the direction of positive charge flow.

C. Explain that current is produced when charge is free to move in an electric field. For most materials, it is the negative electrons that move and their motion is opposite to the direction of the electric field. Current is taken to be in the direction opposite to that of electron drift, in the direction of the field.

D. Distinguish between the drift velocity and the velocities of individual charges. Note that the drift velocity of electrons in an ordinary wire is zero unless an electric field is turned on. Also note that the drift speed is many orders of magnitude smaller than the average electron speed.

E. Explain that current density is a microscopic quantity used to describe current flow at a point. Use the same diagram but now consider a small part of the cross section and state that $J = i/A$ in the limit as the area diminishes to a point. State that current density is a vector in the direction of the drift velocity for positive charge and opposite the drift velocity for negative charge. Explain that $i = \int \vec{J} \cdot d\vec{A}$ is the current through a finite surface, where $d\vec{A}$ is normal to the surface. This reduces to $J = i/A$ for uniform current density and an area that is perpendicular to the current. Unit: A/m^2.

F. Derive $\vec{J} = en\vec{v}_d$ and show how to calculate the drift speed from the free-electron concentration and current in the wire, assuming uniform current density. You may want to go over the calculation of the free-electron concentration from the mass density of the sample and the molar masses of its constituents.

II. Resistance and resistivity.

A. Define resistance by $R = V/i$ and point out that R may depend on V. Unit: $1\,\text{ohm} = 1\,\text{V/A}$; abbreviation: Ω. Also define resistivity ρ and conductivity σ. Point out Table 27–1. Explain that the latter quantities are characteristic of the material while resistance also depends on the sample shape and the positions of the current leads.

B. Make a sketch similar to the one shown here. Indicate that $V_a - V_b = iR$ is algebraically correct, even if i is negative, and effectively defines the resistance of the sample with the leads connected at a and b. Emphasize that the point at which the current enters is iR higher in potential than the point at which it leaves.

C. Show that $R = \rho L/A$ for a conductor with uniform cross section A and length L, carrying a current that is uniformly distributed over the cross section.

D. Point out that for many samples the current is proportional to the potential difference and the resistance is independent of the voltage applied. These materials are said to obey Ohm's law. Also point out that many important materials do not obey Ohm's law. Show Fig. 27–11.

E. Use a variable-voltage power supply and connect, in turn, samples of ohmic (carbon resistor) and non-ohmic (solid state diode) material across the terminals. Use analog meters to display the current and potential difference and vary the supply smoothly and fairly rapidly. For the ohmic material, it will be apparent that i is proportional to V, while for the non-ohmic material, it will be apparent that i is not proportional to V.

F. Give a qualitative description of the mechanism that leads to Ohm's law behavior. Explain that collisions with atoms cause the drift velocity to be proportional to the applied field. Assume the electrons have zero velocity after each collision and that they accelerate for a time τ between collisions. Show that an electron goes the same distance on the average during the first five collisions as it does during the second five so the drift velocity is proportional to the field even though the electron accelerates between collisions. Now consider the quantitative aspects: derive the expression for the drift velocity in terms of $\vec{E}$ and the mean free time τ, then derive $\rho = m/ne^2\tau$. Emphasize that the mean free time is determined by the electron speed and, since drift is an extremely small part of the speed, τ is essentially independent of the electric field. Point out that a long mean free time means a small resistivity because the electrons accelerate for a longer time between collisions and thus have a higher drift speed.

G. Remark that the resistivity of a sample depends on the temperature. Define the temperature coefficient of resistivity and point out the values given in Table 27–1.

III. Energy considerations.

A. Point out that when current flows from the high to the low potential side of any device, energy is transferred from the current to the device at the rate $P = iV$. Reproduce Fig. 27–13 and note that $P = i(V_a - V_b)$ is algebraically correct if P is the power supplied to the device. Note that if P is negative the device is supplying energy at the rate $-P$.

B. Give examples: Energy may be converted to mechanical energy (a motor), to chemical energy (a charging battery), or to internal energy (a resistor). Also note the converse: mechanical energy (a generator), chemical energy (a discharging battery), and internal energy (a thermocouple) may be converted to electrical energy.

C. Explain that in a resistor the electrical potential energy of the free electrons is converted to kinetic energy as the electric field does work on them and that the kinetic energy is lost to atoms in collisions. This increases the thermal motion of the atoms. Show that the rate of energy loss in a resistor is given by $P = i^2R = V^2/R$.

SUPPLEMENTARY TOPICS

1. Semiconductors
2. Superconductors

Both topics are important for modern physics and technology. Say a few words about them if you have time or encourage students to read about them on their own.

SUGGESTIONS

1. Assignments
   a. Discuss question 2 to emphasize the sign convention for current.
   b. Use questions 6 and 7 in a discussion of current density, resistivity, and drift velocity. Definitions are covered in problems 1 (current), 5 (current density), and 4 (drift speed).
   c. Use questions 3, 4, and 5 when you discuss the calculation of resistance. Assign problems 19, 20, and 23. For a greater challenge assign problem 29.
   d. As part of the coverage of energy dissipation by a resistor, assign problems 35 and 38.
2. Demonstrations
   a. Model of resistance: Freier and Anderson Eg1.
   b. Thermal dissipation by resistors: Freier and Anderson Eh3.
   c. Fuses: Freier and Anderson Eh5.
   d. Ohm's law: Freier and Anderson Eg2, Eo1.
   e. Measurement of resistance, values of resistance: Freier and Anderson Eg3, 6.
   f. Temperature dependence of resistance: Freier and Anderson Eg4, 5.
3. Audio/Visual
   a. *Electric Currents*; Side D: Waves (II) & Electricity and Magnetism of Cinema Classics; video disk; available from Ztek Co., PO Box 11768, Lexington, KY 40577–1768.
   b. *Voltage*; VHS video tape (15 min); Films for the Humanities & Sciences, PO Box 2053, Princeton, NJ 08543–2053.
   c. *Resistance*; VHS video tape (10 min); Films for the Humanities & Sciences (see above for address).

d. *Current Electricity*; Electricity; VHS video tape (10 min); Films for the Humanities & Sciences (see above for address).
e.]it Electric Current; VHS video tape; Films for the Humanities & Sciences (see above for address).
f. *Direct Current and Alternating Current*; VHS video tape (14 min); Films for the Humanities & Sciences (see above for address).
g. *Temperature and Resistance*; Physics Demonstrations in Electricity and Magnetism, Part II; VHS video tape; ≈3 min; Physics Curriculum & Instruction, 22585 Woodhill Drive, Lakeville, MN 55044.

4. Laboratory
   a. Meiners Experiment 10–3: *Electrical Resistance.* An ammeter and voltmeter are used to find the resistance of a light bulb and wires of various dimensions, made of various materials. The dependence of resistance on length and cross section is investigated. Resistivities of the substances are calculated and compared.
   b. Bernard Experiment 29: *A Study of the Factors Affecting Resistance.* A Wheatstone bridge and a collection of wire resistors are used to investigate the dependence of resistance on length, cross section, temperature, and resistivity.
   c. Meiners Experiment 10–8: *Temperature Coefficient of Resistors and Thermistors.* A Wheatstone bridge is used to measure the resistances of a resistor and thermistor in a water-filled thermal reservoir. The temperature is changed by an immersion heater. Students see two different behaviors. A voltmeter-ammeter technique can replace the bridge if desired.
   d. Also see Meiners Experiment 9–2 and Bernard Experiment 30, described in the Chapter 20 notes. These experiments can be revised to emphasize the power dissipated by a resistor. In several runs the students measure the power dissipated for different applied voltages.

Chapter 28 CIRCUITS

BASIC TOPICS

I. Emf devices.
   A. Explain that an emf device moves positive charge inside from its negative to its positive terminal or negative charge in the opposite direction and maintains the potential difference between its terminals. Emf devices are used to drive currents in circuits. Example: a battery is an emf device with an internal resistance. Note the symbol used in circuit diagrams to represent an ideal emf device (no internal resistance).
   B. Explain that a direction is associated with an emf and that it is from the negative to the positive terminal, inside the device. This is the direction current would flow if the device acted alone in a completed circuit. Point out that when current flows in this direction the device does positive work on the charge and define the emf of an ideal device as the work per unit positive charge: $\mathcal{E} = \mathrm{d}W/\mathrm{d}q$. Also point out that the positive terminal of an ideal device is $\mathcal{E}$ higher in potential than the negative terminal, regardless of the direction of the current. Unit: volt.
   C. Point out that the rate at which energy is supplied by an ideal device is $i\mathcal{E}$. State that for a battery the energy comes from a store of chemical energy. Mention that a battery is charging if the current and emf are in opposite directions.

II. Single loop circuits.
   A. Consider a circuit containing a single ideal emf and a single resistor. Use energy considerations to derive the steady state loop equation (Kirchhoff's loop rule): equate the power

supplied by the emf to the power loss in the resistor.

B. Derive the loop equation by picking a point on the circuit, selecting the potential to be zero there, then traversing the circuit and writing down expressions for the potential at points between the elements until the zero potential point is reached again. Tell the students that if the current is not known a direction must be chosen for it and used to determine the sign of the potential difference across the resistor. When the circuit equation is solved for i, a negative result will be obtained if the current is actually opposite in direction to the arrow. As you carry out the derivation remind students that current enters a resistor at the high potential end and that the positive terminal of an emf is at a higher potential than the negative terminal.

C. Consider slightly more complicated single loop circuits. Include the internal resistance of the battery and solve for the current. Place two batteries in the circuit, one charging and the other discharging. Once the current is found, calculate the power gained or lost in each element.

D. For the circuits considered, show how to calculate the potential difference between two points on the circuit and point out that the answer is independent of the path used for the calculation. Explain the difference between the closed and open circuit potential difference across a battery.

III. Multiloop circuits.

A. Explain Kirchhoff's junction rule for steady state current flow. State that it follows from the conservation of charge and the fact that charge does not build up anywhere when the steady state is reached.

B. Using an example of a two-loop circuit, go over the steps used to write down the loop and junction equations and to solve for the currents. Explain that if the current directions are unknown an arbitrary choice must be made in order to write the equations and that if the wrong choice is made, the values obtained for the current will be negative.

C. Warn students not to write duplicate junction equations. Define a branch and state that different symbols must be used for currents in different branches. State that the total number of equations will be the same as the number of branches, that the number of independent junction equations equals one less than the number of junctions, and that the remaining equations are loop equations. Also state that each current must appear in at least one loop equation.

D. Derive expressions for the equivalent resistance of two resistors in series and in parallel. Contrast with the expressions for the equivalent capacitance of two capacitors in series and in parallel. Show how to calculate potential differences across resistors in series and currents in resistors in parallel. Show how series and parallel combinations can sometimes be used to solve complicated circuits. Mention that not all circuits can be considered combinations of series and parallel connections.

IV. *RC* circuits.

A. Consider a series circuit consisting of an emf, a resistor, a capacitor, and a switch. Suppose the switch is closed at time $t = 0$ with the capacitor uncharged. Use the loop rule and $i = \mathrm{d}q/\mathrm{d}t$ to show that $R(\mathrm{d}q/\mathrm{d}t) + (q/C) = \mathcal{E}$. By direct substitution, show that $q(t) = C\mathcal{E}[1 - e^{-t/RC}]$ satisfies this equation and yields $q = 0$ for $t = 0$. Also find expressions for the potential differences across the capacitor and across the resistor. Show that $q = C\mathcal{E}$ for times long compared to RC. State that $i = \mathrm{d}q/\mathrm{d}t$ only if the current arrow is into the positive plate of the capacitor. If it is into the negative plate, then $i = -\mathrm{d}q/\mathrm{d}t$.

B. Explain that $\tau = RC$ is called the time constant for the circuit and that it is indicative of the time required to charge the capacitor. If RC is large, the capacitor takes a long time to charge. Show that $q/C\mathcal{E} \approx 0.63$ when $t = \tau$.

C. Show that the current is given by $i(t) = (\mathcal{E}/R)e^{-t/RC}$. Point out that $i = \mathcal{E}/R$ for $t = 0$ and that the potential difference across the capacitor is zero at that time because the capacitor is uncharged. Thus the potential difference across the resistor is $\mathcal{E}$. Also point out that the current tends toward zero for times that are long compared to τ. Then, the potential difference across the resistor is zero and the potential difference across the capacitor is $\mathcal{E}$.

D. Derive the loop equation for a series circuit consisting of a capacitor and resistor. Suppose the capacitor has charge q_0 at time $t = 0$ and show that $q = q_0 e^{-t/RC}$. Again find expressions for the potential differences across the capacitor and resistor. Point out that RC is indicative of the time for discharge.

E. Write the expression for the energy initially stored in the capacitor: $U = \frac{1}{2}q_0^2/C$. Evaluate $\int_0^\infty i^2 R\,dt$ to find the energy dissipated in the resistor as the capacitor discharges. Show that these energies are the same.

SUPPLEMENTARY TOPIC

Electrical measuring instruments (voltmeters and ammeters). This material can be covered as needed in conjunction with the laboratory.

SUGGESTIONS

1. Assignments
   a. Use question 1 in a discussion of emfs and batteries. Problem 2 covers the fundamental idea of emf. Use problems 7 and 16 to discuss the distinction between the emf and terminal potential difference of a battery.
   b. Assign some single-loop problems, such as 5, 8, and 17.
   c. Discuss some of questions 2 through 5, 7, and 9 in connection with parallel and series combinations of resistors. Assign problems 13, 19, and 27.
   d. Assign some problems dealing with multiloop circuits. Consider problems 21, 22, 24, 30, and 31.
   e. Assign problems 39 and 42 if voltmeters and ammeters are discussed in lecture or lab. Also consider problems 40 and 41.
2. Demonstrations
   a. Seats of emf: Freier and Anderson Ee2, 3, 4.
   b. Measurement of emf: Freier and Anderson Eg7.
   c. Resistive circuits: Freier and Anderson Eh1, 2, 4, Eo2 — 8.
3. Audio/Visual
   Series and Parallel Circuits; Demonstrations of Physics: Electricity and Magnetism; VHS video tape (5:16); Media Design Associates, Inc., Box 3189, Boulder, CO 80307–3190.
4. Computer Software
   a. *Electricity & Magnetism*; Macintosh, Windows; Cross Educational Software, Inc., 508 E. Kentucky Avenue, PO Box 1536, Ruston, LA 71270. Contains sections on resistors, basic circuits, and *RC* circuits.
   b. *DC Circuits*; Windows; Miky Ronen, Matzi Eliahu, and Igal Yastrubinezky; Physics Academic Software, North Carolina State University, PO Box 8202, Raleigh, NC 27690–0739. Circuit elements can be put together to form circuits, values of the parameters can be selected, and the circuits can then be analyzed.
5. Computer Projects
   A computer can easily be programmed to solve simultaneous linear equations. Have students use such a program to solve multiloop circuit problems. Details of a program and some specific projects are given in the Computer Projects section of this manual.

6. Laboratory
   a. Meiners Experiment 10–7 (Part A): *Measuring Current with a d'Arsonval Galvanometer.* Students determine the characteristics and sensitivity of a galvanometer. To expand this lab, ask the students to design an ammeter and a voltmeter with full scale deflections prescribed by you. Students practice circuit analysis while trying to understand design considerations.
   b. Meiners Experiment 10–9: *The EMF of a Solar Cell.* Students study a slide wire potentiometer and use it to measure the emf of a solar cell. This is another experiment that gives them practice in circuit analysis.
   c. Bernard Experiment 28: *Measurements of Potential Difference with a Potentiometer.* Students study a slide wire potentiometer and use it to investigate the emf and terminal voltage of a battery and the workings of a voltage divider.
   d. Bernard Experiment 26: *A Study of Series and Parallel Electric Circuits.* Students use ammeters and voltmeters to verify Kirchhoff's laws and investigate energy balance for various circuits. They also experimentally determine equivalent resistances of resistors in series and parallel. This experiment can be extended somewhat by having them consider a network of resistors that cannot be reduced by applying the rules for series and parallel resistors. Also see Bernard Experiment 27: *Methods of Measuring Resistance.* Two voltmeter-ammeter methods and a Wheatstone bridge method are used to measure resistance and to check the equivalent resistance of series and parallel connections.
   e. Bernard Experiment 31: *Circuits Containing More Than One Potential Source.* Similar to Bernard Experiment 26 described above except circuits with more than one battery are considered. The two experiments can be done together, if desired.
   f. Meiners Experiment 10–4: *The R-C Circuit.* Students connect an unknown resistor to a known capacitor, charged by a battery. The battery is disconnected and a voltmeter and timer are used to measure the time constant. The value of the resistance is calculated. In a second part an unknown capacitor is charged by means of a square wave generator and the decay is monitored on an oscilloscope. Again the time constant is measured, then it is used to calculate the capacitance. A third part explains how to use a microprocessor to collect data. Also see Bernard Experiment 32: *A Study of Capacitance and Capacitor Transients.*

Chapter 29 MAGNETIC FIELDS

BASIC TOPICS

I. Definition of the field and the magnetic force on a moving charge.
   A. Explain that moving charges create magnetic fields and that a magnetic field exerts a force on a moving charge. Both the field of a moving charge and the force exerted by a field depend on the velocity of the charge involved. The latter property distinguishes it from an electric field.
   B. Define the magnetic field: the force on a moving test charge is $q_0\vec{v} \times \vec{B}$ after the electric force is taken into account. Review the rules for finding the magnitude and direction of a vector product. Point out that the force must be measured for at least two directions of $\vec{v}$ since the component of $\vec{B}$ along $\vec{v}$ cannot be found from the force. The direction of $\vec{B}$ can be found by trying various directions for $\vec{v}$ until one is found for which the force vanishes. The magnitude of $\vec{v}$ can be found by orienting $\vec{v}$ perpendicular to $\vec{B}$. Units: $1\,\text{tesla} = 1\,\text{N/A}\cdot\text{m}$, $1\,\text{gauss} = 10^{-4}\,\text{T}$. Point out the magnitudes of the fields given in Table 29–1.
   C. Explain that the magnetic force on any moving charge is given by $\vec{F}_B = q\vec{v} \times \vec{B}$. Point out that the force is perpendicular to both $\vec{v}$ and $\vec{B}$ and is zero for $\vec{v}$ parallel or antiparallel to

$\vec{B}$. Also point out that the direction of the force depends on the sign of q. Remark that the field cannot do work on the charge and so cannot change its speed or kinetic energy. A magnetic field can change the direction of travel of a moving charge. It can, for example, be used to produce a centripetal force and can cause a charge to move in a circular orbit.

D. To show a magnetic force qualitatively, slightly defocus an oscilloscope so the central spot is reasonably large. Move a bar magnet at an angle to the face of the scope and note the movement of the beam.

E. Point out that the total force on a charged particle is $q(\vec{E} + \vec{v} \times \vec{B})$ when both an electric and a magnetic field are present.

II. Magnetic field lines.

A. Explain that field lines can be associated with a magnetic field. At any point the field is tangent to the line through that point and the number of lines per unit area that pierce a plane perpendicular to the field is proportional to the magnitude of the field.

B. To show field lines project Fig. 29–4 or place a sheet of clear plastic over a bar magnet and place iron filings on the sheet. Place the arrangement on an overhead projector. Explain that the filings line up along field lines.

C. Point out that magnetic field lines form closed loops; they continue into the interior of the magnet, for example. Contrast with electric field lines and remark that no magnetic charge has yet been found. Mention that magnetic field lines would start and stop at magnetic monopoles, if they exist. Remark that lines enter at the south pole of a magnet and exit at the north pole.

III. Motions of charged particles in magnetic fields.

A. Derive $v = E/B$ for the speed of a charge passing through a velocity selector.

B. Outline the Thompson experiment and derive Eq. 29–8 for the mass-to-charge ratio.

C. Show how the Hall effect can be used to determine the sign and concentration of charge carriers in a conductor. Mention that these measurements are important for the semiconductor industry. Also mention that the Hall effect is used to measure magnetic fields. Show a Hall effect teslameter.

D. Consider a charge with velocity perpendicular to a constant magnetic field. Show that the orbit radius is given by $r = mv/qB$ and the period of the motion is given by $T = 2\pi m/qB$ (independently of v) for non-relativistic speeds. If you are covering modern topics, state that $r = p/qB$ is relativistically correct but $p = mv/\sqrt{1 - v^2/c^2}$, where c is the speed of light, must be used for the momentum. Remark that the orbit is a helix if the velocity of the charge has a component along the field. Show how to calculate the pitch of the helix, given the velocity components parallel and perpendicular to the field. Mention that cyclotron motion is used in cyclotrons and synchrotrons. If you have time, explain how a cyclotron works.

IV. Force on a current-carrying wire.

A. Run a flexible non-magnetic wire near a strong permanent magnet. Observe that the wire does not move. Turn on a power supply so about 1 A flows in the wire and watch the wire move. Remark that magnetic fields exert forces on currents. A car battery and jumper cables can be used. To avoid an explosion, place a heavy-duty switch in the circuit, far from the battery.

B. Consider a thin wire carrying current, with all charges moving with the drift velocity. Start with the force on a single charge and derive $d\vec{F}_B = i\,d\vec{L} \times \vec{B}$ for an infinitesimal segment and $\vec{F}_B = i\vec{L} \times \vec{B}$ for a finite straight segment in a uniform field. Stress that $d\vec{L}$ and $\vec{L}$ are in the direction of the current.

C. Consider an arbitrarily shaped segment of wire in a uniform field. Show that the force on the segment between a and b is $\vec{F}_B = i\vec{L} \times \vec{B}$, where $\vec{L}$ is the vector joining the ends of the segment. This expression is valid only if the field is uniform.

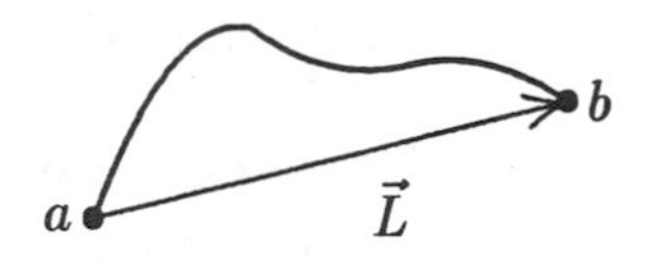

D. Point out that the force on a closed loop in a uniform field is zero since $\vec{L} = 0$.

E. Calculate the force of a uniform field on a semicircular loop of wire, in the plane perpendicular to $\vec{B}$. Do this by evaluating the integral $i \int d\vec{L} \times \vec{B}$ along the wire, then repeat using the result given in C above.

V. Torque on a current loop.

A. Calculate the torque exerted by a uniform field on a rectangular loop of wire arbitrarily oriented with two opposite sides perpendicular to $\vec{B}$. See Fig. 29–21.

B. Define the magnetic dipole moment of a current loop ($\mu = NiA$) and give the right hand rule for determining its direction. For a rectangular loop in a uniform field, show that $\vec{\tau} = \vec{\mu} \times \vec{B}$. State that the result is generally valid for any loop in a uniform field. Mention that other sources of magnetic fields, such as bar magnets and Earth, have dipole moments. Mention that many fundamental particles have intrinsic dipole moments.

C. Note that this is a restoring torque and that if the dipole is free to rotate it will oscillate about the direction of the field. If damping is present, it will line up along the field direction. Remark that this is the basis of magnetic compasses.

D. Explain how analog ammeters and voltmeters work. To demonstrate the torque on a current-carrying coil, remove the case from a galvanometer and wire it to a battery and resistor so that it fully deflects.

E. Remark that a potential energy cannot be associated with a moving charge in a magnetic field but can be associated with a magnetic dipole in a magnetic field. Show that $U = -\vec{\mu} \cdot \vec{B}$. Find the work required to turn a dipole through 90° and 180°, starting with it aligned along the field. Point out that U is a minimum when $\vec{\mu}$ and $\vec{B}$ are parallel and is a maximum when they are antiparallel.

SUGGESTIONS

1. Assignments

a. Use questions 1 and 2 to help in understanding the magnetic force. The dependence of magnetic force on velocity and charge is emphasized in problems 1 and 2.

b. Use questions 5 through 8 to test for understanding of the motion of charges in magnetic fields. Problems 17 and 23 deal with the circular orbit of a charge in a uniform magnetic field. Crossed electric and magnetic fields, used as a velocity filter, are explored in questions 3 and 4 and in problems 7, 9, and 11. Problem 25 deals with a mass spectrometer. Problem 31 deals with cyclotrons. Use some of these problems to include practical applications.

c. Use problem 12 to help students study the Hall effect.

d. Use question 9 to include more detail in the discussion of the magnetic force on a current-carrying wire. Use problems 34 and 36 to stress the importance of the angle between the magnetic field and the current carrying wire on which it exerts a force. Use problems 40 and 44 to emphasize that the force of a uniform magnetic field on a closed loop is zero. Problem 47 asks students about the dynamics of current-carrying wires in magnetic fields. Assign problems 41 in support of the discussion of magnetic torques on current-carrying loops.

e. Magnetic dipoles and the torques exerted on them by magnetic fields are explored in problems 49 and 50. Question 10 and problem 55 deal with the energy of a dipole in a field. Also consider problem 54.

2. Books and Monographs

   Teaching about magnetism; by Robert J. Reiland; available from AAPT, One Physics Ellipse, College Park MD 20740–3845. A PTRA workshop manual. Contains a collection of demonstrations and student activities.

3. Computer Project

   Have students use numerical integration of Newton's second law to investigate the orbits of charges in magnetic and electric fields.

4. Audio/Visual

   a. *The Force on a Current*; from AAPT collection 2 of single-concept films; video tape; available from Ztek Co., PO Box 11768, Lexington, KY 40577–1768.
   b. *Magnetism and Magnetic Fields*; Side D: Waves (II) & Electricity and Magnetism of Cinema Classics; video disk; available from Ztek Co. (see above for address).
   c. *Magnetic Fields*; Demonstrations of Physics: Electricity and Magnetism; VHS video tape (6:14); Media Design Associates, Inc., Box 3189, Boulder, CO 80307–3190.
   d. *Magnetic Effects in Space*; Skylab Demonstrations; VHS video tape (14 min); Media Design Associates, Inc. (see above for address).
   e. *Magnetism in Space*; Skylab Demonstrations; VHS video tape (19 min); Media Design Associates, Inc. (see above for address).
   f. *Magnetic Fields*; VHS video tape (18 min); Films for the Humanities & Sciences, PO Box 2053, Princeton, NJ 08543–2053.
   g. *Magnetic Force*; VHS video tape (9 min); Films for the Humanities & Sciences (see above for address).
   h. *Magnetic Fields*; VHS video tape (15 min); Films for the Humanities & Sciences (see above for address).
   i. *Creating Magnetic Force*; VHS video tape (12 min); Films for the Humanities & Sciences (see above for address).
   j. *Magnetic Fields*; Physics Demonstrations in Electricity and Magnetism, Part III; VHS video tape; ≈3 min; Physics Curriculum & Instruction, 22585 Woodhill Drive, Lakeville, MN 55044.

5. Computer Software

   a. *Dynamic Analyzer*. See Chapter 5 SUGGESTIONS.
   b. *Electricity & Magnetism*; Macintosh, Windows; Cross Educational Software, Inc., 508 E. Kentucky Avenue, PO Box 1536, Ruston, LA 71270. Contains a section on magnetic force.

6. Demonstrations

   a. Force on an electron beam: Freier and Anderson Ei18, Ep8, 11.
   b. Forces and torques on wires: Freier and Anderson Ei7, 12, 13 — 15, 19, 20.
   c. Meters: Freier and Anderson Ej1, 2.
   d. Hall effect: Freier and Anderson Ei16.

7. Laboratory

   a. Bernard Experiment 33: *A Study of Magnetic Fields.* A small magnetic compass is used to map field lines of various permanent magnets, a long straight current-carrying wire, a single loop of current-carrying wire, a solenoid, and Earth. Parts of this experiment might be performed profitably in connection with Chapter 31.
   b. Meiners Experiment 11–3: *Determination of e/m.* Students use the accelerating potential and the radius of the orbit in a magnetic field to calculate the charge-to-mass ratio for the electron.
   c. Meiners Experiment 11–5: *The Hall Effect.* Students measure the Hall voltage and use it to calculate the drift speed and carrier concentration for a bismuth sample. The influence

of the magnetic field on the Hall voltage is also investigated. Values of the magnetic field are given to them by the instructor.

Chapter 30 MAGNETIC FIELDS DUE TO CURRENTS

BASIC TOPICS

I. Magnetic field of a current.

A. Place a magnetic compass near a wire carrying a dc current of several amperes, if possible. Turn the current on and off and reverse the current. Note the deflection of the compass needle and remark that the current produces a magnetic field and that the field reverses when the current reverses.

B. Write the Biot-Savart law for the field produced by an infinitesimal segment of a current-carrying wire. Give the value for μ_0. Draw a diagram to show the direction of the current, the displacement vector from the segment to the field point, and the direction of the field. Explain that $d\vec{B}$ is in the direction of $i\, d\vec{s} \times \vec{r}$. Point out the angle between $\vec{r}$ and $d\vec{s}$. Mention that the integral for the field of a finite segment must be evaluated one component at a time. Point out that the angle between $d\vec{B}$ and a coordinate axis must be used to find the component of $d\vec{B}$.

C. Example: Show how to calculate the magnetic field of a straight finite wire segment. See the text, but use finite limits of integration. State that magnetic fields obey a superposition principle and point out that the result of the calculation can be used to find the field of a circuit composed of straight segments. Specialize the result to an infinite straight wire. Demonstrate the right-hand rule for finding the direction of $\vec{B}$ due to a long straight wire.

D. Explain that the field lines around a straight wire are circles in planes perpendicular to the wire and are centered on the wire. Draw a diagram to illustrate. Use symmetry to argue that the magnitude of the field is uniform on a field line. Point out that for other current configurations B is not necessarily uniform on a field line.

E. Show how to find the force per unit length of one long straight wire on another. Treat currents in the same and opposite directions. Lay two long automobile starter cables on the table. Connect them in parallel to an auto battery, with a $0.5\,\Omega$, 500 W resistor and an "anti-theft" switch or starter relay in each circuit. Close one switch and note that the wires do not move. Close the other switch and note the motion. Show parallel and antiparallel situations. It is better to reconnect the wires or rearrange them rather than to use a reversing switch.

F. Give the definition of the ampere and remind students of the definition of the coulomb.

G. Consider a circular arc of radius R, subtending an angle ϕ, and carrying current i. Use the Biot-Savart law to show that the magnetic field at the center is given by $B = \mu_0 i\phi/4\pi R$. Note that ϕ must be in radians. Specialize to the cases of a semicircle and a full circle.

II. Ampere's law.

A. Write the law in integral form. Explain that the integral is a line integral around a closed contour and interpret it as a sum over segments. Point out that it is the tangential component of $\vec{B}$ that enters. Explain that the current that enters is the net current through the contour. Two currents in opposite directions tend to cancel, for example. Illustrate by considering a contour that encircles five or six wires, with some currents in each direction. Also consider a wire passing through the plane of the contour but outside the contour. Mention that this current produces a magnetic field at all points on the contour but the integral of its tangential component is zero.

B. Explain the right-hand rule that relates the direction of integration around the contour and the direction of positive current through the contour.

C. Pick a functional form for the magnetic field ($B_x = 2xy$, $B_y = -y^2$, $B_z = 0$, for example). Be sure the divergence is zero and the curl is not. Now consider a simple contour, such as a square in the xy plane. Integrate the tangential component of the field around the contour and calculate the net current through it.

D. Use Ampere's law to calculate the magnetic field *outside* a long straight wire. Either use without proof the circular nature of the field lines or give a symmetry argument to show that $\vec{B}$ at any point is tangent to a circle through the point and has constant magnitude around the circle. Point out that the integration contour is taken tangent to $\vec{B}$ in order to evaluate the integral in terms of the unknown magnitude of $\vec{B}$.

E. Use Ampere's law to calculate the field *inside* a long straight wire with a uniform current distribution. Note that the use of Ampere's law to find B has the same limitations as Gauss' law when used to find E: there must be sufficient symmetry.

F. Use Ampere's law to calculate the field inside a solenoid. First argue that, for a long tightly wound solenoid, the field at interior points is along the axis and nearly uniform while the field at exterior points is nearly zero.

G. Similarly, use Ampere's law to calculate the field inside a toroid.

III. Magnetic dipole field.

A. Use the Biot-Savart law to derive an expression for the field of a circular current loop at a point on its axis. Stress the resolution of $d\vec{B}$ into components.

B. Take the limit as the radius becomes much smaller than the distance to the field point and write the result in terms of the dipole moment. Explain that the result is generally true for loops of any shape as long as the field point is far from the loop. Remind students that the dipole moment of a loop is determined by its area and the current it carries.

SUGGESTIONS

1. Assignments

a. Use questions 1 and 2 and problem 1 as part of the discussion of the magnetic field due to a long straight wire. Problems 11 and 13 deal with the field of a finite straight wire. Assign them, then one or more of 12, 14, 15, 16, and 20, which ask students to superpose the fields of finite wires.

b. Question 3 deals with the field of a circular arc. Problems 7 through 10 deal with circuits consisting of straight line and circular segments.

c. Ask questions 4 and 5 in association with the magnetic forces exerted by wires on each other. Assign problems 24 and 28.

d. Use questions 6 through 9 in your discussion of Ampere's law. After discussing line integrals around closed loops, assign problems 31 and 34 to test the fundamentals; problems 32 and 37 give some applications. Assign problem 35 if you want to include the field of a wire with nonuniform current density.

e. Question 10 and problems 40 and 43 can be assigned to support the discussion of solenoids and toroids.

f. Problems 48, 52, and 57 deal with the magnetic fields of coils and dipole loops. Assign problems 50 and 55 if you cover Helmholtz coils or use them in lab.

2. Computer Projects

a. Have students use the Biot-Savart law and numerical integration to calculate the magnetic field due to a circular current loop at off-axis points. They can use a commercial math program or their own programs.

b. Use numerical integration to verify Ampere's law for several long straight wires passing through a square contour. Have them show the result of the integration is independent of the positions of the wires, as long as they are inside the square. Also have them consider a wire outside the square.

3. Demonstrations
   a. Magnetic fields of wires: Freier and Anderson Ei8 — 11.
   b. Magnetic forces between wires: Freier and Anderson Ei1 — 6.
4. Books and Monographs
   The Solenoid; by Carl R. Stannard, Arnold A. Strassenberg, and Gabriel Kousourou; available from AAPT, One Physics Ellipse, College Park MD 20740–3845. Covers the magnetic field of a solenoid and practical applications as a mechanical switch.
5. Audio/Video
   a. *Magnetism and Electron Flow*; VHS video tape (11 min); Films for the Humanities & Sciences, PO Box 2053, Princeton, NJ 08543–2053.
   b. *Creating Magnetic Fields*; VHS video tape (11 min); Films for the Humanities & Sciences (see above for address).
6. Computer Software
   a. *EM Field*; David Trowbridge. See Chapter 23 SUGGESTIONS.
   b. *Electricity & Magnetism*; Macintosh, Windows; Cross Educational Software, Inc., 508 E. Kentucky Avenue, PO Box 1536, Ruston, LA 71270. Contains a section on Ampere's law.
7. Laboratory
   a. Meiners Experiment 11–1: *The Earth's Magnetic Field.* A tangent galvanometer is used to measure Earth's magnetic field. The dip angle is calculated.
   b. Meiners Experiment 11–2: *The Current Balance.* The gravitational force on a current-carrying wire is used to balance the magnetic force due to current in a second wire. The data can be used to find the value of μ_0 or to find the current in the wires. The second version essentially defines the ampere. Part B describes how a microprocessor can be used to collect and analyze the data.
   c. Bernard Experiment 34: *Measurement of the Earth's Magnetic Field.* The oscillation period of a small permanent magnet suspended inside a solenoid is measured with the solenoid and Earth's field aligned. The reciprocal of the period squared is plotted as a function of the current in the solenoid, and the slope, along with calculated values of the solenoid's field, is used to find Earth's field.
   d. Meiners Experiment 11–3: *Determination of e/m.* Students find the speed and orbit radius of an electron in the magnetic field of a pair of Helmholtz coils and use the data to calculate e/m. Information from this chapter is used to compute the field, given the coil radius and current. If you are willing to postulate the field for the students, this experiment can be performed in connection with Chapter 30.

Chapter 31 INDUCTION AND INDUCTANCE

BASIC TOPICS

I. The law of induction.
   A. Connect a coil (50 to 100 turns) to a sensitive galvanometer and move a bar magnet in and out of the coil. Note that a current is induced only when the magnet is moving. Show all possibilities: the north pole entering and exiting the coil and the south pole entering and exiting the coil. In each case point out the direction of the induced current. With a little practice you might also demonstrate effectively that the deflection of the galvanometer depends on the speed of the magnet.

B. To show the current produced by changing the orientation of a loop, align the loop axis with Earth's magnetic field and rapidly rotate the loop once through 180°. Note the deflection of a galvanometer in series with the loop. Explain that this forms the basis of electric generators.

C. Connect a coil to a switchable dc power supply. Connect a voltmeter (digital, if possible) to the supply to show when it is on. Place a second coil, connected to a sensitive galvanometer, near the first. Show that when the switch is opened or closed, current is induced in the second coil, but that none is induced when the current in the first coil is steady.

D. Define the magnetic flux through a surface. Unit: 1 weber = 1 T·m^2. Point out that Φ_B measures the number of magnetic field lines that penetrate the surface. Remark that $\Phi_B = BA\cos\theta$ when $\vec{B}$ is uniform over the surface and makes the angle θ with its normal.

E. Give a qualitative statement of the law: an emf is generated around a closed contour when the magnetic flux through the contour changes. Stress that the law involves the flux through the surface bounded by the contour. Point out the surface and contour for each of the demonstrations done, then remark that the contour may be a conducting wire, the physical boundary of some material, or a purely geometric construction. Remark that if the contour is conducting, then current flows.

F. Give the equations for Faraday's law: $\mathcal{E} = -\mathrm{d}\Phi_B/\mathrm{d}t$ for a single loop and $\mathcal{E} = -N\,\mathrm{d}\Phi_B/\mathrm{d}t$ for N tightly packed loops. Note that the emf's add.

II. Lenz's law.

A. Explain Lenz's law in terms of the magnetic field produced by the current induced if the contour is a conducting wire. Stress that the induced field must re-enforce the external field in the interior of the loop if the flux is decreasing and must tend to cancel it if the flux is increasing. This gives the direction of the induced current, which is the same as the direction of the emf. Review the right-hand rule for finding the direction of the field produced by a loop of current-carrying wire. State that Lenz's law can be used even if the contour is not conducting. The current must then be imagined.

B. Optional: Give the right-hand rule for finding the direction of positive emf. When the thumb points in the direction of $\mathrm{d}\vec{A}$, then the fingers curl in the direction of positive emf. If Faraday's law gives a negative emf, then it is directed opposite to the fingers. Stress that the negative sign in the law is important if the equation, with the right-hand rule, is to describe nature.

C. Consider a rectangular loop of wire placed perpendicular to a magnetic field. Assume a function $B(t)$ and calculate the emf and current. Show how the directions of the emf and current are found. Point out that an *area* integral is evaluated to find Φ_B and a *time* derivative is evaluated to find the emf. Some students confuse the variables and integrate with respect to time.

III. Motional emf.

A. Consider a rectangular loop being pulled with constant velocity past the boundary of a uniform magnetic field. Calculate the emf and current.

B. Consider a rod moving with a constant velocity that is perpendicular to a uniform magnetic field. Show how to complete the loop and calculate the emf. Mention that the emf exists only in the moving rod, regardless of whether the rest of the contour is conducting.

C. Consider a rectangular loop of wire rotating with constant angular velocity about an axis that is in the plane of the loop and through its center. Take the magnetic field to be uniform and point out that now the flux is changing because the angle between the field and the normal to the loop is changing. Derive the expression for the emf and point out it is time dependent.

IV. Energy considerations.

A. Point out that an emf does work at the rate $\mathcal{E}i$, where i is the current. Explain that for a current induced by motion, the energy comes from the work done by an external agent or from the kinetic energy of the moving portion of the loop.

B. Consider four conducting rails that form a rectangle, three fixed and the fourth riding on two of them. Take the magnetic field to be uniform and normal to the loop. Assume that essentially all of the electrical resistance of the loop is associated with the moving rail. First, suppose the moving rail has constant velocity and derive expressions for the emf, current, and magnetic force on the rail. Next, derive expressions for the rate at which an external agent must do work to keep the velocity constant and for the rate at which energy is dissipated by the resistance of the loop. Point out that all the energy supplied by the agent is dissipated.

C. Now suppose the rail is given an initial velocity and, thereafter, it is acted on by the magnetic field alone. Use Newton's second law to derive an expression for the velocity as a function of time. Compare the rate at which the kinetic energy is decreasing with the rate of energy dissipation in the resistance. Remark that this phenomenon finds practical application in magnetic braking.

D. Mention that energy is also dissipated when a current is induced by a changing magnetic field and it comes from the agent that is changing the field.

V. Induced electric fields.

A. Explain that a changing magnetic field produces an electric field, which is responsible for the emf. The emf and electric field are related by $\mathcal{E} = \oint \vec{E} \cdot d\vec{s}$, where the integral is around the contour. Remind students that this integral is the work per unit charge done by the field as a charge goes around the contour. Write Faraday's law as $\oint \vec{E} \cdot d\vec{s} = -\frac{d}{dt} \int \vec{B} \cdot d\vec{A}$. Note that $d\vec{s}$ and $d\vec{A}$ are related by a right-hand rule: fingers along $d\vec{s}$ implies thumb along $d\vec{A}$. This is consistent with Lenz's law.

B. State that the induced electric field is like an electrostatic field in that it exerts a force on a charge but that it is unlike an electrostatic field in that it is not conservative. For an electrostatic field, the integral defining the emf vanishes.

C. Consider a cylindrical region containing a uniform magnetic field along the axis. Assume a time dependence for $\vec{B}$ and derive expressions for the electric field inside the region and outside the region. See Sample Problem 31–4. Point out that the lines of $\vec{E}$ form closed circles concentric with the cylinder and that the magnitude of $\vec{E}$ is uniform around a circle.

VI. Definition of inductance.

A. Connect a light bulb and choke coil in parallel across a switchable dc supply. Close the switch and note that the lamp is initially brighter than when steady state is reached. Open the switch and note that the light brightens before going off. Remark that this behavior is due to the changing magnetic flux through the coil and that the flux is created by the current in the coil itself.

B. Point out that when current flows in a loop, it generates a magnetic field and the loop contains magnetic flux due to its own current. If the current changes, so does the flux and an emf is generated around the loop. The total emf, due to all sources, determines the current. Remark that the self-flux is proportional to the current and the induced emf is proportional to the rate of change of the current.

C. Define the inductance by $L = N\Phi_B/i$, where N is the number of turns, Φ_B is the magnetic flux through each turn, and i is the current in the circuit. Unit: 1 henry = 1 V·s/A.

D. Remark that Faraday's law yields $\mathcal{E} = -L\, di/dt$ for the induced emf.

E. Inductors are denoted by ‿ℓℓℓℓℓ‿ in circuit diagrams. Point out that if the circuit element looks like $a \overset{i \longrightarrow}{\text{ℓℓℓℓℓ}} b$, then $V_b - V_a = -L\,di/dt$ is algebraically correct. As an example, use $i(t) = i_m \sin(\omega t)$. Note that i is positive when it is directed from a to b and negative when it is directed from b to a. Compute $V_a - V_b = Li_m\omega\cos(\omega t)$. Graph i and the potential difference as functions of time to show the phase relationship. Remark that a real inductor can be regarded as a pure inductance in series with a pure resistance.

F. Show how to calculate the inductance of an ideal solenoid. Use the current to calculate the field, then the flux, and finally equate $N\Phi_B$ to Li and solve for L. Point out that L is independent of i but depends on geometric factors such as the cross-sectional area, length, and the number of turns per unit length.

G. Optional: Show how to calculate the inductance of a toroid.

VII. An LR circuit.

A. Derive the loop equation for a single loop containing a source of emf (an ideal battery), a resistor, and an inductor in series: $\mathcal{E} - iR - L\,di/dt = 0$, where the current is positive if it leaves the positive terminal of the seat of emf. Use the prototypes developed earlier:

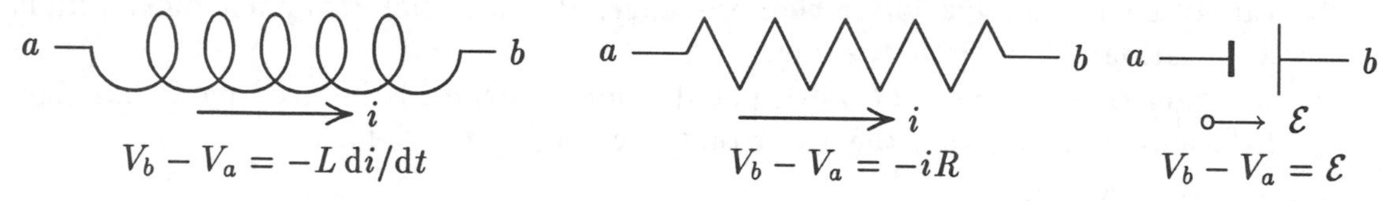

Remark that these are correct no matter if the current is positive or negative or if it is increasing or decreasing. Write down the solution for the current as a function of time for the case $i(0) = 0$: $i = (\mathcal{E}/R)[1 - e^{-Rt/L}]$. Show that the expression satisfies the loop equation and meets the initial conditions. Show a graph of $i(t)$; point out the asymptotic limit $i = \mathcal{E}/R$ and the time constant $\tau_L = L/R$. Remark that if L/R is large, the current approaches its limit more slowly than if L/R is small.

B. Explain the qualitative physics involved. When the battery is turned on and the current increases, the emf of the coil opposes the increase and the current approaches its steady state value more slowly than if there were no inductance. At long times, the current is nearly constant so di/dt and the induced emf are small. The current is nearly the same as it would be in the absence of an inductor. Just after the battery is turned on, the potential difference across the resistor is zero and the potential difference across the inductor is $\mathcal{E}$. After a long time, the potential difference across the resistor is $\mathcal{E}$ and the potential difference across the inductor is zero.

C. Repeat the calculation for a circuit with an inductor and resistor but no battery. Take the initial current to be i_0 and show that $i(t) = i_0 e^{-t/\tau_L}$. Graph the solution and show the position of τ_L on the time axis. Point out that the emf of the coil opposes the decrease in current.

D. Demonstrate the two circuits by connecting a resistor and coil in series to a square-wave generator. Observe the current by placing oscilloscope leads across the resistor. Observe the voltage drop across the coil. Vary the time constant by varying the resistance.

VIII. Energy considerations.

A. Consider a single loop circuit containing an ideal battery, a resistor, and an inductor. Assume the current is increasing. Write down the loop equation, multiply it by i, and identify the power supplied by the battery and the power lost in the resistor. Explain that the remaining term describes the power being stored by the inductor, in its magnetic field. Point out the similarity between $i\mathcal{E}$ and $-iL\,di/dt$ for the rate at which work is being done by an ideal battery and by an inductor (with emf $-Ldi/dt$).

B. Integrate $P = iL\,di/dt$ to obtain $U_B = \frac{1}{2}Li^2$ for the energy stored in the magnetic field (relative to the energy for $i = 0$).

C. Consider the energy stored in a long current-carrying solenoid and show that the energy density is $u_B = B^2/2\mu_0$. Explain that this gives the energy density at a point in any magnetic field and that the energy required to establish a given magnetic field can be calculated by integrating the expression over the volume occupied by the field.

IX. Mutual induction.

A. Repeat the demonstration experiment discussed in note IC. Explain it in terms of the concept of mutual induction. Point out that the flux through the second coil is proportional to the current in the first. Define the mutual induction of the second coil with respect to the first by $M_{12} = N_2\Phi_{12}/i_1$. Show that $\mathcal{E}_2 = -M_{12}\,di_1/dt$ is the emf induced in the second coil when the current in the first changes. State without proof that $M_{12} = M_{21}$.

B. Example: derive the mutual inductance for a small coil placed at the center of a solenoid or for a small, tightly wound coil placed at the center of a larger coil.

C. Show that two inductors connected in series and well separated have an equivalent inductance of $L = L_1 + L_2$. Then, show that if their fluxes are linked, $L = L_1 + L_2 \pm 2M$, where the minus sign is used if the field lines have opposite directions. Also consider inductors in parallel. See problems 43, 44, and 71.

SUGGESTIONS

1. Assignments
   a. Question 1 deals with the magnitudes of induced emf and current. Questions 2, 3, and 4 deal with Lenz's law. Use several as examples and several to test students.
   b. Assign some of problems 1, 2, 3, 4, and 6 to cover the emf's generated by various time dependent magnetic fields. Addition of emf's is covered in problem 15. This is a good problem to test for understanding of the sign of an induced emf.
   c. Motional emf is covered in problems 17, 18, 27, 28 and 30. If you use a flip coil in the lab, assign problems 11 and 13. Problems 29 and 31 deal with energy transfers.
   d. Assign problems 32 and 33 in connection with the discussion of induced electric fields.
   e. Assign problem 36 (coil) or 39 (two parallel wires) as an example of a typical inductance calculation.
   f. Use some of questions 7 through 10 when discussing LR circuits. Assign problem 48. LR time constants are considered in problem 45.
   g. After discussing energy flow in a simple LR circuit with increasing current, assign problems 58 and 60.
   h. Problem 63 deals with energy storage and energy density in an inductor.
2. Demonstrations
   a. As a supplementary demonstration, take a large, long coil, mount it vertically, insert a solid, soft-iron rod with a foot or so sticking out, and connect the coil via a switch to a large dc power supply. Place a solid aluminum ring around the iron rod. The ring should fit closely but be free to move. Close the switch and the ring will jump up, then settle down. Repeat with a ring that has a gap in it. Finally, use an ac power supply. The effect can be enhanced by cooling the ring with liquid nitrogen.

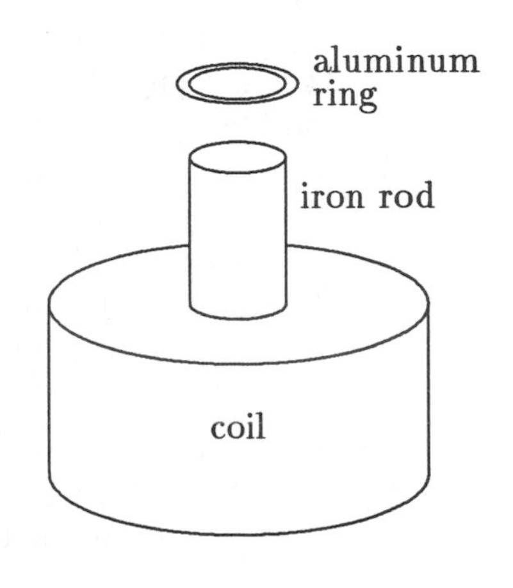

   b. Generation of induced currents: Freier and Anderson Ek1 — 6.
   c. Eddy currents: Freier and Anderson Ei1 — 6.

d. Generators: Freier and Anderson: Eq4 — 7, Er1.
e. Self-inductance: Freier and Anderson Eq1 — 3.
f. LR circuit: Freier and Anderson Eo11, En5 — 7.

3. Audio/Visual
   a. *Electromagnetism*; Side D: Waves (II) & Electricity and Magnetism of Cinema Classics; video disk; available from Ztek Co., PO Box 11768, Lexington, KY 40577–1768.
   b. *Electromagnetic Induction*; Demonstrations of Physics: Electricity and Magnetism; VHS video tape (7:33); Media Design Associates, Inc., Box 3189, Boulder, CO 80307–3190.
   c. *The Generation of Current*; VHS video tape (12 min); Films for the Humanities & Sciences, PO Box 2053, Princeton, NJ 08543–2053.
   d. *Electromagnetic Induction*; VHS video tape (14 min); Films for the Humanities & Sciences (see above for address).
   e. *The Generation of Electricity*; VHS video tape (19 min); Films for the Humanities & Sciences (see above for address).
   f. *Electricity and Magnetism*; *Electromagnetic Effects*; *Induction Application*; *Eddy Currents*; Physics Demonstrations in Electricity and Magnetism, Part III; VHS video tape; ≈3 min each; Physics Curriculum & Instruction, 22585 Woodhill Drive, Lakeville, MN 55044.

4. Laboratory
   a. Bernard Experiment 35: *Electromagnetic Induction.* Students measure the magnitude and observe the direction of current induced by a changing magnetic flux in a simple galvanometer circuit. Changing flux is produced by moving permanent magnets, by moving current-carrying coils, and by changing current in a coil.
   b. Meiners Experiment 11–4: *The Magnetic Field of a Circular Coil.* The emf generated in a small search coil when a low frequency ac current flows in a given circuit (a circular coil in this case) is used to determine the magnetic field produced by the circuit. The field is investigated as a function of position, specified in spherical coordinates.

Chapter 32 MAGNETISM OF MATTER; MAXWELL'S EQUATIONS

BASIC TOPICS

I. Gauss' law for magnetism.
   A. Explain that a magnetic monopole is a particle that produces a magnetic field even while at rest, with magnetic field lines starting or stopping on it. Remark that no magnetic monopole has been observed yet but it is currently being sought. Write down Gauss' law for the magnetic field and state that magnetic field lines form closed contours so the flux through any closed surface vanishes. If monopoles were found to exist, the law would be modified to include them. Compare with Gauss' law for the electric field.
   B. To show that the ends of a magnet are not monopoles, magnetize a piece of hard iron wire. Use a compass to locate and mark the north and south poles. Break the wire into pieces and again use the compass to show that each piece has a north and a south pole. Repeat a few times using smaller pieces each time. Remark that the same results would be obtained if the breaking process were continued to the atomic level. Individual atoms and particles are magnetic dipoles, not monopoles.

II. Magnetic dipoles in matter.
   A. Explain that current loops and bar magnets produce magnetic fields which, for points far away, are dipole fields. Review the expressions for the magnetic field of a dipole and for

the dipole moment of a loop in terms of the current and area. Place a bar magnet under a piece of plastic sheet on an overhead projector. Sprinkle iron filings on the sheet and show the field pattern of the magnet. Remind students that field lines emerge from the north pole and enter at the south pole.

B. Explain that the electron and many other fundamental particles have intrinsic dipole moments, which are related to their intrinsic spin angular momenta. Say that only one component, usually taken to be the z component, can be measured at a time. The z component of the spin angular momentum is $S_z = \pm h/4\pi = \pm 5.2729 \times 10^{-35}\ \text{J}\cdot\text{s}$, where h is the Planck constant. The z component of the associated magnetic dipole moment is $\mu_{S,z} = -(e/m)S_z = \mp(eh/4\pi m) = \mp 9.27 \times 10^{-24}\ \text{J/T}$, where m is the electron mass. Since an electron is negatively charged, the dipole moment and spin angular momentum are in opposite directions. Mention that particle and atomic magnetic moments are often measured in units of the **Bohr magneton** μ_B: $\mu_B = eh/4\pi m$. This is the magnitude of the electron spin dipole moment.

C. Explain that electrons in atoms create magnetic fields by virtue of their orbital motions. Derive Eq. 32–8, which gives the relationship between orbital angular momentum and dipole moment for a negative particle, such as an electron. Say that quantum mechanically the z component of the orbital angular momentum is given by $L_{\text{orb},z} = m_\ell h/2\pi$, where $m_\ell = \pm 1, \pm 2, \ldots, \pm(\text{limit})$ and "limit" is the largest magnitude of m_ℓ. The z component of the dipole moment is $\mu_{\text{orb},z} = -m_\ell(eh/4\pi) = -m_\ell\mu_B$.

D. Say that if an electron is placed in an external magnetic field, in the z direction, its magnetic energy is given by $U = -\mu_z B_{\text{ext}}$. Draw an energy level diagram to show the splitting of the levels in a magnetic field.

E. Remark that it is chiefly the orbital and spin dipole moments of electrons that are responsible for the magnetic properties of materials. Explain how to calculate the dipole moment of an atom: $\vec{\mu} = (-e/2m)\vec{L} + (-e/m)\vec{S}$, where $\vec{L}$ is the total orbital angular momentum and $\vec{S}$ is the total spin angular momentum of the electrons of the atom.

F. Explain that protons and neutrons also have intrinsic dipole moments, but that these are much smaller than the dipole moment of an electron because the masses are so much larger. Remark that nuclear magnetism has found medical applications.

III. Magnetization.

A. Define magnetization as the dipole moment per unit volume. Although only uniformly magnetized objects are considered in the text, you may wish to state the definition as the limiting value as the volume shrinks to zero.

B. State that a magnetized object produces a magnetic field both in its exterior and interior and write $\vec{B} = \vec{B}_0 + \vec{B}_M$ for the total field. Here $\vec{B}_0$ is the applied field and $\vec{B}_M$ is the field due to dipoles in the material. Remark that for some materials $\vec{B}_M$ is in the same direction as $\vec{B}_0$, while for others it is in the opposite direction.

IV. Diamagnetism and paramagnetism.

A. Give a qualitative discussion of diamagnetism. Explain that an external field changes the electron orbits so there is a net dipole moment and that the induced moment is directed opposite to the field. This tends to make the total field weaker than the external field alone. Bismuth is an example of a diamagnetic substance.

B. Give a qualitative discussion of paramagnetism. Explain that paramagnetic substances are composed of atoms with net dipole moments and, in the absence of an external field, the moments have random orientations, so that they produce no net magnetic field. An external field tends to align the moments and the material produces its own field. Since the moments, on average, are aligned with the external field, the total field is stronger than the external field alone. Alignment is opposed by thermal agitation and both the net magnetic moment and magnetic field decrease as the temperature increases.

C. Remind students that the potential energy of a dipole $\vec{\mu}$ in a magnetic field $\vec{B}$ is given by $U = -\vec{\mu} \cdot \vec{B}$ and show that the energy required to turn a dipole end for end, starting with it aligned with the field, is $2\mu B$. Calculate U for $\mu = \mu_B$ and $B = 1\,\text{T}$. Calculate the mean translational kinetic energy for an ideal gas at room temperature ($\frac{3}{2}kT$) and remark that there is sufficient energy for collisions to reorient the dipoles. Calculate the temperature for which $2\mu B = \frac{3}{2}kT$.

D. Give the Curie law for small applied fields. Explain that for small applied fields M is proportional to B and inversely proportional to T. Draw the full graph of magnetization as a function of the applied field and point out the linear region. Describe saturation and explain that there is an upper limit to the magnetization. Point out this region on the graph. The limit occurs when all atomic dipoles are aligned. Use a teslameter or flip coil to measure the magnetic field just outside the end of a large, high-current coil. Put a large quantity of manganese in the coil and again measure the field.

E. Explain that diamagnetic effects are present in all materials but are overshadowed by paramagnetic or ferromagnetic effects if the atoms have dipole moments.

V. Ferromagnetism.

A. Explain that, for iron and other ferromagnetic substances (such as Co, Ni, Gd, and Dy), the atomic dipoles are aligned by an internal mechanism (exchange coupling) so the substance can produce a magnetic field spontaneously, in the absence of an external field. At temperatures above its Curie temperature, a ferromagnetic substance becomes paramagnetic. Gadolinium is ferromagnetic with a Curie temperature of about 20° C. Put a sample in a beaker of cold water ($T < 20°$ C) and use a weak magnet to pick it up from the bottom of the beaker but not out of the water. Add warm water to the beaker and the sample will drop from the magnet.

B. Describe ferromagnetic domains and explain that the dipoles are aligned within any domain but are oriented differently in neighboring domains. The magnetic fields produced by the various domains cancel for an unmagnetized sample. When the sample is placed in a magnetic field, domains with dipoles aligned with the field grow in size while others shrink. The dipoles in a domain may also be reoriented somewhat as a unit.

C. Define hysteresis (see Fig. 32–13) and explain that the growth and shrinkage of domains are not reversible processes. Domain size is dependent not only on the external field but also on the magnetic history of the sample. When the external field is turned off, the material remains magnetized. Draw a hysteresis curve and point out the approach to saturation and the residual field. Explain that to demagnetize an ferromagnet an external field must be applied in the direction opposite to the magnetization.

D. Explain the difference between soft and hard iron in terms of hysteresis. Use a large, high-current coil to magnetize a piece of hard iron and show that it remains magnetized when the current is turned off. Also magnetize a piece of soft iron and show it is magnetized only as long as the current remains on. When the current is turned off, very little permanent magnetization remains. Soft iron is used for transformer coils.

VI. The Maxwell induction law.

A. In the material discussed so far, note the absence of any counterpart to Faraday's law, i.e. the creation of magnetic fields by changing electric flux. Tell students it should be there and you will now discuss its form.

B. Consider the charging of a parallel-plate capacitor. Remind students that in Ampere's law $d\vec{s}$ and $d\vec{A}$ are related by a right-hand rule and the surface integral is over any surface bounded by the closed contour.

C. In the diagram, surfaces A, B, and C are all bounded by the contour that forms the left end of the figure. If we choose surface A or C, then Ampere's law as we have taken it gives $\oint \vec{B} \cdot d\vec{s} = \mu_0 i$, but if we choose surface B, it gives $\oint \vec{B} \cdot d\vec{s} = 0$. Since the integral on the left side is exactly the same in all cases, something is wrong.

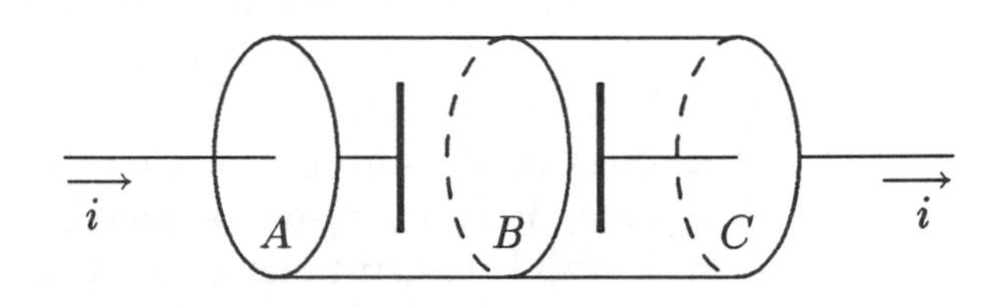

D. Note that the situation discussed and the lack of symmetry in the electromagnetic equations suggests that Ampere's law as used so far must be changed. Experiment confirms this conjecture.

E. Explain that if the electric flux through an open surface changes with time, then there is a magnetic field and the magnetic field has a tangential component at points on the boundary. Write down the Maxwell law of induction: $\oint 8\vec{B} \cdot d\vec{s} = \mu_0\epsilon_0\, d\Phi_E/dt$, where Φ_E is the electric flux through the surface. Compare to Faraday's law and point out the interchange of $\vec{B}$ and $\vec{E}$, the change in sign, and the appearance of the factor $\mu_0\epsilon_0$. State that this law can be combined with Ampere's law and write the complete Ampere-Maxwell law: $\oint \vec{B} \cdot d\vec{s} = \mu_0 i + \mu_0\epsilon_0\, d\Phi_E/dt$.

F. Give the right-hand rule that relates the normal to the surface used to calculate Φ_E and the direction of integration around its boundary. State that the surface may be a purely mathematical construction and that the law holds for any surface.

G. Consider a charging parallel-plate capacitor with circular plates and derive expressions in terms of dE/dt for the magnetic field on a plane between the plates, at points both inside and outside the capacitor. See Sample Problem 32–3.

II. Displacement current.

A. Define the displacement current: $i_d = \epsilon_0\, d\Phi_E/dt$. Explain that it does not represent the flow of charge and is not a true current, but that it enters the Ampere-Maxwell law in the same way as a true current. Discuss the direction of i_d. Consider a region in which the electric field is uniform and is changing. Find the direction for both an increasing and a decreasing field.

B. Refer to the Ampere-Maxwell law. Explain that there are no changing electric fields in the examples of previous chapters so only true currents were considered. Explain that there is no true current in the region between the plates of a charging capacitor, but there is a displacement current.

C. Consider a parallel-plate capacitor with circular plates, for which dE/dt is given. Show that the total displacement current in the interior of the capacitor equals the true current into the capacitor. Explain that the sum of the true and displacement currents is continuous. Optional: discuss a leaky capacitor.

D. Derive expressions in terms of i_d for $\vec{B}$ at various points along the perpendicular bisector of the line joining the plate centers. Consider points between the plates and outside them.

E. Show that the total displacement current between the plates of a capacitor is the same as the true current into or out of the plates.

III. Maxwell's equations.

A. Write down the four equations in integral form and review the physical processes that each describes. See Table 32–1.

B. Carefully distinguish between the line and surface integrals that appear in the equations and give the right-hand rules that relate the direction of integration for the contour integrals and the normal to the surface for the surface integrals.

C. Review typical problems: the electric field of a point charge, the magnetic field of a uniform current in a long straight wire, the magnetic field at points between the plates of a capacitor with circular plates, the electric field accompanying a changing uniform magnetic field with cylindrical symmetry.

D. State that in the absence of dielectric and magnetic materials these equations describe all electromagnetic phenomena to the atomic level and the natural generalizations of them provide valid descriptions of electromagnetic phenomena at the quantum level. They are consistent with modern relativity theory. Optional: for completeness you may want to rewrite the equations and include magnetization and electric polarization terms.

SUPPLEMENTARY TOPIC

Earth's magnetic field. Section 32–3 describes the magnetic field of Earth. The shape, cause, and some of the ramifications of Earth's field are important topics and should be covered if you have the time. If not, you might intersperse some of the information in your other lectures. Explain that the field can be approximated by a magnetic dipole field. Draw a sphere, label the north and south geographic poles, draw a dipole moment vector at the center (pointing roughly from north to south, about 10° away from the axis of rotation), and draw some magnetic field lines. Remark that the north pole of the dipole is near the south geographic pole. Define declination and inclination.

SUGGESTIONS

1. Assignments
   a. To test for understanding of Gauss' law for magnetism, assign problem 1 or 3.
   b. Ask students to think about a permanent bar magnet that pierces the surface of a sphere and explain why the net magnetic flux through the surface is zero. Also ask them about the electric flux as a single charge as it crosses the surface and the magnetic flux of a single magnetic monopole as it crosses the surface.
   c. Questions 1 and 2 deal with the energy of a magnetic dipole in an external magnetic field. They also deal with the intrinsic dipole moments of electrons. Assign problem 9 in connection with the dipole moments of electrons.
   d. Questions 3, 4, and 5 deal with diamagnetism and question 6 deals with paramagnetism. Magnetization in a paramagnetic substance is covered in problems 17 and 18.
   e. The Curie temperature of a ferromagnet is covered in problem 20 and the magnetization of a ferromagnet is covered in problem 22. Use problem 21 to show that magnetic interactions are not responsible for ferromagnetism.
   f. To test for understanding of the direction of the magnetic field induced by a changing electric field, assign questions 8 and 9.
   g. Question 10 helps students think carefully about displacement current. Also consider question 11. Assign problems 31, 34, and 35.
2. Demonstrations
   a. Field of a magnet: Freier and Anderson Er4.
   b. Gauss' law: Freier and Anderson Er12.
   c. Paramagnetism: Freier and Anderson Es3, 4.
   d. Ferromagnetism: Freier and Anderson Es1, 2, 6 — 10.
   e. Levitation: Freier and Anderson Er10, 11.
3. Books and Monographs

   magnetic Monopoles; edited by Alfred S. Goldhaber and W. Peter Trower; available from AAPT, One Physics Ellipse, College Park MD 20740–3845. Reprint collection, with a resource letter.

4. Audio/Visual
   a. *Ferromagnetic Domain Wall Motion*; *Paramagnetism of Liquid Oxygen*; from AAPT Miller collection of single-concept films; video tape; available from Ztek Co., PO Box 11768, Lexington, KY 40577–1768.
   b. *Magnetic Effects in Space*; Skylab Demonstrations; VHS video tape (14 min); Media Design Associates, Inc., Box 3189, Boulder, CO 80307–3190.
   c. *Magnetism in Space*; Skylab Demonstrations; VHS video tape (19 min); Media Design Associates, Inc. (see above for address).
   d. *Earth's Magnetic Field*; VHS video tape (10 min); Films for the Humanities & Sciences, PO Box 2053, Princeton, NJ 08543–2053.
   e. *Domain Theory*; Electromagnetism; VHS video tape (10 min); Films for the Humanities & Sciences (see above for address).
   f. *Magnetism and Static Electricity*; VHS video tape (30 min); Films for the Humanities & Sciences (see above for address).
5. Computer Software
   Electricity & Magnetism; Macintosh, Windows; Cross Educational Software, Inc., 508 E. Kentucky Avenue, PO Box 1536, Ruston, LA 71270. Contains a section on ferromagnetic domains.
6. Laboratory
   Meiners Experiment 11–6: *Magnetization and Hysteresis.* Faraday's law is used to measure the magnetic field inside an iron toroid for various applied fields. A plot of the field as a function of the applied field shows hysteresis. A method for obtaining the hysteresis curve as an oscilloscope trace is also given.

Chapter 33 ELECTROMAGNETIC OSCILLATIONS AND ALTERNATING CURRENT

BASIC TOPICS

I. LC oscillations.
   A. Draw a diagram of an LC series circuit and assume the capacitor is charged. Explain that as charge flows, energy is transferred from the electric field of the capacitor to the magnetic field of the inductor and back again. When the capacitor has maximum charge, the current (dq/dt) vanishes, so no energy is stored in the inductor. When the current is a maximum, the charge on the capacitor vanishes and no energy is stored in that element.
   B. Write down the loop equation, then convert it so the charge q on the capacitor is the dependent variable. If the direction of positive current is into the capacitor plate with positive charge q, then $i = dq/dt$. If it is out of that plate, then $i = -dq/dt$.
   C. Write down the solution: $q(t) = Q\cos(\omega t + \phi)$. Show by direct differentiation that this is a solution if $\omega^2 = 1/LC$. Show that ϕ is determined by the initial conditions and treat the special case for which $q = Q$, $i = 0$ at $t = 0$.
   D. Once the solution is found, derive expressions for the current, the energy stored in the capacitor, and the energy stored in the inductor, all as functions of time. Sketch graphs of these quantities. Show that the total energy is constant.
   E. Derive expressions for the potential differences across the capacitor and the inductor. Draw graphs of them as well. Mention that the charge on the capacitor is proportional to the potential difference across its plates and that the time rate of change of the current is proportional to the potential difference across the terminals of the inductor.

F. Note that the form of the differential equation for q is the same as that for the displacement x of a block oscillating on the end of a spring. Make the analogy concrete by explaining that if q is replaced by x, L is replaced by m, and C is replaced by $1/k$, the equation for q becomes the equation for x. Also point out that the current corresponds to the velocity of the block, that the energy stored in the inductor corresponds to the kinetic energy of the block, and that the energy stored in the capacitor corresponds to the potential energy stored in the spring.

II. Damped oscillations.

A. Write down the loop equation for a single RLC loop, then convert it so q is the dependent variable. State that $q(t) = Q\,e^{-Rt/2L}\cos(\omega' t + \phi)$ satisfies the differential equation. Here ω' is somewhat less than $1/\sqrt{LC}$. If time permits, the expression for ω' can be found by substituting the assumed solution into the differential equation.

B. Draw a graph of $q(t)$ and point out that the envelope decreases exponentially. Each time the capacitor is maximally charged, the charge on the positive plate is less than the previous time. Explain that this does not violate the conservation of charge principle since the total of the charge on both plates of the capacitor is always zero. Energy is dissipated in the resistor.

C. To show the oscillations, wire a resistor, inductor, and capacitor in series with a square-wave generator and connect an oscilloscope across the capacitor. The scope shows a function proportional to the charge. Also connect the oscilloscope across the resistor to show a function that is proportional to the current. If you have a dual-trace scope, show the functions simultaneously. Show the effect of varying C (use a variable capacitor), R (use a decade box), and L (insert an iron rod into the coil). If time permits, show that oscillations occur only if $1/LC > (R/2L)^2$.

III. Elements of ac circuit analysis.

A. Consider a resistor connected to a sinusoidally oscillating emf. State that the potential difference across the resistor has the same angular frequency ω_d as the emf. State that the potential difference across the resistor is in phase with the current and that the amplitudes are related by $I_R = V_R/R$. Emphasize that $v_R(t)$ gives the potential of one end of the resistor relative to the other. Draw a phasor diagram: two arrows along the same line with length proportional to I_R and V_R, respectively. Both make the angle $\omega_d t$ with the horizontal axis and rotate in the counterclockwise direction. Point out that the vertical projections represent $i_R(t)$ and $v_R(t)$ and these vary in proportion to $\sin(\omega_d t)$ as the arrows rotate.

B. Consider a capacitor connected to a sinusoidally oscillating emf. Start with $i_C = dq/dt = C\,dv_C/dt$, substitute $v_C = V_C\sin(\omega_d t)$, and show that v_C lags i_C by 90° and that the amplitudes are related by $I_C = V_C/X_C$, where $X_C = 1/\omega_d C$ is the capacitive reactance. Draw a phasor diagram to show the relationship. Mention that the unit of reactance is the ohm.

C. Consider an inductor connected to a sinusoidally oscillating emf. Start with $v_L = L\,di_L/dt$, substitute $v_L = V_L\sin(\omega_d t)$, and show that v_L leads i_L by 90° and that the amplitudes are related by $I_L = V_L/X_L$, where $X_L = \omega_d L$ is the inductive reactance. Draw a phasor diagram to show the relationship.

D. Wire a small resistor in series with a capacitor and a signal generator. Use a dual trace oscilloscope with one set of leads across the resistor and the other set across the capacitor. Remind students that the potential difference across the resistor is proportional to the current, so the scope shows i_C and v_C. Point out the difference in phase. Repeat with an inductor in place of the capacitor.

IV. Forced oscillations of an RLC series circuit.

A. Draw the circuit. Assume the generator emf is given by $\mathcal{E}(t) = \mathcal{E}_m \sin(\omega_d t)$ and the current is given by $i(t) = I \sin(\omega_d t - \phi)$. Pick consistent directions for positive emf and positive current. Construct a phasor diagram step-by-step (see Fig. 33–11). First draw the current and resistor voltage phasors, in phase. Remind students that the current is the same in every element of the circuit so voltage phasors for the other elements can be drawn using the phase relations between voltage and current developed earlier. Draw the capacitor voltage phasor lagging by 90° and the inductor voltage phasor leading by 90°. Make $V_L > V_C$. Their lengths are IX_C and IX_L, respectively. Draw the projections of the phasors on the vertical axis and remark that the algebraic sum must be $\mathcal{E}(t)$.

B. Draw the impressed emf phasor. Remark that its projection on the inductance phasor must be $V_L - V_C$ and that its projection on the resistance phasor must be V_R. Make the analogy to a vector sum.

C. Use the phasor diagram to derive the expression for the current amplitude: $I = \mathcal{E}_m/Z$, where $Z = \sqrt{R^2 + (X_L - X_C)^2}$ is the impedance of the circuit. Show that the impedance is frequency dependent by substituting the expressions for the reactances.

D. Use the phasor diagram to derive the expression for the phase angle of i relative to $\mathcal{E}$: $\tan\phi = (X_L - X_C)/R$. Point out that $\mathcal{E}$ leads i if $X_L > X_C$, but $\mathcal{E}$ lags i if $X_L < X_C$. For later use, show that $\cos\phi = R/Z$.

V. Resonance

A. Sketch graphs of the current amplitude as a function of the generator frequency for several values of the resistance (see Fig. 33–13). Point out that the current amplitude is greatest when the generator frequency matches the natural frequency of the circuit and that the peak becomes larger as the resistance is reduced. Use the expression derived above for the current amplitude to show that I is greatest for $X_C = X_L$ and that this means $\omega_d = 1/\sqrt{LC}$. Remark that this is the resonance condition. Also show that the phase angle between the current and the generator emf vanishes at resonance.

B. Demonstrate resonance phenomena by wiring an RLC loop in series with a sinusoidal audio oscillator. Look at the current by putting the leads of an oscilloscope across the resistor. Use a decade box for the resistor and measure the current amplitude for various frequencies and for several resistance values. Be sure the amplitude of the oscillator output remains the same. Explain that similar circuits are used to tune radios and TV's.

C. Use a sweep generator to show the current amplitude. Set the oscilloscope sweep rate to accommodate that of the generator and put a small diode in series with the scope leads. Usually this will have enough capacitance that only the envelope will be displayed.

VI. Power considerations.

A. Discuss average values over a cycle. Show that the average of $\sin^2(\omega_d t + \phi)$ is $\frac{1}{2}$ and that the average of $\sin(\omega_d t)\cos(\omega_d t)$ is 0. Define the rms value of a sinusoidal quantity. Point out that ac meters are usually calibrated in terms of rms values.

B. Derive the expression for the power input of the ac source: $P = i\mathcal{E} = i_m \mathcal{E}_m \sin(\omega_d t + \phi)\sin(\omega_d t)$. Show that the average over a cycle is given by $\overline{P} = \mathcal{E}_{\text{rms}} i_{\text{rms}} \cos\phi$. Do the same for the power dissipated in the resistor. In particular, show that its average value can be written $i_{\text{rms}}^2 R$ or $\mathcal{E}_{\text{rms}} i_{\text{rms}} R/Z$. Recall that $R/Z = \cos\phi$ and then use this relationship to show that the average power input equals the average power dissipated in the resistor.

C. Show that the average rate of energy flow into the inductor and capacitor are each zero.

D. Explain that $\cos\phi$ is called the power factor. If it is 1, the source delivers the greatest possible power for a fixed generator amplitude. Remark that the power factor is 1 at resonance.

SUPPLEMENTARY TOPIC

The transformer. Say that ac is in common use because it is efficient to transmit power at high potential and low current but safety considerations require low potential at the user and producer ends of a transmission line. Transformers can be used to change the potential. Use Faraday's law to show how the potential difference across the secondary is related to the potential difference across the primary. Explain what step-up and step-down transformers are. A dual trace oscilloscope can be used to demonstrate transformer voltages. Assume a purely resistive load and show how to find the primary and secondary currents. Show that, as far as the primary current is concerned, the transformer and secondary circuit can be replaced by a resistor with $R_{\text{eq}} = (N_p/N_s)^2 R$, where N_p is the number of turns in the primary coil, N_s is the number of turns in the secondary coil, and R is the load resistance. Explain impedance matching.

SUGGESTIONS

1. Assignments
   a. Questions 1 through 5 can be used to help students think about LC circuit relationships.
   b. If you compare an oscillating LC circuit to an oscillating mass on a spring, assign problem 6 or 7.
   c. Assign problems 1, 3, 4, and 11 to test for understanding of the fundamentals of LC oscillations. The frequency of oscillation is covered in problems 8, 9, and 15. Also ask question 6 in connection with the discussion of energy.
   d. When discussing solutions to the RLC loop equation, include questions 8, 9, and 11.
   e. Assign problem 35 in connection with discussions of the phase and amplitude of separate inductive and capacitive circuits.
   f. Assign problem 42 to have students think about voltages around an LCR circuit.
   g. Resonance is covered in problems 45 and 48.
   h. Power in an RLC circuit is covered in problems 59 and 60 and the power factor in problem 57. Include question 12 in the discussion.
2. Demonstrations
   a. LCR series circuit: Freier and Anderson En12, Eo13.
   b. Measurements of reactance and impedance: Freier and Anderson Eo9.
   c. Transformers: Freier and Anderson Ek7, Em1, 2, 4, 5, 7, 8, 10.
3. Audio/Visual
   Direct Current and Alternating Current; VHS video tape (14 min); Films for the Humanities & Sciences, PO Box 2053, Princeton, NJ 08543–2053.
4. Computer Software
   Electricity & Magnetism; Macintosh, Windows; Cross Educational Software, Inc., 508 E. Kentucky Avenue, PO Box 1536, Ruston, LA 71270. Contains a section on RLC circuits.
5. Laboratory
   a. Meiners Experiment 10–11: *A.C. Series Circuits.* Students use an oscilloscope and ac meters to investigate voltage amplitudes, phases, and power in RC and RLC circuits. Voltage amplitudes and phases are plotted as functions of the driving frequency to show resonance. Reactances and impedances are calculated from the data.
   b. Bernard Experiment 37: *A Study of Alternating Current Circuits.* An ac voltmeter is used to investigate the voltages across circuit elements in R, RC, RL, and RLC circuits, all with 60 Hz sources. Reactances and impedances are computed. If possible, oscilloscopes should be used. A section labelled optional describes their use. This experiment is pedagogically

similar to the text and can be used profitably to reenforce the ideas of the chapter. Warning: the lab book uses the word vector rather than phasor.

Chapter 34 ELECTROMAGNETIC WAVES

BASIC TOPICS

I. Qualitative features of electromagnetic waves.

A. Explain that an electromagnetic wave is composed of electric and magnetic fields. The disturbance, analogous to the string shape that moves on a taut string, is made up of the fields themselves, moving through space or a material medium. Also explain that electromagnetic waves carry energy and momentum.

B. State that the wave speed in a vacuum is given by $c = 1/\sqrt{\mu_0 \epsilon_0}$ and is about 3.00×10^8 m/s. The existence of waves and this expression for the wave speed in vacuum are predicted by Maxwell's equations. Since the values of c and μ_0 are fixed, this fixes ϵ_0.

C. Show the electromagnetic spectrum (Fig. 34–1 of the text) and point out the visible, ultraviolet, infrared, x-ray, microwave, and radio regions. Remark that all the waves are fundamentally the same, differing only in wavelength and frequency. Remind students that ϵ_0 and μ_0 enter electrostatics and magnetostatics, respectively, and were first encountered in situations that had nothing to do with wave propagation.

D. Restate that the visible spectrum extends from just over 400 nm to just under 700 nm. Remark that while color is largely subjective, violet is at the short wavelength end while red is at the high wavelength end. Use a prism to display the spectrum. Show Fig. 34–2 of the text and remark that human eyes are most sensitive in the green–yellow portion of the spectrum and that sensitivity falls off rather rapidly on either side.

E. State that an accelerating charge creates electromagnetic radiation. Show diagrams of an oscillating electric dipole antenna and its fields (see Figs. 34–3, 34–4, and 34–5). Point out that $\vec{E}$ and $\vec{B}$ are perpendicular to each other and to the direction of propagation and that they oscillate in phase with each other at any point. Explain the term polarization.

II. Traveling sinusoidal waves.

A. Take $E(x,t) = E_m \sin(kx - \omega t)$, along the y axis, and $B(x,t) = B_m \sin(kx - \omega t)$, along the z axis. Remark that both fields travel in the positive x direction and that they are in phase. Remind students that the minus sign in the argument becomes a plus sign for a wave traveling in the negative x direction.

B. Consider a rectangular area in the xy plane, with infinitesimal width dx and length h (along y). Evaluate $\oint \vec{E} \cdot d\vec{s}$ and Φ_B, then show that Faraday's law yields $\partial E/\partial x = -\partial B/\partial t$. Substitute the expressions for E and B to show that $E = cB$, where $c = \omega/k$. Stress that the magnitudes of $\vec{E}$ and $\vec{B}$ are related. Remark that $\vec{E}$ is different at different points because $\vec{B}$ changes with time.

C. Consider a rectangular area in the xz plane, with infinitesimal width dx and length h (along z). Evaluate $\oint \vec{B} \cdot d\vec{s}$ and Φ_E, then show that the Ampere-Maxwell law yields $-\partial B/\partial x = \mu_0 \epsilon_0 \, \partial E/\partial t$. Combine this with the result of part B to show that $c = 1/\sqrt{\mu_0 \epsilon_0}$. Remark that $\vec{B}$ is different at different points because $\vec{E}$ changes with time. Emphasize the role played by the displacement current.

III. Energy and momentum transport.

A. Define the Poynting vector $\vec{S} = (1/\mu_0)\vec{E} \times \vec{B}$ and explain that it is in the direction of propagation and that its magnitude gives the electromagnetic energy per unit area that crosses an area perpendicular to the direction of propagation per unit time. Remark that for a plane wave, $S = EB/\mu_0 = E^2/\mu_0 c = cB^2/\mu_0$.

B. Consider the plane wave of Section II, propagating in the positive x direction. Consider a volume of width Δx and cross section A (in the yz plane) and show that the electric and magnetic energies in it are equal and that the total energy is $\Delta U = (EBA/\mu_0 c)\Delta x$, for small Δx. This energy passes through the area A in time $\Delta t = \Delta x/c$ so the rate of energy flow per unit area is EB/μ_0, as previously postulated.

C. Explain that most electromagnetic waves of interest oscillate rapidly and we are not normally interested in the instantaneous values of the energy or energy density. Explain how to find the average over a period of the square of a sinusoidal function. Define the intensity as the time average of the magnitude of the Poynting vector and write expressions for it in terms of the average energy density and in terms of the field amplitudes.

D. Explain that electromagnetic waves transport momentum and that S/c gives the momentum that crosses a unit area per unit time. The momentum is in the direction of $\vec{S}$. Also explain that if an object absorbs energy U, then it receives momentum U/c. If the object reflects energy U, then it receives momentum $2U/c$.

E. Show that if a wave, incident normal to a surface of area A, is completely absorbed, then the force on the surface is IA/c and the radiation pressure is I/c, where I is the intensity. Show that the force and radiation pressure have twice these values if the wave is completely reflected.

F. As an example of radiation pressure, you may wish to consider solar pressure. S can be determined from the solar constant $1.38\,\mathrm{kW/m^2}$ (valid just above Earth's atmosphere).

IV. Polarization.

A. Remind students that a linearly polarized electromagnetic wave is one for which the electric field is everywhere parallel to the same line. As the wave passes by any point, the field oscillates along the line of polarization.

B. Explain that a linearly polarized wave can be resolved into two other linearly polarized waves with mutually orthogonal polarization directions. Take the original polarization direction to be at the angle θ to one of the new directions and show that the amplitudes are given by $E_1 = E_m \cos\theta$ and $E_2 = E_m \sin\theta$, where E_m is the original amplitude.

C. Explain that the electric field associated with unpolarized light does not remain in the same direction for more than about 10^{-8} s and the new direction is unrelated to the old.

D. Shine unpolarized light through crossed Polaroid sheets and note the change in intensity as the second sheet is rotated. Show that the intensity does not change if the first sheet is rotated. Remark that for an ideal polarizing sheet the transmitted intensity is half the incident intensity.

E. Derive the law of Malus. Explain that the light emerging from the first Polaroid sheet is linearly polarized in a direction determined by the orientation of the sheet. Remark that this direction is called the polarizing direction of the sheet. Draw a diagram of the electric field amplitude as the light enters the second sheet, at an angle θ to the polarizing direction of the second sheet. Resolve the amplitude into components along the polarizing direction and perpendicular to it. Explain that the first component is transmitted through the sheet while the second is absorbed. The amplitude of the transmitted wave is proportional to $\cos\theta$ and the intensity is proportional to $\cos^2\theta$.

F. Shine unpolarized light onto two crossed Polaroid sheets and remark that no light is transmitted. Then, slide another sheet between the two and point out the change in transmitted intensity as you rotate the sheet in the middle. The sheets can be taped to ringstands to hold them. Explain the phenomenon by examining the polarization at each stage of the transmission.

V. Wave and geometrical optics.

A. Explain that optical phenomena outside the quantum realm can be understood in terms of Maxwell's equations and that the wave nature of electromagnetic radiation must be taken into account to explain many important phenomena. State that some of these will be discussed later.

B. Explain that if the wavelength of the light is much smaller than any obstacles it meets or any slits through which it passes, then the important property is the direction of motion, not details of the wave nature. This is the realm of geometrical optics.

C. Define a ray as a line that gives the direction of travel of a wave. It is perpendicular to the wave fronts (surfaces of constant phase). Explain that geometrical optics deals largely with tracing rays as light is reflected from surfaces or passes through materials.

VI. Reflection and refraction.

A. Explain that when light traveling in one medium strikes a boundary with another medium, some is reflected and some is transmitted into the second medium. Draw a plane boundary between two media and show an incident, a reflected, and a refracted ray. Label the angles these rays make with the normal to the surface. Use θ_1 to label the angle of incidence, θ_1' to label the angle of reflection, and θ_2 to label the angle of refraction. Emphasize that these angles are measured relative to the normal to the surface.

B. Tell students that the speed of light may be different in different materials and state that the speeds of light in the two media are crucial for determination of the amplitudes of the reflected and refracted light and for determination of the angle of refraction. Define the index of refraction of a medium as the ratio of the wave speed in vacuum to the wave speed in the medium and write $v = c/n$. Remark that the index of refraction is a property of the medium and depends on the wavelength. Point out Table 34–1, which gives the indices of refraction of various materials. Note that the index of refraction for a vacuum is 1 and is nearly 1 for air. State it is wavelength dependent and point out Fig. 34–19.

C. Consider a plane wave incident on a plane surface. Write down the law of reflection: $\theta_1 = \theta_1'$.

D. Write down the law of the law of refraction: $n_1 \sin\theta_1 = n_2 \sin\theta_2$.

E. Explain that light rays are bent toward the normal when light enters a more optically dense medium (higher index of refraction) and are bent away from the normal when it enters a less optically dense medium.

F. Consider light striking a water surface from air and trace a few rays. Consider light from an underwater source and trace a few rays as they enter the air. Consider a slab of glass with parallel sides and show that the emerging ray has the same direction as the entering ray but is displaced along the slab. Optional: derive the expression for the displacement.

G. Trace a ray through a prism and derive the expression for the angle of deviation: $\psi = \theta_1 + \theta_2 + \phi$, where θ_1 is the angle of incidence, θ_2 is the angle of emergence, and ϕ is the prism angle. Explain that ψ is different for different colors because n depends on wavelength.

H. Shine an intense, monochromatic, well-collimated beam on a prism and point out the reflected and refracted beams. A laser works reasonably well but it is difficult for the class to see the beam. Use smoke or chalk dust to make it visible. To avoid the mess, use an arc beam or the beam from a 35 mm projector, filtered by red glass. Make a $\frac{1}{2}$ in. hole in a 2 in. by 2 in. piece of aluminum and insert it in the film gate. Use white light from the projector and the prism to show that different wavelengths are refracted through different angles.

I. Explain total internal reflection. Show that no wave is transmitted when the angle of incidence is greater than the critical angle and derive the expression for the critical angle in terms of the indices of refraction. Stress that the index for the medium of incidence must be greater than the index for the medium of the refracted light. Total internal reflection can be demonstrated with some pieces of solid plastic tubing having a diameter larger than that used for fiber optics. The beam inside is quite visible. If time permits, discuss fiber optics and some of its applications.

VII. Polarization by reflection.

A. Reflect a well collimated beam of unpolarized light from a plane glass surface. A slide projector beam does nicely. Darken the room and obtain a reflection spot on the ceiling. Place a Polaroid sheet in the reflected beam and note the change in intensity of the spot as you rotate it. Remark that the reflected light is partially polarized.

B. Orient the incident beam so the angle of incidence is Brewster's angle and use the Polaroid sheet to show the reflected light is now entirely polarized.

C. Discuss Brewster's law. Explain that unpolarized light incident on a boundary is partially or completely polarized on reflection. When the angle of incidence and the angle of refraction sum to 90°, the reflected light is completely polarized, with $\vec{E}$ perpendicular to the plane of the incident and reflected rays. Show that the angle of incidence θ_B for completely polarized reflected light is given by $\tan\theta_B = n_2/n_1$, where medium 1 is the medium of the reflected ray.

SUGGESTIONS

1. Assignments

   a. Relationships among frequency, wavelength, and speed are explored in problems 2, 3, and 4. These also give some examples of high and low frequency electromagnetic radiation and ask students to interpret Fig. 34–2, which graphs the sensitivity of the human eye as a function of wavelength.

   b. To stress the relationship between $\vec{E}$ and $\vec{B}$ in an electromagnetic wave, assign problem 9. Use questions 1 and 2 in the discussion of the relationships among the directions of the electric field, the magnetic field, and the direction of propagation.

   c. To emphasize the magnitude of the energy and momentum carried by an electromagnetic wave, assign problems 16 and 25. Also consider some problems that deal with point sources: 17 and 19, for example.

   d. Use problem 32 to test for understanding of polarization. Also consider questions 3, 4, and 5. The fundamentals of polarizing sheets are covered in problems 33 and 36. In the first, the incident light is unpolarized while in the second, it is polarized. Also consider problem 41.

   e. Problem 43 covers the law of refraction. Also consider questions 7, 8, and 9.

   f. Assign problems 56 and 57 in connection with total internal reflection. Assign either problem 61 or 62 in connection with polarization by reflection.

2. Demonstrations

   Radiation: Freier and Anderson Ep4, 5.

3. Books and Monographs

   a. *Resource Letters, Book Four* and *Resource Letters, Book Five*; American Association of Physics Teachers, One Physics Ellipse, College Park, MD 20740–3845. Contain resource letters on light.

   b. *Connecting Time and Space*; edited by Harry E. Bates; available from AAPT (see above for address). Reprints dealing with measurements of the speed of light and the redefinition of the meter.

4. Audio/Visual
   a. *Color, Scattering, Polarization*; Side D: Waves (II) & Electricity and Magnetism of Cinema Classics; video disk; available from Ztek Co., PO Box 11768, Lexington, KY 40577–1768.
   b. *Polarized Light*; VHS video tape (10 min); Films for the Humanities & Sciences, PO Box 2053, Princeton, NJ 08543–2053.
   c. *The Determination of the Velocity of Light*; VHS video tape (15 min); Films for the Humanities & Sciences (see above for address).
   d. *The Electromagnetic Spectrum*; VHS video tape (21 min); Films for the Humanities & Sciences (see above for address).
   e. *Reflection*, *Refraction*, and *Dispersion*; Side C: Waves (I) of Cinema Classics; video disk; available from Ztek Co. (see above for address).
   f. *The Wave Model*; VHS video tape (10 min); Films for the Humanities & Sciences (see above for address).
   g. *Reflection*; Demonstrations of Physics: Light; VHS video tape (6:34); Media Design Associates, Inc., Box 3189, Boulder, CO 80307–3190.
   h. *Refraction*; Demonstrations of Physics: Light; VHS video tape (8:07); Media Design Associates, Inc. (see above for address).
   i. *Reflection of Light*; VHS video tape (9 min); Films for the Humanities & Sciences (see above for address).
   j. *Refraction of Light*; VHS video tape (10 min); Films for the Humanities & Sciences (see above for address).
   k. *Propagation of Light*; *Visible and Infrared Spectrum*; Physics Demonstrations in Light, Part I; VHS video tape; ≈3 min each; Physics Curriculum & Instruction, 22585 Woodhill Drive, Lakeville MN 55044.
5. Computer Software
   a. *Physics Simulation Programs.* See Chapter 17 SUGGESTIONS.
   b. *Optics*; Macintosh, Windows; Cross Educational Software, Inc., 508 E. Kentucky Avenue, PO Box 1536, Ruston, LA 71270. Contains a section on polarization.
6. Laboratory
   a. Meiners Experiment 13–7; *Polarization of Light.* Polaroid sheets are first investigated and the law of Malus is verified. Then, a polaroid sheet is used to investigate polarization by reflection, by refraction, and by scattering. Brewster's angle is found. Rotation of the direction of polarization by a sugar solution is also studied and crossed polarizers are used to check various objects for stresses. This is essentially a series of demonstrations performed by students.
   b. Bernard Experiment 46; *Polarized Light.* Similar to Meiners Experiment 13–7 except light transmitted by a calcite crystal is also investigated. A photodetector is used to obtain quantitative data.
   c. Meiners Experiment 13-3: *Prism Spectrometer.* Helium lines are used to determine the index of refraction as a function of wavelength for a glass prism. A good example of dispersion and excellent practice in carrying out a rather complicated derivation involving Snell's law. Also see Bernard Experiment 43: *Index of Refraction with the Prism Spectrometer* and Bernard Experiment 44: *The Wavelength of Light.* In the second of these experiments, students use a prism spectrometer to determine the wavelength of lines from a sodium source.

Chapter 35 IMAGES

BASIC TOPICS

I. Plane mirrors.

A. Consider a plane wave incident on a plane mirror. Remind students of the law of reflection: $\theta_1 = \theta_1'$.

B. Consider a point source in front of a plane mirror. Draw both incident and reflected rays and show that the reflected rays appear to come from a point behind the mirror. Show that the object and image lie on the same normal to the mirror and that they are the same distance from the mirror. Remark that no light comes from the image and that the image is said to be virtual.

C. Define the object distance p and image distance i and explain that the latter is taken to be negative for virtual images. The law of equal distance is written $p = -i$.

D. Give the condition for being able to see an image. Draw a mirror, an eye, a source, and its image. Draw the line from the image to the eye and state that the image can be seen if this line intersects the mirror. Show that length of a wall mirror with its top edge at eye level need reach only halfway to the floor for a person to see his feet. Demonstrate with a mirror resting on the floor and half-covered with a cloth. Have a student stand in front of the mirror. Start with the cloth about shoulder height and lower it until the student can see his feet.

II. Spherical mirrors.

A. Consider a point source in front of a concave spherical mirror. Draw a diagram that shows the central axis, the center of curvature, and the source on the axis, outside the focal point. Show that small-angle rays form an image and that object and image distances are related by $1/p + 1/i = 2/r$. To emphasize the small-angle approximation, consider the case $p = 2r$ and use a full hemispherical concave surface. The small-angle formula predicts all rays cross the axis at $i = (2/3)r$, but the ray that strikes the edge of the mirror crosses at the vertex.

B. Explain that the mirror equation is also valid for convex mirrors and for any position of the object, even virtual objects for which incoming rays converge toward a point behind the mirror. Give the sign convention: p and i are positive for real objects and images (in front of the mirror) and are negative for virtual objects and images (behind the mirror); r is positive for concave mirrors (center of curvature in front of the mirror) and negative for convex mirrors (center of curvature behind the mirror). Remark that a surface is concave or convex according to its shape as seen from a point on the incident ray.

C. Define the focal point as the image point when the incident light is parallel to the axis. By considering a source far away, show that $f = r/2$. Consider a concave mirror and show that for $p > f$, the image is real; for $p < f$, the image is virtual. Also show that for $p = f$, parallel rays emerge after reflection.

D. Describe a geometric construction for finding the image of an extended source. Trace rays from an off-axis point: one through the center of curvature, one through the focal point, and one parallel to the axis. Use both concave and convex mirrors as examples. Explain that the geometric construction gives the same result as the small-angle approximation if reflection is assumed to take place at a plane through the mirror vertex and perpendicular to the optic axis. The law of reflection cannot be applied at this plane, of course.

E. Define lateral magnification and show that $m = -i/p$. Explain the sign: m is positive for erect images and negative for inverted images. Virtual images of real objects are erect and real images of real objects are inverted.

F. Take the limit $r \to \infty$ and show that the mirror equation makes sense for a plane mirror.

III. Spherical refracting surfaces.

A. Draw a convex spherical boundary between two media, use the law of refraction to trace a small-angle ray from a source on the axis, and show that $n_1/p + n_2/i = (n_2 - n_1)/r$, where n_1 is the index of refraction for the region of incident light and n_2 is the index of refraction for the region of refracted light. You can demonstrate the bending of the light using a laser and a round-bottom flask. Use a little smoke or chalk dust to make the beam visible in air and a pinch of powdered milk in the water to make it visible inside the flask.

B. Explain the sign convention. Point out that real images are on the opposite side of the boundary from the incident light and virtual images are on the same side. Explain that p and i are positive for real objects and images, negative for virtual objects and images. r is positive for convex surfaces, negative for concave. With this sign convention, the equation holds for concave or convex surfaces and for $n_2 > n_1$ or $n_1 > n_2$.

C. Consider the limit $r \to \infty$, which yields $i = -pn_2/n_1$. This is the solution to the apparent depth problem. For water four inches deep, a ball on the bottom appears to be at a depth of about three inches. Use an aquarium filled with water and a golf ball to make a hallway display.

IV. Thin lenses.

A. Explain that a lens consists of two refracting surfaces close together in vacuum. State or derive the thin lens equation: $1/p + 1/i = (n - 1)(1/r_1 - 1/r_2)$, where n is the index of refraction for the lens material. Stress that the equation holds for small-angle rays. State that it also holds to a good approximation for a lens in air. r_1 is the radius of the first surface struck by the light and r_2 is the radius of the second. They are positive or negative according to whether the surfaces are convex or concave when viewed from a point on the incident ray. You may wish to generalize the equation by retaining the indices of refraction. The result is $1/p + 1/i = (n_2/n_1 - 1)(1/r_1 - 1/r_2)$. This allows you to consider a thin glass or air lens in water.

B. By considering $p \to \infty$, show that the focal length is given by $1/f = (n - 1)(1/r_1 - 1/r_2)$ or more generally by $1/f = (n_2/n_1 - 1)(1/r_1 - 1/r_2)$. Show that the same value, including sign, is obtained no matter which surface is struck first by light. Then, show that $1/p + 1/i = 1/f$. Point out that there are two focal points, the same distance from the lens but on opposite sides. For a converging lens, rays from a point source at f on one side are parallel on the other side; incident parallel rays converge to f on the other side. For a diverging lens, rays that converge toward f on the other side emerge parallel; rays that are parallel emerge as diverging from f on the incident side.

C. Show how to locate the image of an extended object by tracing a ray parallel to the axis, a ray through the lens center, and a ray along a line through the first focal point (on the incident side for a converging lens and on the other side for a diverging lens).

D. Define lateral magnification and show that $m = -i/p$. Explain that the sign tells whether the image is erect or inverted.

E. Consider all possible situations: converging lens with $p > f$, $p < f$, and $p = f$; diverging lens with $p > f$, $p < f$, and $p = f$. In each case, show whether the image is real or virtual, erect or inverted, and find its position relative to the focal point.

F. Note that most optical instruments are constructed from a combination of two or more lenses. Point out that to analyze them, one considers one lens at a time, with the image of the previous lens as the object of the lens being considered. This sometimes leads to virtual objects. Note that the overall magnification is given by $m = m_1 m_2 m_3 \ldots$ and that the sign of m tells whether the image is erect or inverted. If the image lies on the opposite side of the system from the object and is outside the system, then it is real; otherwise it is virtual.

SUPPLEMENTARY TOPIC

Optical instruments. This section may be studied in the laboratory. Ask students to experiment with the image forming properties of positive and negative lenses, then construct one or more optical instruments. Display several instruments in the lab.

SUGGESTIONS

1. Assignments
   a. Interesting applications of plane mirrors are covered in problems 5 (rotation of mirror) and 6 (can observer see an image?). Problems 3 and 4 and question 3 deal with images in multiple mirrors. Assign at least one of them.
   b. Use questions 4 and 5 to discuss images in spherical mirrors. Problem 10 covers nearly all possibilities. Lateral magnification is covered in problem 9.
   c. Assign problem 13 in connection with spherical refracting surfaces. Problem 14 covers all of the possibilities.
   d. Use questions 6, 7, and 8 to discuss images formed by thin lenses. For comprehensive coverage of nearly all relationships, assign problem 24. Problems 19 and 20 test understanding of the lensmaker's equation. Also assign problem 28, which deals with a compound system and includes a ray tracing exercise.
   e. Consider expanding the course a little by including problem 35, which deals with the human eye.
2. Demonstrations
   a. Plane mirrors: Freier and Anderson Ob1 - 6, Ob8.
   b. Refraction at a plane surface: Freier and Anderson Od1 - 7.
   c. Prisms: Freier and Anderson Of1 - 4.
   d. Total internal reflection: Freier and Anderson Oe1 - 7.
3. Books and Monographs
   a. *Resource Letters, Book Four* and *Resource Letters, Book Five*; American Association of Physics Teachers, One Physics Ellipse, College Park, MD 20740–3845. Contain resource letters on optics.
   b. *The Camera*; by Bill G. Aldridge, Gary S. Waldman, and John Yoder III.; available from AAPT (see above for address). Concepts important for understanding cameras.
4. Audio/Visual
   a. *Optics*; VHS video tape (30 min); Films for the Humanities & Sciences, PO Box 2053, Princeton, NJ 08543–2053.
   b. *The Image*; VHS video tape (8 min); Films for the Humanities & Sciences (see above for address).
   c. *Lenses*; VHS video tape (10 min); Films for the Humanities & Sciences (see above for address).
   d. *Convex and Concave Lenses*; VHS video tape (10 min); Films for the Humanities & Sciences (see above for address).
   e. *Mirror and Image*; VHS video tape (9 min); Films for the Humanities & Sciences (see above for address).
5. Computer Software
   a. *Ray*; Miky Ronen; Macintosh; available from Physics Academic Software, North Carolina State University, PO Box 8202, Raleigh, NC 27690–0739. A ray-tracing program. The user can place reflecting surfaces, refracting surfaces, mirrors, lenses, and prisms on the screen and control their orientation. Rays are traced using either the paraxial approximation or the actual path.

b. Optics; Macintosh, Windows; Cross Educational Software, Inc., 508 E. Kentucky Avenue, PO Box 1536, Ruston, LA 71270. Contains sections on refractions, mirrors, and lenses.

6. Laboratory
   a. Meiners Experiment 13–1: *Laser Ray Tracing.* A laser beam is used to investigate the laws of reflection and refraction and to observe total internal reflection and the formation of images by spherical mirrors. Measurements are used to calculate the index of refraction of several materials, including liquids, and the focal length of mirrors. Tracing is done by arranging the apparatus so the laser beam grazes a piece of white paper on the lab table. Much the same set of activities are described in Bernard Experiment 38: *Reflection and Refraction of Light*, but pins are used as objects rather than a laser source and rays are traced by positioning other pins along them. The technique can be used if you do not have sufficient lasers for the class.
   b. Bernard Experiment 39: *The Focal Length of a Concave Mirror.* Several methods are described, including a technique that involves finding the radius of curvature. Others involve finding the image when the object distance is extremely long, when it is somewhat greater than $2f$, and when it is somewhat less than $2f$. Then, the mirror equation is used to solve for f.
   c. Meiners Experiment 13–2: *Lenses.* A light source and screen on an optical bench are used to find the focal lengths and magnifications of both convex and concave lenses. Chromatic and spherical aberrations are also studied. Also see Bernard Experiment 40: *Properties of Converging and Diverging Lenses*, a compendium of techniques for finding focal lengths.
   d. Bernard Experiment 41: *Optical Instruments Employing Two Lenses.* Students construct simple two-lens telescopes and microscopes on optical benches, then investigate their magnifying powers. By trying various lens combinations, they learn the purposes of the objective and eyepiece lenses.

Chapter 36 INTERFERENCE

BASIC TOPICS

I. Huygens' principle.
   A. Shine monochromatic light through a double slit and project the pattern on the wall. Either use a laser or place a single slit between the source and the double slit. Use a diagram to explain the setup. Point out the appearance of light in the geometric shadow and the occurrence of dark and bright bands. You can make acceptable double slits by coating a microscope slide with lamp black or even black paint. Tape a pair of razor blades together and draw them across the slide. By inserting various thicknesses of paper or shim stock between the blades, you can obtain various slit spacings.
   B. Explain that Huygens' principle will be used to understand the pattern, then state the principle. Describe plane wave propagation in terms of Huygen wavelets: draw a plane wave front, construct spherical wave fronts of the same radius centered at several points along the plane wave front, then draw the plane tangent to these.
   C. Use Huygens' principle to derive the law of refraction. Assume different wave speeds in the two media and show that the wavelengths are different. Consider wavefronts one wavelength apart and show that $\sin\theta_1/\sin\theta_2 = v_1/v_2$. Explain that $n = c/v$ and obtain the law of refraction.
   D. Go back to the double-slit pattern and explain that those parts of an incident wave front that are within the slit produce spherical wavelets that travel to the screen while wavelets from other parts are blocked. Some wavelets reach the geometric shadow. The spreading of

the pattern beyond the shadow is called diffraction and will be studied in the next chapter. Wavelets from different slits arrive at the same point on the screen and interfere to produce the bands. This phenomena will be studied in this chapter.

II. Two-slit interference patterns.

A. Draw a diagram of a plane wave incident normally on a two-slit system and draw a ray from each slit to a screen far away. Remark that the waves are in phase at the slits but they travel different distances to get to the same point on the screen and may have different phases there. The electric fields sum to the total electric field. At some points, the two fields cancel, at other points they reinforce each other. Remind students that the intensity is proportional to the square of the total field, not to the sum of the squares of the individual fields.

B. Point out that if the screen is far away, the two rays are nearly parallel, then show that the difference in distance traveled is $d \sin\theta$, where d is the slit separation and θ is the angle the rays make with the forward direction. Explain the condition $d\sin\theta = m\lambda$ for a maximum of intensity and the condition $d\sin\theta = (m + \frac{1}{2})\lambda$ for a minimum.

C. Show that a lens can be used to obtain the same pattern, even if the screen is not far away.

III. The intensity.

A. Take the two fields to be $E_1 = E_0 \sin(\omega t)$ and $E_2 = E_0 \sin(\omega t + \phi)$, where $\phi = (2\pi/\lambda)d\sin\theta$. This is easily shown by remarking that $\phi = k\Delta d$, where $k = 2\pi/\lambda$ and $\Delta d = d\sin\theta$ (derived earlier).

B. Explain how the fields can be represented on a phasor diagram. Explain that a phasor has a length proportional to the amplitude and makes the angle ωt or $\omega t + \phi$ with the horizontal axis. Its projection on the vertical axis is proportional to the field. Sum the phasors to obtain the total field. Show that the amplitude E_θ of the total field is $2E_0 \cos(\phi/2)$. Plot the intensity $4E_0^2 \cos^2(\phi/2)$ as a function of ϕ. Point out that $\phi = 0$ produces a maximum, that maxima occur at regular intervals, and that the minima are halfway between adjacent maxima.

C. Show that the intensity at a maximum is four times the intensity due to one source alone. Remark that no energy is gained or lost. All energy through the slits arrives at the screen. The presence of the slitted barrier, however, redistributes the energy.

D. Note the half-width of each maximum, at half the peak, is given by $\sin\theta = \lambda/4d$. The smaller λ/d, the sharper the maximum. Near the central maximum, where $\sin\theta \approx \tan\theta \approx \theta$, the linear spread on the screen is $y \approx (\lambda/2d)D$, where D is the distance from the slits to the screen.

E. It is also worth noting that since $\sin\theta = m\lambda/d \leq 1$ for a maximum, the smaller λ/d, the more maxima occur.

F. For completeness, you might mention the amplitude of the wavelets fall off as $1/r$ and are not quite the same at the screen. Show this is a negligible effect for the patterns considered here.

IV. Coherence.

A. Explain that two waves are coherent if their relative phase does not change with time.

B. Explain that the two interfering waves must be coherent to obtain an interference pattern. The phase difference at the observation point must be constant over the observation time. Explain why two incandescent lamps, for example, do not produce a stable interference pattern. The light is from many atoms and the emission time for a single atom is about 10^{-8} s. The phase difference changes in a random way over times that are short compared to the observation time. State that in this case the intensities add.

C. Explain that an extended source can be used to obtain an interference pattern. Light from each atom goes through both slits and forms a pattern, but the patterns of different atoms are displaced from each other, according to the separation of the atoms in the source. No pattern is seen unless the incident light comes only from a small region of the source. If you did not use a laser in the demonstration, explain the role of the single slit in front of the double slit.

D. Explain that a laser produces coherent light even though many atoms are emitting simultaneously. Because emission is stimulated, light from any atom is in phase with light from all other atoms. A laser can be used to form an interference pattern without restricting the incident beam.

V. Thin-film interference.

A. Cut a 1 to 2 mm slit in a 2" square piece of aluminum and insert it in the film gate of a 35 mm projector. Let the beam impinge on a soap bubble to show the effect.

B. Consider normal incidence on a thin film of index n_1 in a medium of index n_2 and suppose the medium behind the film has index n_3. Explain that a wave reflected at the interface with a medium of higher index undergoes a phase change of π. If $n_1 < n_2 < n_3$, waves reflected at both surfaces undergo phase changes of π. Consider all other possibilities and then specialize to a thin film of index n in air. Give the conditions for maxima and minima for both the reflected light and the transmitted light, assuming near normal incidence. Note that the wavelength in the medium must be used to calculate the phase change on traveling through the medium. Define optical path length and point out its importance for thin-film interference.

C. Broaden the discussion qualitatively by including non-normal incidence. Note that for some angles, conditions are right for destructive interference of a particular color while at other angles, conditions are right for constructive interference of the same color. Also note that these angles depend on λ. Hence the soap bubble colors.

D. If time permits, discuss Newton's rings. Use a plano-convex lens and a plane sheet of glass together with a laser. Use a diverging lens to spread the beam.

SUPPLEMENTARY TOPIC

The Michelson interferometer. This is an excellent example of an application of interference effects. Set up a hallway demonstration and give a brief explanation.

SUGGESTIONS

1. Assignments
   a. In the discussion of coherence, give a more detailed explanation of the single slit placed between the source and the double slit.
   b. Questions 1, 2, 3, and 4 and problems 1 through 10 are good tests of the fundamentals. Use them in the discussion or ask students to answer a few of them for homework.
   c. Problems 11 through 18 deal with the basics of interference. Assign one or two. Use problem 21 or 22 to test for understanding of the derivation of the double-slit equation.
   d. Assign problems 26 and 27 in connection with the double-slit interference pattern. Also consider questions 5, 6, and 7.
   e. Use questions 10, 11, and 12 and problems 31 and 38 to help with the discussion of thin films.
   f. Problems 54, 55, and 57 illustrate some applications of a Michelson interferometer.
2. Demonstrations
   a. Double-slit interference: Freier and Anderson O14, 5, 9.
   b. Thin-film interference: Freier and Anderson O115 — 18.
   c. Michelson interferometer: Freier and Anderson O119.

3. Audio/Visual
   a. *Michelson Interferometer*; from AAPT Miller collection of single-concept films; video tape; available from Ztek Co., PO Box 11768, Lexington, KY 40577–1768.
   b. *Interference*; Side D: Waves (II) & Electricity and Magnetism of Cinema Classics; video disk; available from Ztek Co. (see above for address).
   c. *Double Aperture Interference*; VHS video tape (9 min); Films for the Humanities & Sciences, PO Box 2053, Princeton, NJ 08543–2053.
   d. *Interference/Interferometer*; *Thin Film Interference*; Physics Demonstrations in Light, Part II; VHS video tape; ≈3 min each; Physics Curriculum & Instruction, 22585 Woodhill Drive, Lakeville, MN 55044.
4. Computer Software
   Optics; Macintosh, Windows; Cross Educational Software, Inc., 508 E. Kentucky Avenue, PO Box 1536, Ruston, LA 71270. Contains sections on interference.
   a. *Wave Interference*; Mike Moloney; DOS; Physics Academic Software, North Carolina State University, PO Box 8202, Raleigh, NC 27690–0739. Uses phasors to obtain intensity patterns.
5. Laboratory
   a. Meiners Experiment 13–4: *Interference and Diffraction.* Students observe double-slit patterns of water waves in a ripple tank, sound waves, microwaves, and visible light. In each case except water waves, they measure and plot the intensity as a function of angle, then use the data to calculate the wavelength. A microcomputer can be used to take data and plot the intensity of a visible light pattern.
   b. Meiners Experiment 13–6: *The Michelson Interferometer.* An interferometer is used to measure the wavelengths of light from mercury and a laser and to find the index of refraction of a glass pane and air. Good practical applications.

Chapter 37 DIFFRACTION

BASIC TOPICS

I. Qualitative discussion of single-slit diffraction.
   A. Shine coherent monochromatic light on a single slit and project the pattern on the wall. Point out the broad central bright region and the narrower, less bright regions on either side, with dark regions between. Also point out that light is diffracted into the geometric shadow.
   B. Remark that diffraction can be discussed in terms of Huygens wavelets emanating from points in the slit. Explain that they not only spread into the shadow region but that they arrive at any selected point with a distribution of phases and interfere to produce the pattern. Explain that for quantitative work, this chapter deals with Fraunhofer diffraction, with the screen far from the slit.
   C. Draw a single slit with a plane wave incident normal to it. Also draw parallel rays from equally spaced points within the slit, all making the same angle θ with the forward direction. Point out that all wavelets are in phase at the slit. The first minimum can be located by selecting θ so that, at the observation point, the ray from the top of the slit is 180° out of phase with the ray from the middle of the slit. All wavelets then cancel in pairs. Show that this leads to $a\sin\theta = \lambda$, where a is the slit width. Point out that this value of θ determines the width of the central bright region and that this region gets wider as the slit width narrows. Use $\sin\theta \approx \tan\theta \approx \theta$ (in radians) to show that the linear width of the

central region on a screen a distance D away is $2D\lambda/a$. Use a variable width slit or a series of slits to demonstrate the effect.

D. By dividing the slit into fourths, eighths, etc. and showing that in each case the wavelets cancel in pairs if θ is properly selected, find the locations of other minima. Show that $a \sin\theta = m\lambda$ for a minimum.

E. Explain that for $a < \lambda$, the central maximum covers the whole forward direction. No point of zero intensity can be observed. Also remark that the intensity becomes more uniform as a decreases from λ. This was the assumption made in the last chapter when the interference of only one wavelet from each slit was considered.

F. Qualitatively discuss the intensity. Draw a phasor diagram showing ten or so phasors representing wavelets from equally spaced points in the slit. Show that each wavelet at the observation point is out of phase with its neighbor by the same amount. First, show the phasors with zero phase difference ($\theta = 0$), then show them for a larger value of θ. Show that they approximate a circle at the first minimum and then, as θ increases, they wrap around to form another maximum, with less intensity than the central maximum. Point out that as θ increases, the pattern has successive maxima and minima and that the maxima become successively less intense.

II. The intensity.

A. Draw a diagram showing ten or so phasors along the arc of a circle and let ϕ be the phase difference between the first and last. See Fig. 37–8. Explain that you will take the limit as the number of wavelets increases without bound and draw the phasor addition diagram as an arc. Use geometry to show that $E_\theta = E_m(\sin\alpha)/\alpha$, where $\alpha = \phi/2$. Point out that the intensity can be written $I_\theta = I_m(\sin^2\alpha)/\alpha^2$, where I_m is the intensity for $\theta = 0$. By examining the path difference for the rays from the top and bottom of the slit, show that $\alpha = (\pi a/\lambda)\sin\theta$. Explain that these expressions give the intensity as a function of the angle θ.

B. Sketch the intensity as a function of θ (see Fig. 37–7) and show mathematically that the expression just derived predicts the positions of the minima as found earlier.

C. (Optional) Set the derivative of $(\sin\alpha)/\alpha$ equal to 0 and show that $\tan\alpha = \alpha$ at an intensity maximum. State that the first two solutions are $\alpha = 4.493$ rad and 7.725 rad. Use these results to show that the intensity at the first two secondary maxima are 4.72×10^{-2} and 1.65×10^{-2}, relative to the intensity for $\theta = 0$. You might also want to pick a wavelength and slit width, then find the angular positions of the first two secondary maxima. Remark that they are close to but not precisely at midpoints between zeros of intensity.

III. Double-slit diffraction.

A. Consider the double-slit arrangement discussed in the previous chapter. Point out that the electric field for the light from each of the slits obeys the equation developed for single-slit diffraction and these two fields are superposed. They have the same amplitude, $E_m(\sin\alpha)/\alpha$, and differ in phase by $(2\pi d/\lambda)\sin\theta$, where d is the center-to-center slit separation. The result for the intensity is $I_\theta = I_m(\cos^2\beta)(\sin^2\alpha)/\alpha^2$, the product of the single-slit diffraction equation and the double-slit interference equation. Here $\beta = (\pi d/\lambda)\sin\theta$.

B. Sketch I_θ vs. θ for a double slit and point out that the single-slit pattern forms an envelope for the double-slit interference pattern. Remark that this is so because d must be greater than a. See Fig. 37–14.

C. Show how to calculate the number of interference fringes within the central diffraction maximum and remark that the result depends on the ratio d/a but not on the wavelength.

D. Discuss missing maxima. Point out that the first diffraction minimum on either side of the central single-slit diffraction maximum might coincide with a double-slit interference maximum, in which case the maximum would not be seen. Show that the maximum of order m is missing if $d/a = m$.

IV. Diffraction gratings.

A. Make or purchase a set of multiple-slit barriers with 3, 4, and 5 slits, all with the same slit width and spacing. Multiple slits can be made using razor blades and a lamp blackened microscope slide. Use a laser to show the patterns in order of increasing number of slits. Finish with a commercial grating.

B. Qualitatively describe the pattern produced as the number of slits is increased. Point out the principle maxima and, if possible, the secondary maxima. Remark that the principle maxima narrow and that the number of secondary maxima increases as the number of slits increases. Remark that for gratings with a large number of rulings, the principal maxima are called lines. For each barrier, sketch a graph of the intensity as a function of angle. Explain that the single-slit diffraction pattern forms an envelope for the pattern.

C. Remark that you will assume the slits are so narrow that the patterns you will consider lie well within the central maximum of the single-slit diffraction pattern and you need to consider only one wave from each slit. Explain that lines occur whenever the path difference for rays from two adjacent slits is an integer multiple of the wavelength: $d \sin\theta = m\lambda$. Remark that m is called the order of the line. Also remark that the angular positions of the lines depend only on the ratio d/λ and not on the number of slits or their width.

D. Consider N phasors of equal magnitude that form a regular polygon and remark this is the configuration for an interference minimum adjacent to a principal maximum. Show that for one of these minima the phase difference for waves from adjacent slits is $2\pi(m + 1/N)$ and the path difference is $d \sin\theta = \lambda(m + 1/N)$. Replace θ with $\theta + \delta\theta$, where $d \sin\theta = m\lambda$, to derive the expression $\delta\theta = \lambda/Nd \cos\theta$ for the angular half-width of the principal maximum at angle θ. Explain that this predicts narrowing of the principal maxima as the number of slits is increased. Also explain that principal maxima at large angles are wider than those at small angles.

E. Show a commercial transmission grating and tell students a typical grating consists of tens of thousands of lines ruled over a few centimeters. Explain that light is transmitted through both the rulings and the regions between but since these represent different thickness of material, the phase of the waves leaving the rulings is different from that of waves leaving the regions between. As a result, the diffraction pattern is the same as that of a multiple-slit barrier. Say that a diffraction pattern is also produced by lines ruled on a reflecting surface.

F. Put a grating in front of a white-light source and point out the spectrum. Put a grating in front of a discharge tube to display the emission spectrum of hydrogen or mercury. Note the separation of the lines corresponding to the same principal maximum produced by different frequency light. Explain that atoms produce light with certain discrete frequencies and that these are separated by the grating. Remark that measurements of the angles can be used to compute the wavelengths present if the ruling separation is known. Point out the colors of a compact disk or CD ROM.

V. X-ray diffraction (optional).

A. Explain that x rays are electromagnetic radiation with wavelength on the order of 10^{-10} m (1 Å). Point out that crystals are regular arrays of atoms with spacings on that order and so can be used to diffract x rays.

B. Consider a set of parallel crystalline planes and explain that reflection of the incident beam occurs at each plane, with the angle of reflection equal to the angle of incidence. Draw a diagram like Fig. 37–26 and state that x-ray diffraction is conventionally described in terms of the angle between the ray and the plane, rather than the normal to the plane. Show that waves reflected from the planes interfere constructively if $2d \sin\theta = m\lambda$.

C. Explain that for a given set of planes intense diffracted waves are produced only if waves are incident at an angle θ that satisfies the Bragg condition, given above. Measurements of these angles can be used to investigate the crystal structure. Show how to calculate the distance between planes, given the wavelength and the scattering angle. Explain that a crystal with a known structure can be used as a filter to obtain x rays of a given wavelength from a source with a broad range of wavelengths.

SUPPLEMENTARY TOPICS

1. Diffraction from a circular aperture. This topic is important for its application to diffraction patterns of lenses and the diffraction limit to the resolution of objects by a lens system. Show a diagram or picture (like Fig. 37–9) and point out the bright central disk and the secondary rings. Tell students that the angular position of the smallest ring of zero intensity occurs for $\theta = 1.22\lambda/d$, where d is the diameter of the aperture. If you intend to discuss the resolving power of a grating, the Rayleigh criterion for a circular aperture should be covered first since it is easier to present and understand. You can demonstrate the Rayleigh criterion by drilling two small holes, closely spaced, in the bottom of a tin can. Place the can over a light bulb and let students view it from various distances. See problem 37–15. Also use red and blue filters to show the dependence on wavelength.
2. Dispersion and resolving power of a grating. Define the dispersion of a grating and show it is $m/d\cos\theta$ for a line of order m, occurring at angle θ. Note that dispersion can be increased by decreasing the ruling separation but dispersion does not depend on the number of rulings. If you have gratings with different ruling separations, use them to show the hydrogen spectrum and point out the difference. Define the resolving power of a grating and show it is Nm for the line of order m. Remark that the resolving power does depend on the number of rulings and that the greater this number, the greater the resolving power. Show the sodium spectrum with a grating for which the two D lines cannot be resolved, then show it with one for which they can. Explain that dispersion and resolving power measure different aspects of the pattern produced by a grating. The lines produced by two different wavelengths may be fairly well separated in angle (large dispersion) but cannot be resolved because the principal maxima are so wide (small resolving power).

SUGGESTIONS

1. Assignments
   a. Basic relationships for the single-slit pattern are explored in problems 1 through 4 and 7. Assign a few of these. To test for qualitative understanding, consider questions 1, 2, and 5.
   b. Following the discussion of the equation for the double-slit pattern, ask question 7. Characteristics of the pattern are explored in problems 27, 28, and 31.
   c. Diffraction from a circular aperture with application to the Rayleigh criterion for resolution is covered in problems 14 through 26. Assign one or two if you cover this topic.
   d. After discussing diffraction patterns of multiple slits, ask question 10. Problems 33 and 35 cover the fundamental equation for an intensity maximum. Questions 8 and 9 and problem 38 deal with line width.
   e. Ask question 11 in connection with the dispersion and resolving power of a grating. Assign problems 46 and 49.
   f. After discussing x-ray diffraction by crystals, assign problems 53 and 55. Problems 58 and 61 are a little more challenging. Problem 59 deals with the geometry of a square lattice.
2. Demonstrations
   a. Single-slit diffraction: Freier and Anderson Ol2, 3, 6, 7.

b. Multiple-slit diffraction: Freier and Anderson Ol10, 13.
c. Diffraction by circular and other objects: Freier and Anderson Ol21 — 23.
d. Diffraction by crystals: Freier and Anderson Ol14.

3. Audio/Visual
   a. *The Diffraction of Light*; VHS video tape; Films for the Humanities & Sciences, PO Box 2053, Princeton, NJ 08543–2053.
4. Computer Software
   EM Field. See Chapter 23 SUGGESTIONS.
5. Computer Projects
   Assign some of the computer projects described in the Computer Projects section of this manual. A computer is used to plot the intensity pattern for various situations including the case when the screen is not far from the sources.
6. Laboratory
   a. Meiners Experiment 13–4: *Interference and Diffraction.* See Chapter 40 notes.
   b. Meiners Experiment 13–5; *Diffraction Gratings.* Wavelengths of the helium spectrum are found using a grating spectrometer and the influence of the number of grating rulings is investigated.
   c. Bernard Experiment 44; *The Wavelength of Light.* Wavelengths of the sodium spectrum are found using a grating spectrometer. The wavelength of a laser is also found.
   d. Bernard Experiment 45; *A Study of Spectra with the Grating Spectrometer.* Sources used are a sodium lamp, an incandescent bulb, a mercury lamp, and a lamp containing an unknown element. The limits of the visible spectrum are determined and the unknown element is identified.

Chapter 38 RELATIVITY

BASIC TOPICS

I. Introduction.
   A. Consider a wave on a string and remind students that its speed relative to the string is given by $v_w = \sqrt{\tau/\mu}$, where τ is the tension and μ is the linear mass density. Explain that, according to non-relativistic mechanics, an observer running with speed v_0 with the wave measures a wave speed of $v_w - v_o$ and an observer running against the wave measures a wave speed of $v_w + v_o$. Remark that these results are *not* valid for light (or fast moving waves and particles). The speed of light in a vacuum is found to be the same, regardless of the speed of the observer (or the speed of the source).
   B. Remark that this fact has caused us to revise drastically our idea of time. If, for example, two observers moving at high speed with respect to each other both time the interval between two events, they obtain different results.
   C. Explain that special relativity is a theory that relates measurements taken by two observers who are moving with respect to each other. Although it sometimes seems to contradict everyday experience, it is extremely well-supported by experiment.
   D. State the postulates: the laws of physics are the same for observers in all inertial frames; the speed of light in a vacuum is the same for all directions and in all inertial frames. Remind students what an inertial frame is. Explain that the laws of physics are relationships between measured quantities, not the quantities themselves. Newton's laws and Maxwell's equations are examples. State that relativity has forced us to revise Newton's second law but not Maxwell's equations.

II. Time measurements.

A. Explain the term *event* and note that three space coordinates and one time coordinate are associated with each event. Explain that each observer may think of a coordinate system with clocks at all places where events of interest occur and that the clocks are synchronized. Outline the synchronization process involving light. State that the coordinate system and clock used by an observer are at rest with respect to the observer and may be moving from the viewpoint of another observer.

B. State that two observers in relative motion cannot both claim that two events at different places are simultaneous if their motion is not perpendicular to the line joining the coordinates of the events. To illustrate, show Fig. 38–4 and explain that the events are simultaneous in Sam's frame but the Red event occurs before the Blue event in Sally's frame. Show that signals from the events meet at the mid-point of Sam's spaceship but the signal from the Red event gets to the mid-point of Sally's spaceship before the signal from the Blue event. Stress the importance of the second postulate for reaching these conclusions.

C. Explain the light flasher used to measure time, in principle. Consider a flasher at rest in one frame, take two events to be a flash and the subsequent reception of reflected light back at the instrument, then remark that the time interval is $\Delta t_0 = 2D/c$, where D is the separation of the mirror from the flash bulb. Consider the events as viewed in another frame, moving with speed v perpendicularly to the light ray, and show the interval is $\Delta t = 2D/c\sqrt{1 - v^2/c^2} = \Delta t_0/\sqrt{1 - v^2/c^2}$. This is also written $\Delta t = \gamma \Delta t_0$, where γ $(= 1/\sqrt{1 - v^2/c^2})$ is called the Lorentz factor. State that $v/c < 1$ and $\gamma > 1$.

D. Remark that Δt_0 is the *proper time interval* and that both events occur at the same coordinate in the frame in which it is measured. Point out that Δt is larger than Δt_0. Explain that the same result is obtained no matter what clocks are used for the measurement (as long as they are accurate and each is at rest in the appropriate frame). Ask students to identify a frame to estimate the proper time interval for a ball thrown from third to first base. Note that $\Delta t \approx \Delta t_0$ if $v \ll c$.

E. State that time dilation has been observed by comparing clocks carried on airplanes to clocks remaining behind and by comparing the average decay time of fast moving fundamental particles to their decay time when at rest. You might want to discuss the twin paradox here.

III. Length measurements.

A. Point out the problem with measuring the length of an object that is moving relative to the meter stick: the position of both ends must be marked *simultaneously* (in the rest frame of the meter stick) on the meter stick. If the speed v of the object is known, another method can be used to measure its length: put a mark on a coordinate axis along the line of motion of the object, then measure the time Δt_0 taken by the object to pass the mark. The length is given by $L = v\Delta t_0$. Note that Δt_0 is a proper time interval but L is not the proper length.

B. Explain that the length of the object, as measured in its rest frame, is $L_0 = v\Delta t$, where Δt is the time interval measured in that frame. Substitution of $\Delta t = \gamma \Delta t_0$ leads to $L = L_0/\gamma$. State that L_0, the length as measured in the rest frame of the object, is called the *proper length.* Since $\gamma > 1$, all observers moving with respect to the object measure a length that is less than the rest length. The same result is obtained no matter what method is used to measure length. Note that $L \approx L_0$ if $v \ll c$.

IV. The Lorentz transformation.

A. Consider two reference frames: S' moving with speed v in the positive x direction relative to S. Remark that the coordinates of an event as measured in S are written x, y, z, t

while the coordinates as measured in S' are written x', y', z', t'. Write down the Lorentz transformation for the coordinate differences of two events: $\Delta x' = \gamma(\Delta x - v\Delta t)$, $\Delta y' = \Delta y$, $\Delta z' = \Delta z$, $\Delta t' = \gamma(\Delta t - v\Delta x/c^2)$. Remark that these equations reduce to the Galilean transformation if $v \ll c$: $\Delta x' = \Delta x - v\Delta t$, $\Delta y' = \Delta y$, $\Delta z' = \Delta z$, $\Delta t' = \Delta t$.

B. Explain that the transformation equations can be solved for Δx and Δt, with the result $\Delta x = \gamma(\Delta x' + v\Delta t')$, $\Delta t = \gamma(\Delta t' + v\Delta x'/c^2)$. From the viewpoint of an observer in S', S is moving in the negative x' direction, so the two sets of equations are obtained from each other when v is replaced by $-v$ and the primed and unprimed symbols are interchanged.

C. Discuss some consequences of the Lorentz transformation equations:

1. Simultaneity. Take $\Delta t = 0$, $\Delta x \neq 0$ and show that $\Delta t' = -\gamma v\Delta x/c^2$ ($\neq 0$). If two events are simultaneous and occur at different places in S, then they are not simultaneous in S'. Point out that $\Delta t'$ is positive for Δx negative and is negative for Δx positive. Similarly, take $\Delta t' = 0$, $\Delta x' \neq 0$ and show $\Delta t = \gamma v\Delta x'/c^2$ ($\neq 0$).
2. Time dilation. Consider two events that occur at the same place in S and show that $\Delta t' = \gamma\Delta t$. Point out that Δt is the proper time interval. Also show that the events do not occur at the same place in S': $\Delta x' = -\gamma v\Delta t$. Work the same problem for two events that occur at the same place in S'.
3. Length measurement. Suppose the object is at rest in S' and the meter stick is at rest in S. Marks are made simultaneously in S on the meter stick at the ends of the object. Thus, $\Delta t = 0$. Show that $\Delta x' = \gamma\Delta x$ and point out that $\Delta x'$ is the rest length. Work the same problem with the object at rest in S and the meter stick at rest in S'.
4. Causality. Consider two events, the first of which influences the second. For example, a particle is given an initial velocity along the x axis and collides with another particle. Remark that t_2 (the time of the collision) must be greater than t_1 (the time of firing). Take $\Delta t = t_2 - t_1$ and $\Delta x > 0$, then show that the Lorentz transformation predicts $\Delta t'$ is positive for every frame for which $v < c$. The collision cannot happen before the firing in any frame moving at less than the speed of light.
5. Velocity transformation. Tell students that v represents the velocity of frame S' relative to S and that $\vec{u}$ and $\vec{u}'$ represent the velocity of a particle, as measured in S and S', respectively. Now take u and u' to be the x components of the particle velocity. Divide the Lorentz equation for Δx by the Lorentz equation for Δt to show that the x component of the particle velocity in S is $u = (u' + v)/(1 + vu'/c^2)$. Show this reduces to the Galilean transformation $u = u' + v$ for $v \ll c$. Take $u' = c$ and show that $u = c$. If $u' < c$, then $u < c$ for all frames moving at less than the speed of light.

V. Relativistic momentum and energy.

A. Explain that the non-relativistic definition of momentum must be generalized if momentum is to be conserved in collisions involving particles moving at high speeds. State that the proper generalization is $\vec{p} = m\vec{v}/\sqrt{1 - v^2/c^2}$. Remark that $\vec{p}$ is unbounded as the particle speed approaches the speed of light. In this text, m is used for the rest mass and is called simply the mass. The concept of relativistic mass is not used.

B. Remark that the definition of energy must be changed if the work-energy theorem is to hold for particles at high speeds. State that the relativistically correct expression for the energy of a free particle is $E = mc^2/\sqrt{1 - v^2/c^2}$. Take the limit as v/c becomes small and show that E can then be approximated by $mc^2 + \frac{1}{2}mv^2$. Thus, the correct relativistic definition of the kinetic energy is $K = E - mc^2$. Point out that the particle has energy mc^2 when it is at rest and remark that mc^2 is called the rest energy.

C. Explain that mass and rest energy are not conserved in many interactions involving fundamental particles but that total energy E is; rest energy can be converted to kinetic energy and vice versa.

D. Derive $E^2 = (pc)^2 + (mc^2)^2$ and explain that this expression replaces $E = p^2/2m$ ($= mv^2/2$). Remark that $E = pc$ for a massless particle, such as a photon.

SUPPLEMENTARY TOPIC

The Doppler effect for light. The expression for the frequency transformation can be derived easily by considering the measurement of the period in two frames. Suppose an observer in S obtains T for the interval between successive maxima at the same place. This is a proper time interval and the interval in another frame S' is γT. If S' is moving parallel to the wave, however, the two events do not occur at the same place in S' and γT is not the period in that frame. An observer in S' must wait for a time $|\Delta x'|/c$ longer before the next maximum is reached at the place of the first. Thus, $T' = \gamma T + |\Delta x'|/c$ or since $\Delta x' = -\gamma T v$, $T' = \gamma T(1 + v/c) = T\sqrt{(1+\beta)/(1-\beta)}$. Thus, $f' = f\sqrt{(1-\beta)/(1+\beta)}$. If S' is moving perpendicularly to the wave, the two events occur at the same place in both frames and $T' = \gamma T$, so $f' = f/\gamma$.

SUGGESTIONS

1. Assignments
   a. Simultaneity and time measurements are the issues in questions 2 through 6. Ask some of them to test understanding. Also assign problems 5 and 16.
   b. When length contraction is covered, assign problems 8, 11, and 12.
   c. Assign problems 14 and 15 in support of the discussion of the Lorentz transformation.
   d. Discuss questions 7 and 8 and assign problems 22 and 23 in connection with the relativistic velocity transformation.
   e. Discuss question 9 and assign problems 27 and 31 in connection with the relativistic Doppler effect.
   f. Use question 10 to broaden the discussion of mass and rest energy. Assign problems 38, 40, 41, and 42 in connection with relativistic energy and momentum. If you covered cyclotrons in Chapter 29, assign problem 46.
2. Computer Software
   a. *RelLab*; Paul Horwitz, Edwin F. Taylor, and Kerry Shetline; Macintosh; available from Physics Academic Software, North Carolina State University, PO Box 8202, Raleigh, NC 27690–0739. Show the coordinates and times of events, as measured in user-selected reference frames. Presents some relativity paradoxes that can be understood with the aid of the program.
   b. *Spacetime*; Edwin F. Taylor; Windows, Macintosh; available from Physics Academic Software (see above for address). Shows a "spacetime highway", on which objects in different lanes move with different speeds. Shows the corresponding spacetime diagram, on which events are identified. All the clocks and rulers are also shown so the user can compare readings in different frames.
3. Computer Project
   Have students write a computer program or design a spreadsheet to evaluate the Lorentz transformation equations. Then, have them use it to investigate simultaneity, length contraction, and time dilation. Specific projects are given in the Computer Projects Section of this manual.

Chapter 39 PHOTONS AND MATTER WAVES

BASIC TOPICS

I. Introduction.

A. Explain that this chapter deals with some of the fundamental results of quantum physics. The first few sections describe experimental results that can be understood only if light is regarded as made up of particles. Remark that interference and diffraction phenomena require waves for their explanation. Reconciliation of these opposing views will be discussed later.

B. Explain that the energy of a photon is related to the frequency of the wave through $E = hf$ and the momentum of a photon is related to the wavelength of the wave through $p = h/\lambda$. Show these equations predict $p = E/c$, the classical relationship. Also explain that the energy density is nhf, where n is the photon concentration, and that the intensity is Rhf, where R is the rate per unit area with which photons cross a plane perpendicular to their direction of motion. Recall the discussion of the Poynting vector in Chapter 34. Explain that the Planck constant is a constant of nature and pervades quantum physics. Give its value (6.63×10^{-34} J·s) and calculate the photon energy and momentum for visible light, radio waves, and x rays.

C. Point out that classically monochromatic electromagnetic radiation can have any value of energy. Quantum mechanically, this is not true, but since h is so small, the discreteness of the energy values is important only at the atomic level.

II. The photoelectric effect.

A. Sketch a schematic of the experimental setup. Explain that monochromatic light is incident on a sample. It is absorbed and part of the energy goes to electrons, some of which are emitted. The energy of the most energetic electron is found by measuring the stopping potential V_0.

B. Point out that the stopping potential is independent of the light intensity. As the intensity is increased, more electrons are emitted but they are not more energetic. Show a plot of the stopping potential as a function of frequency and point out that the relationship is linear and that as the frequency is increased the electrons emitted are more energetic. Also state that electrons are emitted promptly when the light is turned on. If the radiation energy were distributed throughout the region of a wave, it would take a noticeable amount of time for an electron to accumulate sufficient energy to be emitted, since an electron has a small surface area. This argument can be made quantitative (see Sample Problem 39–2).

C. Give the Einstein theory. Electromagnetic radiation is concentrated in photons, with each photon having energy hf. The most energetic electrons after emission are those with the greatest energy while in the material and, in the interaction with a photon, receive energy hf. If the light intensity is increased without changing the frequency, there are more photons and, hence, more electrons emitted, but no single electron can receive more energy. Furthermore, the electron receives energy immediately and need not wait to absorb the proper amount.

D. Show that this analysis leads to $hf = \Phi + K_m$, where Φ is the work function, the energy needed to remove the most energetic electron from the material. It is characteristic of the material. Remark that $K_m = eV_0$ and that the Einstein theory predicts a linear relationship between V_0 and f and predicts a minimum frequency for emission: $hf = \Phi$. Remark that the emitted electrons have a distribution of speeds if $hf > \Phi$ because they come from states with different energies.

III. The Compton effect.

A. Note that in the explanation of the photoelectric effect, a photon is assumed to give up all its energy to an individual electron. The photon then ceases to exist. Explain that a photon might transfer only part of its original energy in an interaction with an electron. Since a lower energy means a lower frequency, the scattered light has a longer wavelength than the incident light. State that a photon also carries momentum and part of it is transferred to a target electron.

B. Discuss the experiment. Light is scattered from electrons in matter and the intensity of the scattered light is measured as a function of wavelength for various scattering angles. Show Figs. 39–3 and 39–4. Stress that the experimental data can be explained by considering the interaction to be a collision between two particles, with energy and momentum conserved. Relativistic expressions, however, must be used for energy and momentum.

C. Remark that the situation is exactly like a two-dimensional collision between two particles. Write down the relativistic expressions for the momentum and energy of a particle with mass (the electron) and remind students of the rest energy. Assume the electron is initially at rest and that the photon is scattered through the angle ϕ. The electron leaves the interaction at an angle θ to the direction of the incident photon. Write down the equations for the conservation of energy and the conservation of momentum in two dimensions. Write down the momentum and energy of the photon in terms of the wavelength and solve for the change on scattering of the wavelength: $\Delta\lambda = (h/mc)(1 - \cos\phi)$. Emphasize that agreement with experiment strongly supports the conclusion that the momentum of a photon is $p = h/\lambda$.

D. Note that the change in wavelength is independent of wavelength and that the change is significant only for short wavelength light, in the x-ray and gamma ray regions. Also state that the theoretical results successfully predict experimental data. The widths of the curves are due chiefly to moving electrons, for which $\Delta\lambda$ is slightly different, and the peak near $\Delta\lambda = 0$ is due to scattering from more massive particles (atoms as a whole). Stress that the particle picture of light accounts for experimental data.

IV. Matter waves.

A. Explain that electrons and all other particles have waves associated with them, just as photons have electromagnetic waves associated with them. State that the waves exhibit interference and diffraction effects. Draw a diagram of a single-slit barrier with a beam of monoenergetic electrons incident on it and a fluorescent screen or other mechanism for detecting electrons behind it. Explain that an intense central maximum is obtained and that many electrons arrive in this region. Secondary maxima are also obtained.

B. State that the width of the central maximum depends on the speed of the electrons and narrows if the speed is increased. The maximum also narrows if more massive particles are used at the same speed. Remind students that when they studied the single-slit diffraction of electromagnetic waves, they found the width of the central maximum narrowed as the wavelength decreased. Conclude that the momentum of the particle is related to the wavelength of the wave and that one is proportional to the reciprocal of the other.

C. State that the particle energy and the wave frequency are related by $E = hf$ and that the particle momentum and the wavelength are related by $p = h/\lambda$, just as for photons. Calculate the wavelengths of a 1-eV electron and a 35-m/s baseball.

D. By way of example, state that crystals diffract electrons of appropriate wavelength ($\approx 10^{-10}$ m) and the angular positions of the scattering maxima can be found using Bragg's law, suitably modified to account for changes in the propagation direction that occur when matter waves enter the crystal.

E. Explain that, at the atomic and particle level, physics deals with probabilities. What can be analyzed is the probability for finding a particle, not its certain position. State that a one-dimensional matter wave is denoted by $\psi(x)$ and that $|\psi|^2$ gives the probability density for finding the particle near x. That is, the probability that the particle is in the region between x and $x + dx$ is given by $|\psi(x)|^2\, dx$. Similarly, if E is the electric field amplitude for an electromagnetic wave, then E^2 is proportional to the probability density for finding a photon. In the limit of a large number of particles, $|\psi|^2$ is proportional to the particle concentration.

F. State that space-dependent part of a particle wave function obeys the Schrödinger equation:

$$\frac{d^2\psi}{dx^2} + \frac{8\pi^2 m}{h^2}\left[E - U(x)\right]\psi = 0\,,$$

where E is the energy of the particle and $U(x)$ is its potential energy function. Explain that, for a free particle, we may take $U(x) = 0$ and write

$$\frac{d^2\psi}{dx^2} + k^2\psi = 0\,,$$

where $k^2 = (8\pi^2 m/h^2)E$. The most general solution is

$$\psi(x) = Ae^{ikx} + Be^{-ikx}\,,$$

where A and B are arbitrary constants. The first term represents a particle moving in the positive x direction with momentum $hk/2\pi$ and the second represents a particle moving in the negative x direction with the same magnitude momentum.

G. You may want to review some properties of complex numbers. Explain that a complex number can be written $\psi = \psi_R + i\psi_I$, where ψ_R is the real part and ψ_I is the imaginary part. Say that $i = \sqrt{-1}$. Define the complex conjugate: state that $|\psi|^2 = \psi^*\psi$, where ψ^* is the complex conjugate of ψ, and show that $|\psi|^2 = \psi_R^2 + \psi_I^2$. Also show that $|e^{ikx}|^2 = 1$.

V. The uncertainty principle.

A. Because a different answer might result each time the position of the electron is measured, there is an uncertainty in the position. It can be defined similarly to the standard deviation of a large collection of experimental results. Similar statements can be made about momentum measurements. Explain that the uncertainties in position and momentum are both determined by the particle wave function. Explain that if the electron is placed in a state for which the uncertainty in position is small then the uncertainty in momentum is large and vice versa.

B. Give the Heisenberg uncertainty relations: $\Delta x \cdot \Delta p_x \geq \hbar$, $\Delta y \cdot \Delta p_y \geq \hbar$, and $\Delta z \cdot \Delta p_z \geq \hbar$. Note that it is impossible to reduce both the uncertainty in position and the uncertainty in momentum simultaneously to zero. Compare this conclusion with the classical result by setting $\hbar = 0$.

VI. Barrier tunneling.

A. Show Fig. 39–12 and explain that the wave function penetrates a finite barrier. It is oscillatory (in position) outside the barrier, where $E > U_0$, and exponential inside, where $E < U_0$. The figure shows the probability density.

B. Explain that the particle has a probability of being found on either side of the barrier. Contrast to the behavior of a classical particle.

C. Write down Eqs. 39–21 and 39–22 for the transmission coefficient and explain that this measures the probability of transmission through the barrier. Remark that transmission is small for high, wide barriers and becomes larger as the barrier height decreases and as the barrier width narrows. Also define the reflection coefficient R by $R = 1 - T$.

SUGGESTIONS

1. Assignments
   a. Ask questions 1, 9, and 10 and assign one or two of problems 3, 6, and 11 as part of a discussion of photon properties. Emphasize that the energy in a light beam is the product of the number of photons and the energy of each photon. Assign some of problems 5, 6, 10, 11, 12, 13, 14, and 15.
   b. After discussing the photoelectric effect, ask some of questions 2 through 6. Assign problems 26 and 29.
   c. After discussing the Compton effect, ask questions 7, 8, and 11 and assign problem 34. Also consider problems 42 and 48.
   d. In the discussion of the properties of matter waves, include questions 12 through 16. Assign problems 57 and 64.
   e. Following the discussion of the uncertainty principle, assign problems 76 and 78.
   f. Ask some of questions 17 through 21 in connection with tunneling. Also assign problems 79 and 80.

2. Demonstrations
   Photoelectric effect: Freier and Anderson MPb1

3. Audio/Visual
   a. *Photons and X-rays*, *Electrons*, and *Particles and Waves*; Side F: Angular Momentum and Modern Physics of Cinema Classics; video disk; available from Ztek Co., PO Box 11768, Lexington, KY 40577–1768.
   b. *Spectra*; Demonstrations of Physics: Light; VHS video tape (6:02); Media Design Associates, Inc., Box 3189, Boulder, CO 80307–3190.
   c. *The Dual Nature of Light*; Demonstrations of Physics: Light; VHS video tape (8:16); Media Design Associates, Inc. (see above for address).
   d. *Photons*; VHS video tape (10 min); Films for the Humanities & Sciences, PO Box 2053, Princeton, NJ 08543–2053.
   e. *The Bohr Model*; VHS video tape (10 min); Films for the Humanities & Sciences (see above for address).
   f. *Spectra*; VHS video tape (10 min); Films for the Humanities & Sciences (see above for address).
   g. *Electron Diffraction*; Physics Experiments; VHS video tape (15 min); Films for the Humanities & Sciences (see above for address).

4. Computer Software
   a. *Atomic Physics*; Macintosh, Windows; Cross Educational Software, Inc., 508 E. Kentucky Avenue, PO Box 1536, Ruston, LA 71270. Macintosh. Contains sections on experimental foundations of quantum physics and photons.

5. Computer Project
   A commercial math program or a student-generated root-finding program can be used to solve the equations for the photoelectric and Compton effects. Students may be interested, for example, in seeing how the Compton lines broaden when the electrons are not initially at rest. Assign some exercises as homework or set aside some laboratory time for a more detailed investigation.

6. Laboratory
   a. Meiners Experiment 14–2: *The Photoelectric Effect.* Students investigate the characteristics of various photocells, then use a plot of stopping potential vs. frequency to determine the Planck constant. A mercury source and optical filters are used to obtain monochromatic light of various frequencies.

b. Meiners Experiment 14–5: *Electron Diffraction.* The Sargent-Welch electron diffraction apparatus is used to investigate the diffraction of electrons by aluminum and graphite. Since powder patterns (rings) are obtained, you will need to explain their origin.

Chapter 40 MORE ABOUT MATTER WAVES

BASIC TOPICS

I. One-dimensional particle traps

A. Explain that, for a particle confined by infinite potential energy barriers to the region between 0 and L on the x axis, possible wave functions are given by $\psi_n(x) = A\sin(n\pi x/L)$, where $n = 1, 2, \ldots$. Show that these satisfy the Schrödinger equation

$$\frac{d^2\psi}{dx^2} + \frac{8\pi^2 m}{h^2}E\psi = 0$$

inside the trap and that ψ goes to zero at the boundaries. Explain that a condition for the given function to be a solution is that the energy of the particle must be $E_n = n^2h^2/8mL^2$. You might want to include the time dependence by writing $\Psi = A\sin(n\pi x/L)f_n(t)$ and explaining that $f_n(t)$ is a function of time with magnitude 1.

B. Explain that confinement of the particle leads to energy quantization and that energy is quantized for any bound particle. Plot the allowed values of the energy, as in Fig. 40–3. Point out that the particle has kinetic energy even in the ground state and mention that this energy is called its zero-point energy.

C. Explain that the particle can certainly be found between $x = 0$ and $x = L$, so $\int_0^L |\psi_n|^2\,dx = 1$. The wave function is said to be *normalized* if it obeys this condition. Show that the normalization condition leads to $A = \sqrt{2/L}$.

D. Use the particle confined to a one-dimensional trap as an example and explain that $\psi_n^2\,dx = (2/L)\sin^2(n\pi x/L)\,dx$ gives the probability that the particle can be found between x and $x+dx$ when it is in the state with the given wave function. Sketch several of the probability density functions and point out that there are several places where the probability density vanishes. See Fig. 40–6.

E. Explain that experimentally the probability can be found, in principle, by performing a large number of position measurements and calculating the fraction for which the particle is found in the designated segment of the x axis. Since a position measurement changes the state of the particle, it must be restarted in the same state each time.

F. Draw a diagram of a one-dimensional trap with finite potential energy barriers at the ends and state that the particle wave function now extends into the barriers, although it decreases exponentially there. Show Fig. 40–8. Mention that the allowed values of the energy are different from those for infinite barriers, but that the energy is still quantized.

II. Two- and three-dimensional particle traps

A. Describe a two-dimensional rectangular trap with sides of length L_x and L_y, such that the particle has infinite potential energy at the boundaries and zero potential energy within. Give the expression for the energies:

$$E_{n_x,n_y} = \frac{h^2}{8m}\left[\frac{n_x^2}{L_x^2} + \frac{n_y^2}{L_y^2}\right],$$

where n_x and n_y are integers. Explain that neither n_x or n_y can be zero since either of those values would make the wave function zero everywhere.

B. Repeat the discussion for a three-dimensional trap in the form of a rectangular solid with sides of lengths L_x, L_y, and L_z. Show that the energies are given by

$$E_{n_x,n_y,n_z} = \frac{h^2}{8m}\left[\frac{n_x^2}{L_x^2} + \frac{n_y^2}{L_y^2} + \frac{n_z^2}{L - z^2}\right],$$

III. The hydrogen atom and line spectra.

A. Use a commercial hydrogen tube to show the visible hydrogen spectrum. Since the intensity is low, you will not be able to project this but you can purchase inexpensive $8'' \times 10''$ sheets of plastic grating material, which can be cut into pieces and passed out to the students. Point out Fig. 40–17.

B. Give the expression for the hydrogen energy levels in terms of the principal quantum number: $E_n = -(me^4/8\epsilon_0^2 h^2)(1/n^2)$. State that quantum physics predicts these allowed values. Say that a photon is emitted when a hydrogen atom changes state and derive $f = (me^4/8\epsilon_0^2 h^3)[(1/n_2^2) - (1/n_1^2)]$ for the frequency of the emitted electromagnetic wave.

C. Explain that the Schrödinger equation is a differential equation for the wave function of a particle and that the main ingredient that causes two identical particles to have different wave functions is their potential energy function. For an electron in a hydrogen atom, the potential energy function is $U(r) = -e^2/4\pi\epsilon_0 r$, where r is the distance from the proton to the electron. Mention that when this potential energy function is used in the Schrödinger equation and the reasonable condition that the wave functions remain finite everywhere is applied, then the allowed energy values are predicted. Draw a graph of $U(r)$ and draw lines across it to indicate the values of the first few energy levels.

D. Explain that states for hydrogen are classified using three quantum numbers:

1. The principal quantum number n, which determines the energy.
2. The orbital quantum number ℓ, which determines the magnitude of the orbital angular momentum.
3. The magnetic quantum number m_ℓ, which determines the z component of the orbital angular momentum.

E. Explain that traditionally each value of n is said to label a shell. Remark that a shell may consist of many states, but each is associated with the same value of the energy. Tell students that for a given shell, ℓ may take on the values 0, 1, 2, ..., $n - 1$. There are n different values in all. Explain that all the states with given values of n and ℓ are said to form a subshell. Say that for a given value of ℓ, m_ℓ may take on any integer value from $-\ell$ to $+\ell$ and there are $2\ell + 1$ values in all. As examples, list all the states for $n = 1$, 2, and 3. Group them according to n and remark that all states with the same n have the same energy, all states with the same ℓ have the same magnitude of orbital angular momentum, and all states with the same m_ℓ have the same z component of orbital angular momentum. Remark that states with different values of n, ℓ, or m_ℓ have different wave functions.

F. Give the ground state wave function and obtain the expression for the probability density. Define the Bohr radius ($a = h^2\epsilon_0/\pi m e^2 = 52.9\,\text{pm}$). Remark that ψ has spherical symmetry and explain that this is true of all $\ell = 0$ wave functions. Remind students that the volume of a spherical shell with thickness dr is $4\pi r^2\,dr$ and define the radial probability density as $P(r) = 4\pi r^2|\psi(r)|^2$. Sketch $P(r)$ for the hydrogen atom ground state (Fig. 40–18) and point out there is a range of radial distances at which the electron might be found. Locate the most probable radius and the average radius.

G. Show a dot plots (Figs. 40–20 and 40–22) for the $n = 2$ states and write the expressions for the wave functions (see problems 58 and 59). Remark that the individual probability densities are not spherically symmetric but their sum is.

SUGGESTIONS

1. Assignments
   a. Use questions 1 through 8 in your discussion of a particle in a one-dimensional infinite well. Assign problems 4, 7, and 9.
   b. Ask questions 9 through 12 in connection with a particle trapped in a one-dimensional finite well. Consider problems 19, 21, and 22.
   c. Assign problems 26 and 27 in connection with two-dimensional traps and problems 28 and 29 in connection with three dimensional traps.
   d. After discussing the hydrogen spectrum, ask questions 13, 15, and 16. Also assign problems 33, 35, 37, and 38. If you have discussed the terms *binding energy* and *excitation energy*, assign problem 46.
   e. When you discuss the enumeration of hydrogen atom states, assign problem 54.
   f. Discuss problems 49 and 52 in connection with the ground state of a hydrogen atom. You might also assign problems 51 and 53 if you did not show the given ground state wave function is a solution to the Schrödinger equation and is normalized. Give problem 57 to students who are math oriented.
   g. The $n = 2$ hydrogen wave functions are covered in problems 58 and 59.
2. Demonstrations
   Thompson and Bohr models of the atom.
3. Audio/Visual
   a. *Absorption Spectra*; from AAPT Miller collection of single-concept films; video tape; available from Ztek Co., PO Box 11768, Lexington, KY 40577–1768.
   b. *Spectra*; Demonstrations of Physics: Light; VHS video tape (6:02); Media Design Associates, Inc., Box 3189, Boulder, CO 80307–3190.
   c. *The Quantum Idea*; VHS video tape (10 min); Films for the Humanities & Sciences, PO Box 2053, Princeton, NJ 08543–2053.
   d. *Photons*; VHS video tape (10 min); Films for the Humanities & Sciences (see above for address).
   e. *The Wave-Mechanical Model*; VHS video tape (10 min); Films for the Humanities & Sciences (see above for address).
   f. *Matter Waves*; VHS video tape (10 min); Films for the Humanities & Sciences (see above for address).
   g. *Electron Distribution in the Hydrogen Atom*; A.F. Burr and Robert Fisher; American Association of Physics Teachers slide set; Publications Sales, AAPT Executive Office, 5112 Berwyn Road, College Park, MD 20740-4100. Probability distributions for n =1 to n = 6.
   h. *Atoms, Molecules, and Models*; Side F: Angular Momentum and Modern Physics of Cinema Classics; video disk; available from Ztek Co., PO Box 11768, Lexington, KY 40577–1768.
4. Computer Software
   a. *Atomic Physics*; Macintosh, Windows; Cross Educational Software, Inc., 508 E. Kentucky Avenue, PO Box 1536, Ruston, LA 71270. Macintosh. Contains sections on particles and waves and on the Schrödinger equation. equation, and the hydrogen atom.
   b. *Quantum Numbers*; Macintosh, Windows; Cross Educational Software, Inc. (see above for address).
   c. *Bellbox*; Physics Academic Software; North Carolina State University, Raleigh, NC 27695–8202. This simulation program allows students to experiment with the Einstein-Podolsky-Rosen paradox.

5. Laboratory

   Meiners Experiment 14–3: *Analysis of Spectra.* A spectroscope is used to obtain the wavelengths of hydrogen and helium lines. Hydrogen lines are compared with predictions of the Balmer equation.

Chapter 41 ALL ABOUT ATOMS

BASIC TOPICS

I. Orbital and spin angular momentum.

   A. Remark that orbital angular momentum is quantized and that the allowed values of its magnitude are given by $L = \sqrt{\ell(\ell+1)}\hbar$, where $\hbar = h/2\pi$. The orbital quantum number ℓ can take on the values 0, 1, 2, ..., $n-1$ for a given value of n. Emphasize that for a hydrogen atom, the $n = 2$, $\ell = 0$ and $n = 2$, $\ell = 1$ states, for example, have the same energy but different angular momenta. For other atoms, the energy depends on ℓ as well as on n.

   B. State that the z component of the angular momentum is given by $L_z = m_\ell \hbar$, where $m_\ell = 0, \pm 1, \pm 2, \ldots, \pm \ell$. m_ℓ is called the magnetic quantum number. The z axis can be in any direction, perhaps defined by an external magnetic field. Point out that the angle θ between the angular momentum vector and the z axis is given by $\cos\theta = m_\ell/\sqrt{\ell(\ell+1)}$. The smallest value of θ occurs when $m_\ell = \ell$ and it is not zero. Explain that the angles $\vec{L}$ makes with the x and y axes cannot be known if the angle between $\vec{L}$ and the z axis is known. Discuss this in terms of the precession of $\vec{L}$ about the z axis.

   C. Explain that the electron and some other particles have intrinsic angular momentum, as if they were spinning. The magnitude of the electron spin angular momentum is $S = \sqrt{s(s+1)}\hbar = \sqrt{3/4}\hbar$ and the z component is either $m_s = -\frac{1}{2}\hbar$ or $+\frac{1}{2}\hbar$ (there are two possible states). You might want to remark that spin is not predicted by the Schrödinger equation but that it is predicted by relativistic modifications to quantum physics.

   D. Say that the total angular momentum of an atom is the vector sum of the orbital angular momenta and the spin angular momenta of its electrons.

II. Magnetic dipole moments.

   A. Explain that the electron has a magnetic dipole moment because of its orbital motion and write $\vec{\mu}_{\text{orb}} = -(e/2m)\vec{L}$ and $\mu_{\text{orb},z} = -(e/2m)L_z = -(e\hbar/2m)m_\ell$. Give the value of the Bohr magneton ($\mu_B = e\hbar/2m = 9.28 \times 10^{-24}$ J/T). Remind students that because of its motion, the electron experiences a torque in an external magnetic field and produces its own magnetic field (provided $\vec{\mu}_{\text{orb}} \neq 0$).

   B. State that the spin magnetic moment is $\mu_{sz} = -2m_s\mu_B$. Stress the appearance of the factor 2. The electron produces a magnetic field and experiences a torque in a magnetic field because of this moment.

   C. Remark that the energy of an electron is changed by$-\mu_z B$ when an external field $\vec{B}$ is applied in the positive z direction. Thus, states with the same n but different m_ℓ have different energies in a magnetic field. This is called the Zeeman effect. Photons with an energy equal to the energy difference of the two spin states cause the spin to flip. The phenomenon can be detected by measuring the absorption of the beam.

   D. Briefly describe the Stern-Gerlach experiment. Explain that a magnetic dipole in a *non-uniform* magnetic field experiences a force and that $F_z = \mu_z \, dB/dz$ for a field in the z direction that varies along the z axis. Atoms with different values of m_ℓ experience different forces and arrive at different places on a screen. That discrete regions of the screen receive atoms is experimental evidence for the quantization of the z component of angular momentum.

E. To emphasize the practical, qualitatively explain NMR and its use in diagnostic medicine. You might also explain how local magnetic fields in solids, for example, can be measured using magnetic resonance techniques.

IV. Pauli exclusion principle.

A. State the principle. For any two electrons in the same trap at least one of their quantum numbers must be different. State that this is a principle that holds for electrons, protons, neutrons, and many other particles. Also state that it does not hold for all particles and give the photon as an example of a particle for which it does not hold.

B. As an example consider a small group of electrons in a square trap. State that the single-particle energy levels are given by $(h^2/8mL^2)(n_x^2 + n_y^2)$, where L is the length of an edge of the trap and n_x and n_y are quantum numbers that may take on any integer value greater than zero. For each of the (n_x, n_y) pairs of values $(1,1)$, $(1,2)$, $(2,1)$, $(1,3)$, $(3,1)$, $(2,3)$, and $3,1)$ calculate the energy in units of $h^2/8mL^2$, order the states according to energy, and point out the degeneracies, including spin. Note how many electrons can have each value of the energy. Assume there are five electrons, give the ground state configuration, and calculate the ground state energy of the system in units of $h^2/8mL^2$. Emphasize the role of the Pauli exclusion principle.

C. Repeat for the first excited state of the system.

IV. Atomic states.

A. Explain that quantum mechanical states for an electron in an atom are classified using four quantum numbers:

1. The principal quantum number n, which determines the energy.
2. The orbital quantum number ℓ, which determines the magnitude of the orbital angular momentum and, to a lesser extent, the energy.
3. The magnetic quantum number m_ℓ, which determines the z component of the orbital angular momentum.
4. The spin quantum number m_s, which determines the z component of the spin angular momentum.

B. Explain that traditionally each value of n is said to label a shell and the shells are named $K, L, M, N, \ldots$, in order of increasing n. Remark that a shell may consist of many states.

C. Remind students that for a given shell, ℓ may take on the values $0, 1, 2, \ldots, n-1$. There are n different values in all. Explain that all the states with given values of n and ℓ are said to form a subshell. Remind students that for a given value of ℓ, m_ℓ may take on any integer value from $-\ell$ to $+\ell$. Since m_s can have either of two values, a subshell consists of $2(2\ell+1)$ states. Either state or prove that the shell with principal quantum number n has $2n^2$ states.

D. Give the spectroscopic notation: s labels an $\ell = 0$ subshell, p labels an $\ell = 1$ subshell, d labels an $\ell = 2$ subshell, etc. Explain that the value of n is placed in front of the letter and the number of electrons in the subshell is given as a superscript: $3d^2$ indicates two electrons in the $n = 3$, $\ell = 2$ subshell.

E. As you may have done for the last chapter, list all the states for $n = 1$, 2, and 3. Group them according to n and remark that all states with the same ℓ have the same magnitude of orbital angular momentum and all states with the same m_ℓ have the same z component of orbital angular momentum. Remark that states with different values of n, ℓ, and m_ℓ have different wave functions.

V. Atom building and the periodic table.

A. Give the "rules" for atom building:

1. The four quantum numbers n, ℓ, m_ℓ, m_s can be used to label states. Remark that wave functions and energies are different for electrons with the same quantum numbers

in different atoms.

2. The electrons in an atom obey the Pauli exclusion principle: No more than one electron can have any given set of quantum numbers.

B. Explain that as more protons are added to the nucleus, the electron wave functions pull in toward regions of low potential energy. This and the dependence of the energy on ℓ means that states associated with one principal quantum number may not be filled before states associated with the next principal quantum number are started. For example, a $5s$ state is lower in energy than a $4d$ state, in different atoms. It also accounts for the fact that all atoms are nearly the same size.

C. Show a periodic table. Point out the inert gas atoms and explain they all have filled shells. Point out the alkali metal and alkaline earth atoms and state they have one and two electrons, respectively, outside closed shells. Remark that electrons in partially filled shells are chiefly responsible for chemical activity. Point out the atoms in which d and f states are being filled and finally those in which p states are being filled.

VI. X rays and the numbering of the elements.

A. Explain that x rays are produced by firing energetic electrons into a solid target. Show Fig. 41–14 and point out the continuous part of the spectrum and the peaks. Also point out that there is a sharply defined minimum wavelength to the x-ray spectrum. Explain that the continuous spectrum results because the electrons lose some or all of their kinetic energy in close (decelerating) encounters with nuclei. This energy appears as photons and $\Delta K = hf$. Explain that a photon of minimum wavelength is produced when an electron loses all its kinetic energy in a single encounter. Derive the expression for the minimum wavelength in terms of the original accelerating potential and point out it is independent of the target material.

B. Explain that the line spectrum in Fig. 41–14 appears because incident electrons interact with atomic electrons and knock some of the deep-lying electrons out of the atoms. Electrons in higher levels drop to fill the holes, emitting photons with energy equal to the difference in energy of the initial and final atomic levels. The K_α line is produced when electrons drop from the L ($n = 2$) shell to the K ($n = 1$) shell and the K_β line is produced when electrons drop from the M ($n = 3$) shell to the K shell. Explain Fig. 41–16.

C. Show Fig. 41–17 and state that when the square root of the frequency for any given line is plotted as a function of the atomic number of the target atom, the result is nearly a straight line. Argue that the innermost electrons have an energy level scheme close to that of hydrogen but with an effective nuclear charge of $(Z - 1)e$, where the 1 accounts for screening by electrons close to the nucleus. Z is the number of protons in the nucleus, the atomic number. Use the expression for hydrogen energy levels. For K_α, put $n = 2$ for the initial state and $n = 1$ for the final state, then show that $\sqrt{f}$ is proportional to $(Z - 1)$.

D. Remark that this relationship was used to position the chemical elements in the periodic table independently of their chemical properties. This technique was particularly important for elements in the long rows of the periodic table, which contain many elements with similar chemical properties. Today the technique is used to identify trace amounts of impurities in materials.

VII. The laser.

A. List the characteristics of laser light: monochromatic, coherent, directional, can be sharply focused. See the text for quantitative comparisons with light from other sources.

B. Explain the mechanism of light absorption: an incident photon is absorbed if hf corresponds to the energy difference of two electron states of the material and the upper state is initially empty. An electron jumps from the lower to the upper state. Explain spontaneous emission: an electron spontaneously (without the aid of external radiation) makes the transition from one state to a lower state (if that state is empty) and a photon with hf equal to the energy difference is emitted. Emphasize that in most cases the electron remains in the upper state for a time on the order of 10^{-9} s but that there are metastable states in which the electron remains for a longer time ($\approx 10^{-3}$ s). Explain stimulated emission: with the electron in an upper state, an incident photon with the proper energy can cause it to make the jump to a lower state. The result is two photons of the same energy, moving in the same direction, with waves having the same phase and polarization. Remark that laser light is produced by a large number of such events, each triggered by a photon from a previous event. Hence, all laser photons are identical. Explain that metastable states are important since the electron must remain in the upper state until its transition is induced. Compare with light produced by random spontaneous transitions.

C. Explain that, in thermodynamic equilibrium, upper levels are extremely sparsely populated compared to the ground state. To obtain laser light, the population of an upper level must be increased; otherwise absorption events would equal or exceed stimulated emission events. A laser must be pumped. Write down the expression for the thermal equilibrium number of atoms in the state with energy E: $n(E) = Ce^{-E/kT}$. Explain that C is independent of energy but depends on the number of atoms present. State that the temperature T is on the Kelvin scale.

D. Discuss the helium-neon laser, paying particular attention to the role played by helium atoms in maintaining population inversion in the neon atoms. Also explain the roles played by the walls and mirror ends. Go over the four characteristics of laser light discussed earlier and tell how each is achieved.

SUGGESTIONS

1. Assignments
   a. To test for understanding of the angular momentum quantum numbers, go over questions 1 through 5 and assign problems 7, 8, and 9. To stress the connection between angular momentum and magnetic dipole moment, assign problem 11.
   b. To discuss the Stern-Gerlach experiment in more detail, include question 10. Also assign problems 16 and 18.
   c. Use problems 23 and 25 (two-dimensional trap) or problems 27 and 28 (three-dimensional trap) to test for understanding of the Pauli exclusion principle. To emphasize the role played by spin in the building of the periodic table, ask problem 31. To help in the discussion of the periodic table, assign problems 30 and 33.
   d. The existence of a minimum wavelength in the continuous x-ray spectrum provides an argument for the particle nature of light. Either discuss this or see if the students can devise the argument. Assign problems 39 and 40. After discussing characteristic x-ray lines and Moseley plots, ask questions 11 and 12. Assign problems 43, 45, and 47.
   e. Ask questions 13 and 14 to see if students understand how lasers work. Also assign problems 59, 62, and 63. Populations of states are covered in problems 57, 58, and 66.
2. Demonstrations
   Zeeman effect: Freier and Anderson MPc1.
3. Books and Monographs
   Resource Letters, Book Five; American Association of Physics Teachers, One Physics Ellipse, College Park, MD 20740–3845. Contains a resource letter on atomic physics.
4. Audio/Visual
   a. *The Atom revealed*; VHS video tape (50 min); Films for the Humanities & Sciences, PO Box 2053, Princeton, NJ 08543–2053.

b. *Applications of Lasers*; VHS video tape (58 min); Films for the Humanities & Sciences (see above for address).
c. *Constructing a Laser*; VHS video tape (58 min); Films for the Humanities & Sciences (see above for address).
d. *Viewing a Hologram*; Demonstrations of Physics: Light; VHS video tape (3:36); Media Design Associates, Inc., Box 3189, Boulder, CO 80307–3190.

5. Computer Software
   Quantum Numbers; Macintosh, Windows; Cross Educational Software, Inc., 508 E. Kentucky Avenue, PO Box 1536, Ruston, LA 71270. Macintosh.

Chapter 42 THE CONDUCTION OF ELECTRICITY IN SOLIDS

BASIC TOPICS

I. Electron energy bands.
   A. Explain that a crystalline solid is a periodic arrangement of atoms and show some ball and stick models or Fig. 42–1.
   B. Explain that energy levels for electrons in crystalline solids are grouped into bands with the levels in any band being nearly continuous and with gaps of unallowed energies between. Remark that bands are produced when atoms are brought close together. Wave functions then overlap and extend throughout the solid. Show Fig. 42–3 and remark that low energy bands are narrow since the wave functions are highly localized around nuclei and overlap is small. High energy bands are wide because overlap is large. When the atoms are close together, outer-shell electrons are influenced by many atoms rather than just one.
   C. Remind students that since the Pauli exclusion principle holds, the lowest total energy is achieved when electrons fill the lowest states with one electron in each state. Thus, at $T = 0\,\mathrm{K}$, all states are filled up to a maximum energy.
   D. Remark that for a metal at $T = 0\,\mathrm{K}$, the highest occupied state is near the middle of a band, while for an insulator or semiconductor, it is at the top of a band.
   E. Write down the Fermi-Dirac occupancy probability $P(E)$, given by Eq. 42–6, and state that it gives the thermodynamic probability that a state with energy E is occupied. State that E_F is a parameter, called the Fermi energy, that is different for different materials. Show that for $T = 0\,\mathrm{K}$, $P(E) = 1$ for $E < E_F$ and $P(E) = 0$ for $E > E_F$. To give a numerical example, calculate the probabilities of occupation for states 0.1 and 1 eV above the Fermi energy, then 0.1 and 1 eV below, at room temperature. Graph $P(E)$ vs. E for $T = 0$ and for $T > 0$. See Fig. 42–6. Also show the graph for a still higher temperature and point out that the central region (from $P = 0.9$ to $P = 0.1$, say) widens. This quantitatively describes the thermal excitation of electrons to higher energy states. Remark that the Fermi-Dirac occupancy probability is valid for any large collection of electrons, including the collections in metals, insulators, and semiconductors.

II. Metallic conduction.
   A. Write down Eq. 27–20 for the resistivity and remark that n is the concentration of conduction electrons and τ is the mean time between collisions of electrons with atoms. Ask students to review Section 27–6. Remark that a low resistivity results if the electron concentration is large or the mean free time is long. In a rough way, if there are few collisions per unit time, then the mean free time is long and the electrons are accelerated by the electric field for a long time before colliding, so the drift velocity is large. Remark that quantum physics must be used to determine n and τ.

B. Explain that for metals, the energies of conduction electrons (those in partially filled bands) are primarily kinetic and to a first approximation, we may take the electrons to be trapped in a box the size of the sample. The so-called free electron model of a metal takes the potential energy to be zero in the box.

C. Define the density of states function $N(E)$ and the density of occupied states function $N_o(E)$. Explain that $N_o(E) = N(E)P(E)$ and that the total electron concentration in a metal is given by $n = \int N(E)P(E)\,dE$. In principle, this equation can be solved for the Fermi energy as a function of temperature. State that for nearly free electrons in a metal, $N(E)$ is given by Eq. 42–5, and that the Fermi energy is given by Eq. 42–9. Evaluate the expression for copper and show that E_F is about 7 eV above the lowest free electron energy. Strictly, this is the result for $T = 0$ but the variation of E_F and n with temperature is not important in a first approximation for metals.

D. Explain that the electric current is zero when no electric field is present because states for which the velocities are $+\vec{v}$ and $-\vec{v}$, for example, have the same energy. If one is filled, then so is the other. Thus, the average velocity of the electrons vanishes. A current arises in an electric field because the electrons accelerate: they tend to make transitions within their band to other states such that the changes in their velocities are opposite to the field.

E. Explain that the acceleration caused by an electric field does not continue indefinitely because the electrons are scattered by atoms of the solid. As a result, the electron distribution distorts only slightly. Some states with energy slightly greater than E_F and velocity opposite the field become occupied while some states with energy slightly less than E_F and velocity in the direction of the field become vacant. Electrons with energy E_F have speeds v_F given by $E_F = \frac{1}{2}mv_F^2$ but the average speed (the drift speed) is considerably less because most electrons can be paired with others moving with the same speed in the opposite direction.

F. Explain that a steady state is reached and that the drift velocity is then proportional to the applied electric field. Only electrons near the Fermi energy suffer collisions and the additional velocities they obtain from the field between collisions are insignificant compared to their velocities in the absence of the field. Thus, the mean free time is essentially independent of the field and Ohm's law is valid.

G. State that electrons in a perfectly periodic lattice do not suffer collisions, a result that is predicted by quantum physics. Collisions with the atoms occur because they are vibrating. Collisions also occur if the solid contains impurities or other imperfections. As the temperature increases, vibrational amplitudes of the atoms increase and so does the number of collisions per unit time. As a result, the mean free time becomes smaller. This explains the increase with temperature in the resistivity of a metal.

III. Insulators and semiconductors.

A. Explain that a filled band cannot contribute to an electric current because the average electron velocity is always zero, even in an electric field. State that insulators and semiconductors have just the right number of electrons to completely fill an integer number of bands and that, in the lowest energy state, all bands are either completely filled or completely empty. For metals, on the other hand, the highest occupied state is near the middle of a band. Metals always have partially filled bands. Show Fig. 42–4 and identify the valence and conduction bands for an insulator.

B. Explain that as the temperature is raised from $T = 0\,\mathrm{K}$, a small fraction of the electrons in the valence band of an insulator or semiconductor are thermally excited across the gap into the conduction band. For a semiconductor, the gap is small (about 1 eV) and at room temperature, both bands can contribute to the current. The conductivity, however, is still small compared to that of a metal. For an insulator, the gap is large (more than 5 eV), so the number of promoted electrons is extremely small and the current is insignificant for laboratory fields. Explain that silicon and germanium are the only elemental semiconductors although there are many semiconducting compounds. Carbon is a prototype insulator, with a gap of 5.5 eV. Compare with silicon, which has a gap of 1.1 eV. Resistivities of metals and semiconductors are compared in Table 42–1.

C. When electrons are promoted across the gap, they contribute to the current in an electric field. The valence band becomes partially filled and electrons there also contribute. It is usually convenient to think about the few empty states in this band rather than the large number of electrons there. That is, the electrons in a nearly filled band are replaced by a collection of fictitious particles, called holes, so that the properties of the hole system are identical to the properties of the electron system they replace. Holes behave as if they were positive charges. In contrast to electrons, holes drift in the direction of the electric field. Compare the carrier concentrations of metals and semiconductors at room temperature. See Table 42–1.

D. Explain the different signs for the temperature coefficients of resistivity, also given in Table 42–1. Explain that for both metals and semiconductors near room temperature, the mean free time decreases with increasing temperature. For metals, the electron concentration is essentially constant but for semiconductors, n increases dramatically with temperature as electrons are thermally promoted across the gap. This effect dominates and the resistivity of an intrinsic semiconductor decreases with increasing temperature.

E. Explain that the proper kind of replacement atoms (donors) can increase the number of electrons in the conduction band and another kind (acceptors) can increase the number of holes in the valence band. They produce n and p type semiconductors, respectively. By considering the number of electrons in their outer shells, explain why phosphorus is a donor and aluminum is an acceptor. Point out that wave functions for impurity states are highly localized around the impurity and so do not contribute to the conductivity. Go over Sample Problem 42–6, which shows that only a relatively small dopant concentration can increase the carrier concentration enormously. Doped semiconductors are used in nearly all semiconducting devices.

IV. Semiconducting devices.

A. Show a commercial junction diode and draw a graph of current vs. potential difference (Fig. 42–12). Include both forward and back bias. Explain that it is a rectifier, with high resistance for current in one direction and low resistance for current in the other direction. Demonstrate the i-V characteristics by placing a diode across a variable power supply and measuring the current for various values of the potential. Reverse the potential to show the rectification.

B. Describe a p-n junction and remark that the diffusion of carriers leaves a small depletion region, nearly devoid of carriers, straddling the metallurgical junction. Explain the origin of the electric field in the depletion region and the origin of the contact potential. Stress that the field is due to uncovered replacement atoms, positive donors on the n side and negative acceptors on the p side.

C. Describe a diffusion current as one that arises because particles diffuse from regions of high concentration toward regions of low concentration. Explain that this motion results from the random motion of the particles. More particles leave a high concentration region simply because there are more particles there, not because they are driven by any applied force. State that the diffusion current for both electrons and holes in an unbiased p-n junction is from the p to the n side, against the contact electric field. Point out that the drift current is from the n toward the p side and that the diffusion and drift currents cancel when no external field is applied. Point out the depletion zone and the currents on Fig. 42–11.

D. Draw a circuit with a battery across a p-n junction, the positive terminal attached to the n side. Explain that this is a back bias. The internal electric field is now larger, the barrier to diffusion is higher, and the reverse current is extremely small. Also explain that the width of the depletion zone is increased by application of a reverse bias.

E. Draw the circuit for forward bias. The internal electric field is now smaller, the barrier to diffusion is lower, and the current increases dramatically. The depletion zone narrows.

F. Explain how diodes are used for rectification and how light-emitting diodes work.

G. Optional. Explain how a field-effect transistor works. Explain the mechanism by which the gate voltage of a MOSFET controls current through the channel. Remove the covers from a few chips and pass them around with magnifying glasses for student inspection.

SUGGESTIONS

1. Assignments
   a. Questions 1 and 2 deal with crystal structure. Ask them if you include more than a passing mention of this topic. Also consider problem 37.
   b. Questions 4 and 5 deal with electrons in solids and questions 3 and 8 deal with energy bands.
   c. Assign problems 4, 8, and 11 in connection with the Fermi energy of a metal. Assign problem 10 in connection with the density of states for a metal. Assign some of problems 13, 14, and 21 in connection with the occupancy probability. Also assign problem 23 and one of problems 25 or 26. The justification for the free electron model of a metal is covered in problem 17.
   d. Problem 39 should be assigned or covered in class when you discuss intrinsic semiconductors.
   e. Doped semiconductors are considered in questions 10, 11, and 12. Discuss them and then assign problems 40 and 41. Also consider problem 43.
   f. p-n junctions are the subject of questions 13, 14, and 16. Be sure to assign or discuss problem 44. If you include LED's, assign either problem 46 or 47 and if you discuss field-effect transistors, assign problem 49.
2. Audio/Visual
   Condensed Matter; Side F: Chapter 70 of Cinema Classics; video disk. Ztek Co., PO Box 1055, Louisville, KY 40201.
3. Computer Project
   Ask students to use a root finding program to carry out calculations of the electron concentration in the conduction band and hole concentration in the valence band of both intrinsic and doped semiconductors. Then, ask them to calculate the contact potential for a p-n junction with given dopant concentrations.

Chapter 43 NUCLEAR PHYSICS

BASIC TOPICS

I. Nuclear properties.

A. Explain that the nucleus of an atom consists of a collection of tightly bound neutrons, which are neutral, and protons, which are positively charged. A proton has the same magnitude charge as an electron. Define the term nucleon and state that the number of nucleons is called the mass number and is denoted by A, the number of protons is called the atomic number and is denoted by Z, and the number of neutrons is denoted by N. Point out that $A = Z + N$. Remark that nuclei with the same Z but different N are called *isotopes.*

The atoms have the same chemical properties and the same chemical symbol. Show a wall chart of the nuclides. Refer to Table 43–1 when discussing properties of nuclides.

B. Explain that one nucleon attracts another by means of the strong nuclear force and that this force is different from the electromagnetic force. It does not depend on electrical charge and is apparently the same for all pairs of nucleons. It is basically attractive; at short distances (a few fm), it is much stronger than the electrostatic force between protons, but it becomes very weak at larger distances. Two protons exert attractive strong forces on each other only at small separations but they exert repulsive electric forces at all separations. Because of the short range, a nucleon interacts only with its nearest neighbors via the strong force. Because the nucleus is small, the much stronger nuclear force dominates and both protons and neutrons can be bound in stable nuclei. Explain that the force is thought to be a manifestation of the strong force that binds quarks together to form nucleons.

C. Show Fig. 43–4 and point out the $Z = N$ line and the stability zone. Explain why heavy nuclei have more neutrons than protons. Also explain that unstable nuclei are said to be radioactive and convert to more stable ones with the emission of one or more particles. Show Fig. 43–12 and point out the stable and unstable nuclei.

D. Explain that although the surface of a nucleus is not sharply defined, nuclei can be characterized by their mean radii and these are given by $r = r_0 A^{1/3}$, where $r_0 \approx 1.2\,\text{fm}$ ($1\,\text{fm} = 10^{-15}\,\text{m}$). Stress how small this is compared to atomic radii. Show that this relationship between r and A leads to the conclusion that the mass densities of all nuclei are nearly the same. Show that the density of nuclear matter is about $2 \times 10^{17}\,\text{kg/m}^3$.

E. Explain that the mass of a nucleus is less than the sum of the masses of its constituent nucleons, well separated. The difference in mass is accounted for by the binding energy through $\Delta E_{\text{be}} = \Delta m\, c^2$, where Δm is the magnitude of the mass difference. The binding energy is the energy that must be supplied to separate the nucleus into well separated particles, at rest. Generalize this equation to the case of a nucleus with Z protons and N neutrons: $\Delta E_{\text{be}} = Zm_p c^2 + Nm_n c^2 - mc^2$. Also define the binding energy per nucleon ΔE_{ben}. Show Fig. 43–6 and point out that there is a region of greatest stability, near iron. For heavier nuclei, the binding energy per nucleon falls slowly but nevertheless does fall. For lighter nuclei, the binding energy per nucleon rises rapidly with increasing mass number. Explain the terms fission and fusion, then remark that the high mass number region is important for fission processes, the low mass number region is important for fusion processes.

F. State that nuclear masses are difficult to measure with precision, so masses are usually expressed in atomic mass units: $1\,\text{u} = 1.6605 \times 10^{-27}\,\text{kg}$. Also state that tables usually give atomic rather than nuclear masses and so include the mass of the atomic electrons. Show that the electron masses cancel in the expression for the binding energy. Give the mass-energy conversion factor: 931.5 MeV/u.

G. Explain that nuclei have discrete energy levels, with separations on the order of MeV. An excited nucleus can make a transition to a lower energy state with the emission of a photon, typically in the gamma ray region of the spectrum. Explain that a nucleus may have intrinsic angular momentum and a magnetic moment. Spins are on the order of $\hbar$, like atomic electrons, but moments are much less than electron moments because the mass of a nucleon is much greater than the mass of an electron.

II. Radioactive decay.

A. Explain that nuclei may be either stable or unstable and those that are unstable ultimately decay to stable nuclei. Decay occurs by spontaneous emission of an electron (e^-), a positron (e^+), a helium nucleus (α), or larger fragments. The resulting nucleus has a different complement of neutrons and protons than the original nucleus.

B. Explain that decay is energetically favorable if the total mass of the products is less than the original mass. Define a decay symbolically as $X \rightarrow Y + b$, where X is the original nucleus, Y is the daughter nucleus, and b is everything else. Point out that charge, number of nucleons, and energy are all conserved. Define the disintegration energy by $Q = (m_X - m_Y - m_b)c^2$. Note that an appropriate number of electron rest energies must be added or subtracted so that atomic masses may be used. Note also that Q must be positive for spontaneous decays and Q appears as the kinetic energy of the decay products or as an excitation energy if the daughter nucleus is left in an excited state.

C. Explain that each radioactive nucleus in a sample has the same chance of decaying and that the decay rate or activity ($R = -\mathrm{d}N/\mathrm{d}t$) is proportional to the number of undecayed nuclei present at time t: $-\mathrm{d}N/\mathrm{d}t = \lambda N$. This has the solution $N = N_0 \exp(-\lambda t)$, so the decay rate is given by $R = R_0 \exp(-\lambda t)$. Define the term half-life and show that it is related to λ by $T_{1/2} = (\ln 2)/\lambda$. Go over Sample Problems 43–4 and 43–5, show Fig. 43–8, and point out the half-life. Emphasize that R decreases by a factor of two in every half-life interval. Define the becquerel unit.

D. Discuss α decay. Write down Eq. 43–21 and explain that the daughter nucleus has two fewer neutrons and two fewer protons than the parent. Go over Sample Problem 43–6 to show that α decay is energetically favorable for ^{238}U. Show Fig. 43–9 and explain that the deep potential well is due to the strong attraction of the residual nucleus for the nucleons in the α particle, while the positive potential is due to Coulomb repulsion. The two forces form a barrier to decay. Explain that the α particle can tunnel through the barrier. Its wave function does not go to zero at the inside edge, but rather has a finite amplitude in the barrier and on the outside. There is a non-zero probability of finding the α particle on the outside. High, wide barriers produce a small probability of tunneling and a long half-life while low, narrow barriers produce the opposite effect. Note the wide range of half-lives that occur in nature (Table 43–2).

E. Discuss β decay. Explain that a neutron can transform into a proton with the emission of an electron and a neutrino (strictly, an antineutrino) and that a proton can transform into a neutron with the emission of a positron and a neutrino. Mention the properties of a neutrino: massless, neutral, weakly interacting. Only protons bound in nuclei can undergo β decay but both free and bound neutrons can decay. These transformations lead to decays such as the ones given in Eqs. 43–23 and 43–24. Explain that the energy is shared by the decay products and that the electrons or positrons show a continuous spectrum of energy up to some maximum amount (see Fig. 43–10). Explain that neutron rich nuclides generally undergo β^- decay while proton rich nuclides generally undergo β^+ decay. This is a mechanism for bringing the nucleus closer to stability. Carefully discuss the inclusion of electron rest energies in the equation for Q so that atomic masses can be used. In particular, show that in β^- decay there is no excess electron mass but in β^+ decay there is an excess of two electron masses.

F. Define the units used to describe radiation dosage: grey and sievert.

SUPPLEMENTARY TOPICS

1. Radioactive dating. If time permits, cover this topic as an application of radioactive decay processes.
2. Nuclear models. This topic adds a little breadth to the nuclear physics section and helps students understand nuclear processes a little better.

SUGGESTIONS

1. Assignments
   a. Nuclear constitution is covered in problems 7, 8, and 9. Nuclear radius and density are covered in problems 4, 6, and 12.
   b. Include questions 2, 4, 5, 7, and 8 in the discussion of nuclear stability and nuclear binding. These ideas are illustrated in problems 10, 11, 13, 19, and 20. Be sure to include problem 10 if you intend to discuss fission (Chapter 44).
   c. Questions 9 through 14 deal with the decay law, activity, and half-life. Discuss a few. Problems 27 and 28 cover basic half-life calculations. Problems 32, 33, 36, and 45 involve half-life calculations drawn from many interesting applications. Assign some of them.
   d. Following the discussion of α decay, students should be able to answer question 16. The disintegration energy and barrier height are covered in problems 46 and 51. Problem 49 asks students to take into account the recoil of the residual nucleus. Problem 48 shows why alphas are emitted rather than well-separated nucleons.
   e. After discussing β decay, assign one or more of problems 53, 55, and 57. Problem 54 shows that β particles do not exist inside nuclei before decay occurs. The β decay discussion can be broadened somewhat by including the recoil of the nucleus. See problem 61.
2. Demonstrations
   Geiger counter: Freier and Anderson MPa2.
3. Books and Monographs
   Resource Letters, Book Four; American Association of Physics Teachers, One Physics Ellipse, College Park, MD 20740–3845. Contains a resource letter on nuclear physics.
4. Audio/Visual
   a. *Radioactive Decay*; *Scintillation Spectrometry*; from AAPT Miller collection of single-concept films; video tape; available from Ztek Co., PO Box 11768, Lexington, KY 40577–1768.
   b. *Nuclear Physics*; Side F: Angular Momentum and Modern Physics of Cinema Classics; video disk; available from Ztek Co. (see above for address).
   c. *Rutherford Scattering*, *Thomson Model of the Atom*; from AAPT collection 1 of single-concept films; video tape; available from Ztek Co. (see above for address).
   d. *The Determination of a Radioactive Half-Life*; VHS video tape (25 min); Films for the Humanities and Sciences, Inc., Box 2053, Princeton, NJ 08543.
   e. *The Rutherford Scattering of Alpha Particles*; VHS video tape (15 min); Films for the Humanities & Sciences (see above for address).
   f. *Radioactivity*; VHS video tape (23 min); Films for the Humanities & Sciences (see above for address).
   g. *The determination of a Radioactive Half-Life*; VHS video tape (15 min); Films for the Humanities & Sciences (see above for address).
   h. *The Rutherford Model*; VHS video tape (10 min); Films for the Humanities & Sciences (see above for address).
   i. *Natural Transmutations*; VHS video tape (10 min); Films for the Humanities & Sciences (see above for address).
5. Computer Software
   a. *Conservation Laws*; Macintosh, Windows; Cross Educational Software, Inc., 508 E. Kentucky Avenue, PO Box 1536, Ruston, LA 71270. Macintosh, Windows. Includes a section on nuclear conservation laws.
   b. *Atomic Physics*; Macintosh; Macintosh, Windows; Cross Educational Software, Inc. (see above for address). Contains sections on Rutherford scattering, nuclear structure, and

radioactive decay.

c. *Chart of the Nuclides: A Tutorial*; Philip DiLavore; Physics Academic Software, North Carolina State University, PO Box 8202, Raleigh, NC 27690–0739. The chart is shown and a click on any nuclide produces information about that nuclide.

6. Laboratory

Many of the following experiments make use of a Geiger tube and scalar.

a. Bernard Experiment 48: *The Characteristics of a Geiger Tube* describes how students can systematically investigate the plateau and resolving time of a Geiger tube. They also learn how to operate a scalar. Consider prefacing the other experiments either with this experiment or with a demonstration of the same material.

b. Meiners Experiment 14–7: *Half-Life of Radioactive Sources.* A Geiger counter and scalar are used to measure the decay rate as a function of time for indium, cesium 137, and barium 137. For the first and last, the data is used to compute the half-life. Other sections explain how to use a microcomputer to collect data and make the calculation and how to use a emanation electroscope to collect data. A neutron howitzer or minigenerator is required to produce radioactive sources.

c. Bernard Experiment 52: *Measurement of Radioactive Half-Life.* Nearly the same as Meiners 14–7. The generation of sources with short half-lives is discussed.

d. Meiners Experiment 14–6: *Absorption of Gamma and Beta Rays.* The particles are incident on sheets of aluminum and the number that pass through per unit time is counted. Students make a logarithmic plot of the counting rate as a function of the thickness of the aluminum and determine the range of the particles.

Chapter 44 ENERGY FROM THE NUCLEUS

BASIC TOPICS

I. The fission process.

A. Refer back to the binding energy per nucleon vs. A curve (Fig. 43–6). It suggests that a massive nucleus might split into two or more fragments nearer to iron, thereby increasing the total binding energy. Each fragment is more stable than the original nucleus. This is the fission process.

B. Remark that many massive nuclei can be rendered fissionable by the absorption of a thermal neutron. Such nuclei are called fissile. Give the example $^{235}\text{U} + \text{n} \rightarrow {}^{236}\text{U}^* \rightarrow \text{X} + \text{Y} + b\text{n}$. Explain that a thermal neutron ($\approx 0.04\,\text{eV}$) is absorbed by a ^{235}U nucleus and together they form the intermediate fissionable $^{236}\text{U}^*$ nucleus. This nucleus splits into two fragments (X and Y) and several neutrons. The sequence of events is illustrated in Fig. 44–2. Point out ^{236}U on Fig. 43–6. The disintegration energy for one possible fission event is calculated in Sample Problem 44–1.

C. Explain that different fission events, starting with the same nucleus, might produce different fragments. The fraction of events that produce a fragment of a given mass number A is graphed in Fig. 44–1. Point out that fragments of equal mass occur only rarely. Explain that the parent nucleus is neutron rich, the initial fragments are neutron rich, and that the initial fragments expel neutrons to produce the fragments X and Y. These generally decay further by β emission and some may emit delayed neutrons following β decay.

D. Show Fig. 44–3 and explain that the parent nucleus starts in the energy well near $r = 0$. The incoming neutron must supply energy to start the fission process. The required energy is slightly less than E_b since tunneling can occur. Point out the energy Q released by the process. Point out Table 44–2 and explain that E_n is the actual energy supplied by an incoming thermal neutron. Point out nuclides in the table for which fission does not occur.

E. Write out several fission modes for ^{235}U and note that on average more than one neutron is emitted. Explain that some neutrons come promptly while others come from later decays (the delayed neutrons). Point out that the average mode yields $Q \approx 200\,\text{MeV}$, of which 190 MeV or so appears as the kinetic energy of the fission fragments and 10 MeV goes to the neutrons.

II. Fission reactors.

A. Note that to have a practical reactor, the fission process must be self sustaining, once started. Also, there must be a way to control the rate of the process and to stop it, if desired.

B. To be self-sustaining, a chain reaction must occur: neutrons from one fission event trigger another. The neutrons emitted from a typical fission event share about 5 to 10 MeV energy and they must be slowed to thermal speeds to be useful. Some sort of moderator, often water, is used.

C. Explain that on average about 2.5 neutrons are produced per fission event. Describe in detail what happens to them. Some leak out of the system, some of the slowed neutrons are captured by ^{238}U, some are captured by fission fragments, and the rest start fission in ^{235}U. Fig. 44–4 gives some typical numbers.

D. Explain the terms critical, subcritical, and supercritical. Note that the control rods, which absorb slow neutrons, are used to achieve criticality. Point out that without the delayed neutrons, control would not be possible since time is needed to move the rods into or out of the reactor.

E. Define the multiplication factor k as the ratio of the number of neutrons present at one time that participate in fission to the number present in the previous generation. Remark that $k = 1$ for critical operation, $k < 1$ for subcritical operation, and $k > 1$ for supercritical operation. Explain that k is determined by the positions of the control rods. The rods are pulled out to increase k and thereby increase power output. They are pushed in to decrease k and thereby decrease power output. When the desired power level is obtained, the rods are positioned so $k = 1$.

F. Use Fig. 44–5 to describe the essential features of a nuclear power plant. Apart from the fact that the fission process is used to heat water or generate steam, this schematic could apply to any power plant. Remark on the special problems attendant on nuclear plants.

III. Fusion.

A. Return to Fig. 43–6 and remark that if two low-mass nuclei are combined to form a nucleus with greater mass, the binding energy is increased considerably. The energy is transformed to the kinetic energy of the resulting nucleus and any particles emitted. In order to carry out the fusion process, the nuclei must be given sufficient energy to overcome the electrostatic repulsion of their protons. They can then approach each other closely enough for the attraction of the strong force to bind them. For ^{3}He, the height of the barrier is about 1 MeV. Since tunneling is possible, fusion can occur at slightly smaller energies.

B. To achieve a large number of fusion events, hydrogen or helium gases must be raised to high temperatures. Even at the temperature of the Sun, only a small fraction of the nuclei have sufficient energy to overcome the Coulomb barrier. Go over Fig. 44–10.

C. Discuss fusion in the Sun. Remark that the core of the Sun is 35% hydrogen and 65% helium by mass. Outline the principal proton-proton cycle: two protons fuse to form a deuteron, a positron, and a neutrino. A deuteron fuses with a proton to form ^{3}He and two ^{3}He nuclei fuse to form ^{4}He and two protons. Remark that six protons are consumed and two are produced for a net loss of four. The two positrons are annihilated with electrons to produce photons. Note that the process can be simplified to $4\text{p} + 2\text{e}^- \to \alpha + 2\nu + 6\gamma$ and the Q value is computed from the mass difference between the alpha particle and the four protons.

D. Calculate the energy released. Show that $Q = 26.7\,\text{MeV}$ and note that the neutrinos take about $0.5\,\text{MeV}$ with them when they leave the Sun. Point out that the fusion process produces about 20 million times as much energy per kg of fuel as the burning of coal.

E. If time permits, discuss helium burning. Use the solar constant to calculate the rate at which the Sun converts mass to energy. Speculate on the future of the Sun. Also mention the carbon cycle, which is essentially the same as the proton-proton cycle. Carbon acts as a catalyst.

F. Discuss controlled thermonuclear fusion. Explain that deuteron-deuteron and deuteron-triton fusion events are being studied. Point out that high particle concentrations at high temperatures must be maintained for sufficiently long times in order to make the process work. Discuss some means for doing this: the tokamak for plasma confinement by magnetic fields, inertial confinement, and laser fusion. State that the right combination has not yet been achieved but work continues.

SUGGESTIONS

1. Assignments
   a. After explaining the basic fission process, test for understanding with questions 2, 3, 4, and 5. Also assign problems 2, 12, and 13.
   b. Following the discussion of the fission reactor, ask questions 7 and 8. To help students understand the role of a moderator, assign problem 27. To illustrate the role of the control rods, assign problems 22 and 25.
   c. Following the discussion of the basic fusion process, assign problems 33 and 34. Also ask question 11.
   d. To help students understand the fusion process as an energy source, assign problems 42 and 43. The carbon cycle is covered in problem 46.

6. Demonstrations
   Chain reaction: Freier and Anderson MPa1.

2. Books and Monographs
   a. *Fission Reactors*; edited by Melvin M. Levine.; available from AAPT, One Physics Ellipse, College Park MD 20740–3845. Covers both physics and engineering aspects.
   b. *Introduction to Nuclear Fusion Power and the Design of Fusion Reactors*; edited by J.A. Fillo and P. Lindenfeld; available from AAPT (see above for address). Covers both physics and engineering aspects.

3. Audio/Visual
   a. *Nuclear Energy*; VHS video tape (23 min); Films for the Humanities & Sciences, PO Box 2053, Princeton, NJ 08543–2053.
   b. *Energy Alternatives: Fusion*; VHS video tape (26 min); Films for the Humanities & Sciences (see above for address).
   c. *Energy from the Nucleus*; VHS video tape (10 min); Films for the Humanities & Sciences (see above for address).
   d. *Electrical Energy from Fission*; VHS video tape (10 min); Films for the Humanities & Sciences (see above for address).

4. Computer Software
   Atomic Physics; Macintosh; Macintosh, Windows; Cross Educational Software, Inc., 508 E. Kentucky Avenue, PO Box 1536, Ruston, LA 71270. Contains a section on nuclear reactors.

Chapter 45 QUARKS, LEPTONS, AND THE BIG BANG

BASIC TOPICS

I. The particle "zoo".

A. Show a list of particles already familiar to students. Include the electron, proton, neutron, and neutrino, then add the muon and pion. Explain that many other particles have been discovered in cosmic ray and accelerator experiments. To impress students with the vast array of particles and the enormous collection of data, make available to them a Review of Particle Properties paper, published roughly every two years in Reviews of Modern Physics.

B. Explain that many new particles are discovered by bombarding protons or neutrons with electrons or protons and show a picture of a detector, such as Fig. 45–1 or a bubble chamber picture, such as Fig. 45–3. State that the picture shows tracks of charged particles in a strong magnetic field, hence the curvature. Remind students that the radius of curvature can be used to find the momentum of a particle if the charge is known. Indicate the collision point and emphasize that the new particles were not present before the collision: the original particles disappear and new particles appear. In most cases, the total rest energy after the collision is much greater than the total rest energy before. Kinetic energy was converted to rest energy.

C. Mention that a few particles seem to be stable (electron, proton, neutrino) but most decay spontaneously to other particles. Point out decays on a bubble chamber picture. Explain the statistical nature of decays and remind students of the meaning of half-life. Examples: $n \rightarrow p + e^- + \nu$, $\pi^+ \rightarrow \mu^+ + \nu$.

D. Explain that for each particle there is an antiparticle with the same mass. A charged particle and its antiparticle have charge of the same magnitude but opposite sign. Their magnetic moments are also opposite. A particle and its antiparticle can annihilate each other, the energy (including rest energy) being carried by photons or other particles produced in the annihilation. Example: $e^+ + e^- \rightarrow \gamma + \gamma$. Antiparticles (except the positron) are denoted by a bar over the particle symbol. Some uncharged particles (such as the photon and π^0) are their own antiparticles. The universe seems to be made of particles, not antiparticles.

II. Particle properties.

A. Spin angular momentum. Remind students that many particles have intrinsic angular momentum. Explain that the magnitude is always an integer or half integer times $\hbar$. Remark that particles with half integer spins are called fermions while particles with integer spins are called bosons. Remind students of the Pauli exclusion principle and its significance, then state that fermions obey the principle while bosons do not. Give examples: electrons, protons, neutrons, and neutrinos are fermions; photons, pions, and muons are bosons. Remark that spin angular momentum is conserved in particle decays and interactions. An odd number of fermions, for example, cannot interact to yield bosons only.

B. Charge. Remind students of charge quantization and charge conservation. Even if the character and number of particles change in an interaction, the total charge before is the same as the total charge after. Example: $n \rightarrow p + e^- + \nu$.

C. Momentum and energy. Explain that energy and momentum are conserved in decays and interactions. Give masses and rest energies for the particles in the list of part I. Give the expressions for relativistic energy and momentum in terms of particle velocity.

D. Forces. Remark that all particles interact via the force of gravity and all charged particles interact via the electromagnetic force. The force of gravity is too weak to have observable influence at energies presently of interest. Remark that there are two additional forces, called strong and weak, respectively. Remind students of the role played by the strong force in holding a nucleus together and the role played by the weak force in beta decay. These topics were covered in Chapter 43. Note that lifetimes for strong decays are about 10^{-23} s, lifetimes for electromagnetic decays are about 10^{-14} to 10^{-20} s, and lifetimes for weak decays are about 10^{-8} to 10^{-13} s.

E. Leptons and hadrons. State that particles that interact via the strong force (as well as the weak) are called hadrons and that particles that interact via the weak force but not the strong are called leptons. List the leptons (electron, muon, tauon, and their neutrinos) and explain that a different neutrino is associated with each of the leptons. Remark that the neutrino that appears following muon decay is not the same as the neutrino that appears following beta decay. Neutrinos are labelled with subscripts giving the associated lepton: ν_e, ν_μ, and ν_τ.

F. Lepton numbers. State that a lepton number is associated with each lepton family, with particles in the family having a lepton number of +1, antiparticles in the family having a lepton number of −1, and all other particles having a lepton number of 0. Explicitly give the electron lepton numbers and the muon lepton numbers for members of the electron and muon families. Explain that each lepton number is conserved in all decays and interactions. Give some beta and muon decay examples.

G. Baryons and mesons. Remark that some strongly interacting particles (proton, neutron) are fermions and are called baryons while others (pion, kaon) are bosons and are called mesons. Explain that a baryon number of +1 is assigned to each baryon particle, a baryon number of −1 is assigned to each baryon antiparticle, and a baryon number of 0 is assigned to each meson. Then, baryon number is conserved in exactly the same way charge is conserved: the total baryon number before a collision or decay is the same as the total baryon number after. This conservation law (and conservation of energy) accounts for the stability of the proton, the baryon with the smallest mass. There is some speculation that baryon number is not strictly conserved and that protons may decay to other particles, but the half-life is much longer than the age of the universe. Some physicists are trying to observe proton decay.

H. Strangeness. Explain that another quantity, called strangeness, is conserved in strong interactions. Neutrons and protons have $S = 0$, K^- and Σ^+ have $S = -1$. A particle and its antiparticle have strangeness of opposite sign. Conservation of strangeness allows $\pi^+ + \mathrm{p} \rightarrow \mathrm{K}^+ + \Sigma^+$ but prohibits $\pi^+ + \mathrm{p} \rightarrow \pi^+ + \Sigma^+$, for example.

III. Quarks and the eight-fold way.

A. Show the eight-fold way patterns (Fig. 45–4) and point out the oblique axes. Remark that these patterns are to fundamental particles as the periodic table of chemistry is to atoms and that they have provided clues to the existence of particles not previously observed.

B. Remark that the properties of strongly interacting particles can be explained if we assume they are made up of more fundamental particles (called quarks). State that there are six quarks, not including the antiquarks, and list them and their properties (Table 45–5). Particularly note the fractional charge and baryon number. Baryons are constructed of three quarks, antibaryons of three antiquarks, and mesons of a quark and antiquark. Show that uud has the charge, spin, and baryon number of a proton and udd has the charge, spin, and baryon number of a neutron. Give the quark content of the spin 1/2 baryons (Fig. 45–5*a*) and the quark content of the spin 0 mesons (Fig. 45–5*b*). Point out that the strange quark accounts for the strangeness quantum number. Mention the charm, bottom, and top quarks and point out they lead to other particles.

C. Explain that the existence of internal structure allows for excited states: there are other particles with exactly the same quark content as those in the figures but they are different particles because the quarks have different motions. The additional energy results in greater mass. Contrast this with the leptons, which have no internal structure. Quarks and leptons are believed to be truly fundamental.

D. Messenger particles. Explain that particles interact by exchanging other particles. Electromagnetic interactions proceed by exchange of photons, for example. Also explain that energy may not be conserved over short periods of time but this is consistent with the uncertainty principle. State that the strong interaction proceeds by the exchange of gluons by quarks and the weak interaction proceeds by the exchange of Z and W particles by quarks and leptons. The interaction that binds nucleons in a nucleus is the same as the interaction that binds quarks in a baryon or meson. In the former case, gluons are exchanged between quarks of different nucleons; in the latter, they are exchanged between quarks of the same baryon or meson.

E. Explain that quarks are conserved in strong interactions. Either the original quarks are rearranged to form new particles or quark-antiquark pairs are created, then both the original and the new quarks are rearranged. This accounts for conservation of strangeness. Example: $K^+ \rightarrow K^0 + \pi^+$ ($u\bar{s} \rightarrow d\bar{s} + u\bar{d}$). A $d\bar{d}$ pair is formed. The d quark couples to the $\bar{s}$ quark to form a K^0 and the $\bar{d}$ quark couples to the u quark to form a π^+. Contrast this with the weak interaction, which can change one type quark into another. Illustrate with beta decay, in which a d quark is converted to a u quark.

F. Explain that quarks have another property, called color. Color produces the gluon field, much as charge produces the electromagnetic field: baryons interact via the strong interaction because quarks have color. Be sure students understand that "color" in this context has nothing to do with the frequency of light. Mention that gluons carry color. The emission or absorption of a gluon changes the color of a quark. Contrast this with the electromagnetic interaction: a photon does not carry charge.

IV. The Big Bang and cosmology.

A. Remind students of the doppler shift for light and state that spectroscopic evidence convinces us that on a large scale, matter in the universe is receding from us and we are led to conclude that the universe is expanding. Write down Hubble's law and give the approximate value of the Hubble parameter: $63.0\,\text{km}/(\text{s} \cdot \text{Mpc}$ $(= 19.3\,\text{mm/s} \cdot \text{ly})$. Define the parsec ($3.084 \times 10^{13}$ km). Show that this implies a minimum age for the universe of about 1.5×10^{10} y.

B. State that the future expansion (or contraction) of the universe depends on its mass density and that the density of matter that radiates is too small to prevent expansion forever. Explain that there is evidence for the existence of matter that does not radiate (dark matter). Explain how the rotational period of a star in a galaxy, as a function of its distance from the galactic center, provides such evidence. The nature of the dark matter is not presently known.

C. Discuss the microwave background radiation and state that physicists believe it was generated about 300,000 years after the big bang, when the universe became tenuous enough to allow photons to exist without being quickly absorbed.

D. Remark that in the early universe the temperature was sufficiently high that the exotic particles now being discovered (and others) existed naturally. We need the results of high energy physics to understand the early universe.

E. Go over the chronological record given at the end of Section 45–14.

SUGGESTIONS

1. Assignments

a. To test for understanding of the conservation laws and the stability of particles, ask questions 4, 5, and 6. Problems 3, 9 (or 11), 12, 13, 15, 16, 17, and 21 each deal with one or more of the conservation laws. Assign several.

b. To help clarify particle properties and classifications, ask questions 10, 11, 12, and 13.

c. Problems 23, 24, 25, 27, and 28 provide excellent illustrations of the quark model.

d. Assign problem 30 in connection with Hubble's law. Assign problem 31 or 32 in connection with the red shift. If you discussed the relativistic Doppler shift in connection with Chapter 38, assign problem 34. Problem 35 deals with the cosmic background radiation. Dark matter and the future of the universe are the subjects of problem 38.

2. Demonstrations

Show nuclear emulsion plates, available from Brookhaven National Laboratory, Fermilab, and other high energy laboratories.

3. Books and Monographs

a. *Resource Letters, Book Four* and *Resource Letters, Book Five*; American Association of Physics Teachers, One Physics Ellipse, College Park, MD 20740–3845. Contain resource letters on high energy and particle physics and on cosmology.

b. *Quarks*; edited by O.W. Greenberg; available from AAPT (see above for address). Reprints covering important aspects of the quark model.

c. *Quarks, Quasars, and Quandaries*; edited by Gordon Aubrecht; available from AAPT (see above for address). Summaries of particle physics and cosmology.

d. *Cosmology and Particle Physics*; edited by David Lindley, Edward W. Kolb, and David N. Schramm. Reprint collection dealing the evolution of the universe from the big bang.

e. *Black Holes*; edited by Steven Detweiler; available from AAPT (see above for address). Reprints dealing with structure and dynamics of black holes.

4. Audio/Visual

a. *Matter and the Universe*; VHS video tape (26 min); Films for the Humanities & Sciences, PO Box 2053, Princeton, NJ 08543–2053.

b. *Particle Physics*; VHS video tape (26 min); Films for the Humanities & Sciences (see above for address).

c. *The Unification Theory*; VHS video tape (26 min); Films for the Humanities & Sciences (see above for address).

d. *Quarks and the Universe: Murray Gell-Mann*; VHS video tape (30 min); Films for the Humanities & Sciences (see above for address).

e. *The Grand Design*; VHS video tape (58 min); Films for the Humanities & Sciences (see above for address).

f. *The Birth of the Stars and the Great Cosmic Cycle*; VHS video tape (60 min); Films for the Humanities & Sciences (see above for address).

g. *The Origin of Quasars*; VHS video tape (59 min); Films for the Humanities & Sciences (see above for address).

h. *The Origin of Galaxies*; VHS video tape (58 min); Films for the Humanities & Sciences (see above for address).

i. *The Origin of the Universe*; VHS video tape (59 min); Films for the Humanities & Sciences (see above for address).

j. *Black Holes, Dark Matter*; VHS video tape (11 min); Films for the Humanities & Sciences (see above for address).

k. *The Expanding Universe: From Big Bang to Big Crunch?*; VHS video tape (20 min); Films for the Humanities & Sciences (see above for address).

5. Computer Software

a. *Atomic Physics*; Macintosh, Windows; Cross Educational Software, Inc., 508 E. Kentucky Avenue, PO Box 1536, Ruston, LA 71270. Contains sections on elementary particles.

b. *Chamber Works* by Robert Estes; OnScreen Science, Inc., 46 Wallace Street, Suite 205, Sommerville, MA 02144–1807; Macintosh. Shows particle decays in a magnetic field. Student controls the field and can measure the radii of orbits. Can be used as a demonstration or as a source of problems. Reviewed TPT **33**, 408 (September 1995).

6. Laboratory
Meiners Experiment 14–8: *Nuclear and High Energy Particles.* A dry ice and alcohol cloud chamber is used to observe the tracks of alpha and beta particles as well as the tracks produced by cosmic rays. A magnet is used to make circular tracks.

SECTION THREE
PROBLEMS IN THE STUDENT SOLUTION MANUAL AND ON THE STUDENT'S COMPANION WEBSITE

The *Student Solution Manual* contains fully worked out solutions to many end-of-chapter problems and the Wiley Website contains guidelines and strategies for working many others. The *Student Solution Manual* is available to students as a print supplement. A password to the Wiley Website is given to students who purchase the study guide (*A Student's Companion*). The problems included in the *Student Solution Manual* and on the Wiley Website are listed here. A few complete solutions for each chapter are included in a free website, with the address given in the text. They are underlined here.

As part of each assignment you may wish to have students study a few of the solutions in the *Student Solution Manual* before attempting their own solutions to other problems. You may also wish to include in the assignment several of the problems discussed on the Wiley Website. These will help students develop problem-solving strategies and learn effective problem-solving techniques.

Chapter 1

Solution Manual: 1E, 3E, 5E, 7P, 9P, 11E, 13P, 15P, 17P, 19E, 21P, 23P
Website: 2E, 6E, 8P, 10E, 14P, 20P, 24P

Chapter 2

Solution Manual: 1E, 3E, 7P, 17E, 19P, 21P, 23E, 25E, 27E, 29E, 31E, 33P, 35P, 37P, 41E, 43E, 45E, 47P, 49P, 51P, 53P, 57P, 59P, 61P
Website: 5P, 8P, 11E, 12P, 18P, 20P, 26E, 28E, 32E, 36P, 38P, 44E, 52P

Chapter 3

Solution Manual: 3E, 5E, 7P, 9P, 11E, 13E, 15E, 17E, 21P, 23P, 25P, 27P, 29E, 31P, 33P, 35P
Website: 1E, 8P, 12E, 18P, 22P, 28E, 34P, 38P

Chapter 4

Solution Manual: 5E, 9E, 13P, 17E, 21E, 23E, 25P, 27P, 31P, 35P, 37P, 39P, 41P, 43E, 45E, 47P, 49P, 51P, 55P, 57E, 59P, 61P
Website: 4P, 8P, 11E, 14P, 16P, 20E, 28E, 30P, 36P, 42E, 44E, 48P, 52P 56E

Chapter 5

Solution Manual: 7P, 9E, 11E, 13E, 15E, 19E, 21E, 25P, 29P, 31P, 33P, 35P, 37P, 41P, 43P, 45P, 47P, 49P, 51P, 53P
Website: 2E, 6P, 12E, 16E, 22E, 26P, 34P, 36P, 40P, 44P, 48P, 50P, 56P

Chapter 6

Solution Manual: 1E, 3E, 5E, 7E, 9P, 11P, 13P, 15P, 19P, 21P, 23P, 27P, 29P, 31P, 33E, 37E, 41E, 43P, 45P, 47P
Website: 8E, 14P, 18P, 22P, 25P, 28P, 30P, 34E, 40P, 42P, 44P, 46P

Chapter 7

Solution Manual: 1E, 5P, 7E, 11P, 13P, 17P, 19P, 21E, 23P, 25E, 27P, 29P, 31E, 33P, 35P, 37P
Website: 6E, 8E, 10P, 15E, 16E, 18P, 22P, 24E, 26P, 28P, 32E, 34P, 36P, 40P

Chapter 8

Solution Manual: 1E, 3E, 5E, 7P, 9E, 11E, 13E, 15P, 17P, 21P, 25P, 27P, 29P, 31P, 33P*, 35P*, 37P, 41P, 43E, 51P, 53P, 55P, 61P
Website: 4E, 6P, 8P, 14P, 19P, 26P, 30P, 32P, 36P, 40E, 42P, 46E, 50P, 58P

Chapter 9

Solution Manual: 1E, 3E, 9P*, 11E, 15P, 17P, 19P, 27E, 29E, 33P, 37P, 39P, 41E, 43E, 47P, 55E
Website: 6P, 7P, 8P, 10E, 13E, 18P, 24P, 25P, 28E, 32E, 36P, 40P, 42E, 46E, 54P

Chapter 10

Solution Manual: 1E, 5E, 9P, 11P, 13P, 15P, 19P, 21E, 23E, 27P, 29P, 31P, 35E, 37E, 41P, 43P, 45P, 49E, 51P, 53P, 55P
Website: 2E, 8P, 14P, 16P, 20E, 26P, 33P, 34P, 40P, 48E, 50P, 52P

Chapter 11

Solution Manual: 3E, 5E, 9E, 13P, 19E, 21E, 25P, 27P, 29P, 31P, 33E, 35E, 39E, 41P, 43P, 45E, 47P, 49E, 51E, 55P, 63P, 65P, 67P
Website: 4E, 6P, 8E, 15P, 16P, 23E, 26P, 34P, 38E, 42P, 44P, 46E, 48P, 52E, 58E

Chapter 12

Solution Manual: 3E, 5E, 9P, 15E, 21P, 25E, 27P, 29E, 33E, 35E, 37P*, 39E, 41E, 43E, 45P, 51P*
Website: 2P, 7E, 8P, 12P, 14P, 20P, 24E, 26P, 32P, 36P, 38P, 42E, 46P, 48P, 52P, 58P

Chapter 13

Solution Manual: 3E, 7E, 9E, 11E, 13E, 15P, 17P, 21P, 23P, 27P, 29P, 31P, 33P, 35P, 37E, 39P
Website: 6E, 10E, 12E, 16P, 18P, 20P, 22P, 24P, 28P, 30P, 35P, 38P, 40P

Chapter 14

Solution Manual: 1E, 3E, 7E, 13P, 15E, 17P, 19P, 25P, 29E, 31P, 33P, 35P, 37P, 41E, 43E, 45E, 47E, 55P*, 57E, 59P, 63P
Website: 6E, 8P, 14E, 18P, 21P, 22E, 23P, 26E, 36P, 42P, 46E, 52P, 54P, 58P

Chapter 15

Solution Manual: 1E, 3E, 6P, 7P, 9E, 13E, 15P, 17P, 19P, 21P, 22E, 23E, 25E, 27E, 29P, 33P, 35P, 37P, 39E, 41P, 43E, 47E, 49E, 55P
Website: 1E, 5E, 10E, 11E, 14E, 18P, 22E, 24E, 26E, 32P, 38P, 40E, 44E, 46E, 48E, 54P, 58P

Chapter 16

Solution Manual: 3E, 5E, 7E, 9E, 13E, 17P, 21P, 23P, 27P, 29P, 31E, 35E, 37E, 39P, 41P, 43E, 47E, 49E, 53P, 55P, 59E, 61E

Website: 1E, 6E, 12E, 14P, 18P, 19P, 22P, 25P, 26P, 33E, 40E, 48E, 52P, 58P, 62P, 64P

Chapter 17

Solution Manual: 3E, 5E, 7P, 11E, 13E, 15P, 17P, 19P, 21P, 23P*, 25P, 27E, 29E, 31P, 35E, 37E, 39P, 41P, 43P, 45P, 47P, 49P, 51P

Website: 4E, 6P, 8P, 9P, 14E. 16P, 20P, 22P, 24E, 26E, 28P, 30P, 33E, 38E, 42P, 46P,

---

Chapter 18

Solution Manual: 3E, 5P, 7P, 9E, 13P, 17E, 19E, 21E, 23E, 25P, 27P, 29P, 33E, 35P, 37P, 39P, 41P, 43E, 45P, 47E, 55P, 59P

Website: 4E, 6P, 10P, 12P, 15P, 24P, 26P, 28P, 34P, 38P, 40P, 42E, 50P, 57P, 60P

Chapter 19

Solution Manual: 1E, 3E, 5E, 7P, 11E, 15E, 17E, 19P, 21P, 25P, 27E, 29E, 33E, 35E, 37P, 45P, 47P, 49E, 51E, 53E, 57P, 65P

Website: 2E, 6E, 9P, 12E, 13E, 18P, 22P, 28E, 36P, 41P, 46P, 48E, 50E, 54E, 61P, 66P

Chapter 20

Solution Manual: 1E, 5E, 7E, 11P, 13P, 15P, 19E, 23P, 25E, 27P, 29P, 31E, 33E, 35P, 37E, 39P, 41P, 43P, 45E, 47P, 53P, 55E, 57E, 61P

Website: 6E, 10E, 12P, 14P, 16P, 22E, 24P, 28P, 34P, 42P, 44E, 46P, 48P, 52E, 54E, 60P, 63P, 64P

Chapter 21

Solution Manual: 3E, 9P, 15P, 17P, 19P, 23E, 25E, 27P, 31P, 33P, 35E, 37E, 41P, 45P

Website: 2E, 6E, 8P, 12P, 13P, 15P, 16P, 20P, 24E, 28P, 30P, 32P, 34E, 40P, 44P

Chapter 22

Solution Manual: 1E, 7P, 9P, 11P, 13P, 15P, 17P, 19E, 21E, 27P, 29E

Website: 3E, 4E, 8P, 10P, 12P, 14P, 20E, 25P, 28P

Chapter 23

Solution Manual: 3E, 5E, 7E, 9P, 13P, 15E, 17P*, 19P, 23P, 25P*, 27P, 29E, 33E, 35E, 37E, 41P, 43P, 47P

Website: 1E, 4E, 11P, 16P, 21P, 24P, 26E, 30E, 36E, 39P, 45E, 46P

Chapter 24

Solution Manual: 2E, 5E, 9P, 11P, 13E, 15P, 17E, 19P, 23P, 25P, 27E, 29P, 31P, 33P*, 35E, 37E, 39P, 43P, 45P

Website: 3E, 7P, 14E, 18P, 20P, 22P, 26E, 28E, 32P, 34E, 41P, 46P

Chapter 25

Solution Manual: 1E, 3P, 5E, 7P, 9P*, 11P*, 17P, 19P, 21P, 23P, 25E, 27E, 33P, 35P, 39P, 41P, 45P, 47P, 51E, 53P

Website: 8P, 10E, 13E, 16E, 18P, 20P, 24E, 28P, 32P, 34P, 38E, 46P, 52E

Chapter 26

Solution Manual: 3E, 5E, 7E, 9P, 11E, 15P, 17P, 19P, 21P, 23E, 25E, 29P, 31P, 33P, 35E, 37E, 39P, 41P, 43E, 45P, 47P

Website: 2E, 8E, 10E, 14P, 16P, 20P, 24E, 32P, 34E, 38P, 40P, 44E, 46P

Chapter 27

Solution Manual: 1E, 3P, 5E, 7E, 9P, 13E, 15E, 17E, 19E, 21P, 23P, 27P, 29P, 31P, 33E, 37P, 39P, 41P

Website: 2P, 4E, 8P, 11P, 12E, 16E, 18E, 22P, 24P, 26P, 30P, 35E, 38P

Chapter 28

Solution Manual: 1E, 3E, 5E, 9E, 15P, 17P, 19E, 21E, 23E, 25E, 27P, 29P, 31P, 35P, 39P, 43P, 45E, 49P, 51P, 53P, 55P*

Website: 4E, 7E, 12P, 14P, 16P, 20E, 26P, 30P, 32P, 33P, 38P, 40P, 47E, 50P, 52P

Chapter 29

Solution Manual: 3E, 5P, 7E, 11P, 15E, 17E, 21E, 25P, 27P, 29P, 31P, 33E, 35E, 37P, 39E, 41E, 43P, 45P, 47P, 49E, 51E, 53E, 55P

Website: 2E, 4P, 10P, 14P, 16E, 20E, 22P, 28P, 32P, 36P, 38P, 40E, 46P, 50E, 54P

Chapter 30

Solution Manual: 1E, 3E, 7P, 9P, 11P, 13P, 15P, 17P, 21E, 25P, 31E, 33P, 37P, 39P, 41E, 43E, 45P, 47P, 49E, 51E, 55P

Website: 4E, 5E, 8P, 14P, 20P, 23E, 28P, 29P, 30E, 35P, 38P, 46P, 52E, 57P

Chapter 31

Solution Manual: 1E, 7P, 9P, 11P, 15P, 17P, 21P, 23P*, 25E, 27E, 29P, 31P, 33E, 35P, 37E, 39P, 41E, 45E, 49E, 53P, 55P*, 59P, 61P, 63E, 67P, 69E, 71P, 73P

Website: 10P, 13P, 16P, 22P, 24P, 30P, 34P, 36E, 38P, 40E, 43P, 44P, 46E, 50P, 54P, 58E, 60P, 66E, 72P, 74P

Chapter 32

Solution Manual: 3P, 5E, 7P, 9E, 11E, 13P*, 15E, 17E, 19P, 21E, 23E, 25P, 27E, 29E, 31E, 33P, 35P, 37P

Website: 6P, 10E, 14E, 16E, 18E, 22E, 28P, 32E, 34P, 36P, 38P

Chapter 33

Solution Manual: 3E, 7P, 9E, 13P, 15P, 17P, 19P, 23P*, 25E, 27P, 29P*, 31E, 33E, 35P, 39E, 41P, 43P, 45P, 47P, 49E, 55E, 57P, 59P, 63E, 65P

Website: 4E, 5P, 11P, 12P, 20P, 26P, 30E, 34P, 36P, 44P, 46P, 52E, 56P, 61P

Chapter 34

Solution Manual: 1E, 5P, 7E, 11E, 13E, 19P, 21E, 23E, 25P, 27P, 29P, 31P, 33E, 35E, 37P, 39P, 41P, 43E, 45E, 47P, 49P, 53E, 55E, 57P, 59P, 61E

Website: 6E, 9E, 16P, 18P, 24P, 28P, 30P, 32E, 36P, 40P, 46P, 51P, 56P, 62E

Chapter 35

Solution Manual: 1E, 3E, 5P, 7P, 11P, 17E, 19E, 21E, 25P, 29P, 31P, 33E, 35P

Website: 2E, 6P, 10P, 12P, 13P, 18E, 23P, 27P, 30P, 32E, 36P

Chapter 36

Solution Manual: 7P, 13E, 15E, 17E, 19P, 21P, 23E, 27P, 29P*, 31E, 35E, 37E, 41P, 43P, 45P, 49P, 53P, 55E, 57P

Website: 1E, 8P, 10P, 11E, 14E, 16E, 20P, 24E, 28P, 34E, 40P, 44P, 48P, 54E, 58P

Chapter 37

Solution Manual: 1E, 3E, 5E, 7P, 9E, 11P, 13P, 15E, 19E, 23P, 27E, 31P, 35E, 37P, 39P, 43P, 45P*, 47E, 49E, 51P, 53E, 57P, 59P, 61P

Website: 2E, 6P, 12P, 16E, 22P, 28E, 32P, 33E, 38P, 40P, 44P, 48E, 52P, 58P, 60P

Chapter 38

Solution Manual: 1E, 3E, 7E, 11E, 13P, 15E, 17E, 19P, 21E, 23E, 25P, 27E, 31P, 35E, 37P, 39P, 43P, 45P, 47P

Website: 5P, 10E, 12P, 16E, 18P, 22E, 26P, 29E, 34E, 38P, 40P, 42P, 44P, 48P

Chapter 39

Solution Manual: 3E, 13P, 15P, 19E, 23P, 27P, 31E, 37P, 39P, 49E, 51E, 53P, 57P, 59P, 63P, 65P, 69P, 71P, 75E, 77P, 79P, 81P

Website: 14P, 16E, 20E, 22P, 29P, 34P, 42P, 48P, 54P, 56P, 61P, 64P, 68P, 73P, 76E, 78P, 80P, 82P

Chapter 40

Solution Manual: 3E, 11P, 15E, 17P, 19E, 23P, 27P, 33E, 35E, 37E, 41P, 45P, 49P, 51P, 53P, 55P, 57P*, 59P

Website: 4E, 7E, 10P, 14E, 23P, 26P, 32E, 34E, 39E, 40E, 43P, 47P, 52P, 54P, 56P

Chapter 41

Solution Manual: 3E, 7E, 11P, 13P, 15E, 17E, 19E, 27P, 33P, 35P, 39P, 41P, 43P, 45P, 49E, 55E, 59E, 65P, 67P

Website: 4E, 9E, 16E, 18P, 22E, 26P, 30E, 34P, 37E, 40P, 42P, 50P, 51P, 58E, 63E, 66P, 68P

Chapter 42

Solution Manual: 1E, 7E, 9E, 11E, 15P, <u>21P</u>, 23P, 25P, 29P, 33P, 35E, 41P, <u>43P</u>, 45E, 47P
Website: 10E, 14P, 16P, 17P, 18P, 19P, 24P, 26P, 27P, 39P, 40P, 42P, 44E

Chapter 43

Solution Manual: 13E, 15E, 19P, 21P, <u>25P</u>, 29E, 31E, 33E, <u>35P</u>, 39P, <u>41P</u>, 43P, 47E, 49P, 53E, 55E, 57P, <u>61P</u>*, 63E, 67E, 69P, <u>75P</u>
Website: 3P, 4E, 7E, 12E, 20P, 24P, 36P, 37P, 51P, 56P, 59P, 62E, 65P, 71P, 76P, 79P

Chapter 44

Solution Manual: 1E, 3E, 7E, <u>11P</u>, 13P, 15E, 19P, 23P, <u>25P</u>, 27P, <u>31P</u>, 33E, 39E, 43P, 47P, 49P, 53P
Website: 2E, 8E, 12P, 21P, 22P, 24P, 29E, 34E, 37P, 44P, 46P, 50P, 54E

Chapter 45

Solution Manual: 3E, 5E, 9P, 13E, 15E, 17E, <u>21P</u>, 25E, <u>31E</u>, <u>37P</u>, 39E
Website: 8P, 11P, 12E, 14P, 16E, 22P, 24E, 28P, 29P, 33P, 34P, 35P, 38P, 40E

SECTION FOUR
ANSWERS TO CHECKPOINTS

The following are the answers to the Checkpoints that appear throughout the text.

Chapter 2

1. b and c
2. zero (zero displacement for the entire trip)
3. (check the derivative dx/dt) (a) 1 and 4; (b) 2 and 3
4. (see Tactic 5) (a) plus; (b) minus; (c) minus; (d) plus
5. 1 and 4 ($a = d^2x/dt^2$ must be a constant)
6. (a) plus (upward displacement on y axis); (b) minus (downward displacement on y axis); (c) $a = -g = -9.8\,\text{m/s}^2$

Chapter 3

1. (a) 7 m ($\vec{a}$ and $\vec{b}$ are in the same direction); (b) 1 m ($\vec{a}$ and $\vec{b}$ are in opposite directions)
2. c, d, f (components must be head-to-tail; $\vec{a}$ must extend from the tail of one component to the head of the other)
3. (a) +, +; (b) +, −; (c) +, + (draw vector from the tail of $\vec{d_1}$ to the head of $\vec{d_2}$)
4. (a) 90°; (b) 0 (vectors are parallel, in the same direction); (c) 180° (vectors are antiparallel, in opposite directions)
5. (a) 0° or 180°; (b) 90°

Chapter 4

1. (a) $(8\,\text{m})\hat{\imath} - (6\,\text{m})\hat{\jmath}$; (b) yes, the xy plane (no z component)
2. Draw $\vec{v}$ tangent to the path with its tail on the path. (a) first; (b) third
3. Take the second derivative with respect to time. (1) and (3): a_x and a_y are both constant and thus $\vec{a}$ is constant; (2) and (4): a_y is constant but a_x is not, thus $\vec{a}$ is not
4. $4\,\text{m/s}^3$, $-2\,\text{m/s}$, 3 m
5. (a) v_x constant; (b) v_y initially positive, decreases to zero, and then becomes progressively more negative; (c) $a_x = 0$ throughout; (d) $a_y = -g$ throughout
6. (a) $-(4\,\text{m/s})\hat{\imath}$; (b) $-(8\,\text{m/s}^2)\hat{\jmath}$
7. (a) 0, distance not changing; (b) +70 km/h, distance increasing; (c) +80 km/h, distance decreasing
8. (a) — (c) increase

Chapter 5

1. c, d, and e ($\vec{F_1}$ and $\vec{F_2}$ must be head-to-tail, $\vec{F}_{\text{net}}$ must be from the tail of one of them to the head of the other)
2. (a) and (b) 2 N, leftward (acceleration is zero in each situation)
3. (a) and (b) 1, 4, 3, 2
4. (a) equal; (b) greater (acceleration is upward, thus net force on body must be upward)
5. (a) equal; (b) greater; (c) less
6. (a) increase; (b) yes; (c) same; (d) yes
7. (a) $F \sin\theta$; (b) increase
8. 0 (because now $a = -g$)

Chapter 6

1. (a) zero (because there is no attempt at sliding); (b) 5 N; (c) no; (d) yes; (e) 8 N
2. (a) same (10 N); (b) decreases; (c) decreases (because N decreases)
3. greater (from Sample Problem 6–5, v_t depends on $\sqrt{R}$)
4. ($\vec{a}$ is directed toward the center of the circular path) (a) $\vec{a}$ downward, $\vec{N}$ upward; (b) $\vec{a}$ and $\vec{N}$ upward
5. (a) same (must still match the gravitational force on the rider); (b) increases ($N = mv^2/R$); (c) increases ($f_{s\,\max} = \mu_s N$)
6. (a) $4R_1$; (b) $4R_1$

Chapter 7

1. (a) decrease; (b) same; (c) negative, zero
2. d, c, b, a
3. (a) same; (b) smaller
4. (a) positive; (b) negative; (c) zero
5. zero

Chapter 8

1. no
2. 3, 1, 2
3. (a) all tie; (b) all tie
4. (a) CD, AB, BC (zero); (b) positive direction of x
5. all tie

Chapter 9

1. (a) origin; (b) fourth quadrant; (c) on y axis below origin; (d) origin; (e) third quadrant; (f) origin
2. (a) to (c) at the center of mass, still at the origin (their forces are internal to the system and cannot move the center of mass)
3. (a) 1, 3, then 2 and 4 tie (zero force); (b) 3
4. (a) 0; (b) no; (c) negative
5. (a) 500 km/h; (b) 2600 km/h; (c) 1600 km/h
6. (a) yes; (b) no

Chapter 10

1. (a) unchanged; (b) unchanged; (c) decreased
2. (a) zero; (b) positive; (c) positive direction of y
3. (a) 10 kg·m/s; (b) 14 kg·m/s; (c) 6 kg·m/s
4. (a) 4 kg·m/s; (b) 8 kg·m/s; (c) 3 J
5. (a) 2 kg·m/s; (b) 3 kg·m/s

Chapter 11

1. b and c
2. a and d
3. (a) yes; (b) no; (c) yes; (d) yes
4. all tie
5. 1, 2, 4, 3
6. 1 and 3 tie, 4, then 2 and 5 tie (zero)
7. (a) downward in the figure; (b) less

Chapter 12

1. (a) same; (b) less
2. less
3. (a) $\pm z$; (b) $+y$; (c) $-x$
4. (a) 1 and 3 tie, then 2 and 4 tie, then 5 (zero); (b) 2 and 3
5. (a) 3, 1, and then 2 and 4 tie (zero); (b) 3
6. (a) all tie (same r, same t, thus same ΔL); (b) sphere, disk, hoop (reverse order of I)
7. (a) decreases; (b) same; (c) increases

Chapter 13

1. c, e, f
2. directly below the rod (torque due to $\vec{F}_g$ on the apple, about the suspension point, is zero)
3. (a) no; (b) at site of $\vec{F}_1$, perpendicular to plane of figure; (c) 45 N
4. (a) at C (to eliminate forces there from a torque equation); (b) plus; (c) minus; (d) equal
5. d
6. (a) equal; (b) B; (c) B

Chapter 14

1. all tie
2. (a) 1, tie of 2 and 4, then 3; (b) line d
3. negative y direction
4. (a) increase; (b) negative
5. (a) 2; (b) 1
6. (a) path 1 [decreased E (more negative) gives decreased a]; (b) less than (decreased a gives decreased T)

Chapter 15

1. all tie
2. (a) all tie (penguin's weight is the same); (b) $0.95\rho_0$, ρ_0, $1.1\rho_0$
3. 13 cm^3/s, outward

4. (a) all tie; (b) 1, then 2 and 3 tie, 4 (wider means slower);
   (c) 4, 3, 2, 1 (wider and lower mean more pressure)

Chapter 16

1. (sketch x versus t) (a) $-x_m$; (b) $+x_m$; (c) 0
2. a (must have the form of Eq. 6–10)
3. (a) 5 J; (b) 2 J; (c) 5 J
4. all tie (in Eq. 16–29, m is included in I)
5. 1, 2, 3 (the ratio m/b matters; k does not)

Chapter 17

1. a, 2; b, 3; c, 1 (compare with phase in Eq. 17–2, then see Eq. 17–5)
2. (a) 2, 3, 1 (see Eq. 17–12); (b) 3, then 1 and 2 tie (find amplitude of dy/dt)
3. same (independent of f); (b) decrease ($\lambda = v/f$); (c) increase; (d) increase
4. (a) increase; (b) increase; (c) increase
5. 0.20 and 0.80 tie, then 0.60, 0.45
6. (a) 1; (b) 3; (c) 2
7. (a) 75 Hz; (b) 525 Hz

Chapter 18

1. beginning to decrease (example: mentally move the curves of Fig. 18–7 rightward past the point at $x = 42$ m)
2. (a) fully constructive, 0; (b) fully constructive, 4
3. (a) 1 and 2 tie, then 3 (see Eq. 18–28); (b) 3, then 1 and 2 tie (see Eq. 18–26)
4. second (see Eqs. 18–39 and 18–41)
5. loosen
6. (a) greater; (b) less; (c) can't tell; (d) can't tell; (e) greater; (f) less
7. (measure speeds relative to the air) (a) 222 m/s; (b) 222 m/s

Chapter 19

1. (a) all tie; (b) 50° X, 50° Y, 50°W
2. (a) 2 and 3 tie, then 1, then 4; (b) 3, 2, then 1 and 4 tie (from Eqs. 19–9 and 19–10, assume that change in area is proportional to initial area)
3. A (see Eq. 19–14)
4. c and e (maximize area enclosed by a clockwise cycle)
5. (a) all tie (ΔE_{int} depends on i and f, not on path); (b) 4, 3, 2, 1 (compare areas under curves); (c) 4, 3, 2, 1 (see Eq. 19–26)
6. (a) zero (closed cycle); (b) negative (W_{net} is negative; see Eq. 19–26)
7. b and d tie, then a, c (P_{cond} identical; see Eq. 19–32)

Chapter 20

1. all but c
2. (a) all tie; (b) 3, 2, 1
3. gas A
4. 5 (greatest change in T), then tie of 1, 2, 3, and 4
5. 1, 2, 3 ($Q_3 = 0$, Q_2 goes into work W_2, but Q_1 goes into greater work W_1 and increases gas temperature

Chapter 21

1. a, b, c
2. smaller (Q is smaller)
3. c, b, a
4. a, d, c, b
5. b

Chapter 22

1. C and D attract; B and D attract
2. (a) leftward; (b) leftward; (c) leftward
3. (a) a, c, b; (b) less than
4. $-15e$ (net charge of $-30e$ is equally shared)

Chapter 23

1. (a) rightward; (b) leftward; (c) leftward; (d) rightward; (p and e have same charge magnitude and p is farther)
2. all tie

3. (a) toward positive y; (b) toward positive x; (c) toward negative y
4. (a) leftward; (b) leftward; (c) decrease
5. (a) all tie; (b) 1 and 3 tie, then 2 and 4 tie

Chapter 24

1. (a) $+EA$; (b) $-EA$; (c) 0; (d) 0
2. (a) 2; (b) 3; (c) 1
3. (a) equal; (b) equal; (c) equal
4. $+50e$; (b) $-150e$
5. 3 and 4 tie, then 2, 1

Chapter 25

1. (a) negative; (b) increase
2. (a) positive; (b) higher
3. (a) rightward; (b) 1, 2, 3, 5: positive; 4: negative; (c) 3, then 1, 2, and 5 tie, then 4
4. all tie
5. a, c (zero), b
6. (a) 2, then 1 and 3 tie; (b) 3; (c) accelerate leftward

Chapter 26

1. (a) same; (b) same
2. (a) decreases; (b) increases; (c) decreases
3. (a) V, $q/2$; (b) $V/2$, q
4. (a) $q_0 = q_1 + q_{34}$; (b) equal (C_3 and C_4 are in series)
5. (a) same; (b) – (d) increase; (e) same (same potential difference across same plate separation)
6. (a) same; (b) decrease; (c) increase

Chapter 27

1. 8 A, rightward
2. (a) – (c) rightward
3. a and c tie, then b
4. device 2
5. a and b tie, then d, then c

Chapter 28

1. (a) rightward; (b) all tie; (c) b, then a and c tie; (d) b, then a and c tie
2. (a) all tie; (b) R_1, R_2, R_3
3. (a) less; (b) greater; (c) equal
4. (a) $V/2$, i; (b) V, $i/2$
5. (a) 1, 2, 4, 3; (b) 4, tie of 1 and 2, then 3

Chapter 29

1. (a) $+z$; (b) $-x$; (c) $\vec{F}_B = 0$
2. (a) 2, then tie of 1 and 3 (zero); (b) 4
3. (a) $+z$ and $-z$ tie, then $+y$ and $-y$ tie, then $+x$ and $-x$ tie (zero); (b) $+y$
4. (a) electron; (b) clockwise
5. $-y$
6. (a) all tie; (b) 1 and 4 tie , then 2 and 3 tie

Chapter 30

1. a, c, b
2. b, c, a
3. d, tie of a and c, then b
4. d, a, tie of b and c (zero)

Chapter 31

1. b, then d and e tie, and then a and c tie (zero)
2. a and b tie, then c (zero)
3. c and d tie, then a and b tie
4. b, out; c, out; d, into; e, into
5. d and e
6. (a) 2, 3, 1 (zero); (b) 2, 3, 1
7. a and b tie, then c

Chapter 32

1. d, b, c, a (zero)
2. (a) 2; (b) 1
3. (a) away; (b) away; (c) less
4. (a) toward; (b) toward; (c) less
5. a, c, b, d (zero)
6. tie of b, c, and d, then a

Chapter 33

1. (a) $T/2$; (b) T; (c) $T/2$; (d) $T/4$
2. (a) 5 V; (b) 150 μJ
3. (a) same; (b) same
4. (a) C, B, A; (b) 1, A; 2, B; 3, S; 4, C; (c) A
5. (a) same; (b) increases
6. (a) same; (b) decreases
7. (a) 1, lags; 2, leads; 3, in phase; (b) 3 ($\omega_d = \omega$ when $X_L = X_C$)
8. (a) increase (circuit is mainly capacitive; increase C to decrease X_C to be closer to resonance for maximum P_{avg}); (b) closer
9. (a) greater; (b) step-up

Chapter 34

1. (a) (Use Fig. 34–5.) On the right side of rectangle, $\vec{E}$ is in the negative y direction; on the left side, $\vec{E}+d\vec{E}$ is greater and in the same direction; (b) $\vec{E}$ is downward. On the right side, $\vec{B}$ is in the negative z direction; on the left side $\vec{B}+d\vec{B}$ is greater and in the same direction.
2. positive direction of x
3. (a) same; (b) decrease
4. a, d, b, c (zero)
5. a
6. (a) no; (b) yes

Chapter 35

1. $0.2d$, $1.8d$, $2.2d$
2. (a) real; (b) inverted; (c) same
3. (a) e; (b) virtual, same
4. virtual, same as object, diverging

Chapter 36

1. b (least n), c, a
2. (a) top; (b) bright intermediate illumination (phase difference is 2.1 wavelengths
3. (a) 3λ, 3; (b) 2.5λ, 2.5
4. a and d tie (amplitude of resultant wave is $4E_0$), then b and c tie (amplitude of resultant wave is $2E_0$)
5. (a) 1 and 4; (b) 1 and 4

Chapter 37

1. (a) expand; (b) expand
2. (a) second side maximum; (b) 2.5
3. (a) red; (b) violet
4. diminish
5. (a) increase; (b) same
6. (a) left; (b) less

Chapter 38

1. (a) same (speed of light postulate); (b) no (the start and end of the flight are spatially separated); (c) no (because this measurement is not a proper time)
2. (a) Sally's; (b) Sally's
3. (a) positive; (b) negative; (c) positive
4. (a) right; (b) more
5. (a) equal; (b) less

Chapter 39

1. b, a, d, c
2. (a) lithium, sodium, potassium, cesium; (b) all tie
3. (a) same; (b) – (d) x rays
4. (a) proton; (b) same; (c) proton
5. same

Chapter 40

1. b, a, c
2. (a) all tie; (b) a, b, c
3. a, b, c, d
4. $E_{1,1}$
5. (a) 5; (b) 7

Chapter 41

1. 7
2. (a) decrease; (b) – (c) same
3. less
4. A, C, B

Chapter 42

1. (a) larger; (b) same
2. Cleveland: metal; Boca Raton: none; Seattle: semiconductor
3. a, b, and c
4. b

Chapter 43

1. ^{90}As and ^{158}Nd
2. a little more than 75 Bq (elapsed time is a little less than three half-lives)
3. ^{206}Pb

Chapter 44

1. c and d
2. (a) no; (b) yes; (c) no
3. e

Chapter 45

1. (a) the muon family; (b) a particle; (c) $L_\mu = +1$
2. b and c
3. c

SECTION FIVE
ANSWERS TO QUESTIONS

The following are the answers to the end-of-chapter questions.

Chapter 2

1. (a) all tie; (b) 4, tie of 1 and 2, then 3
2. (a) negative direction of x; (b) positive direction of x; (c) yes (when graphed line crosses the t axis); (d) positive; (e) constant
3. E
4. (a) 2, 3; (b) 1, 3; (c) 4
5. a and c
6. 60 km/h, not 0
7. $x = t^2$ and $x = 8(t-2) + (1.5)(t-2)^2$
8. (a) 3, 2, 1; (b) 1, 2, 3; (c) all tie; (d) 1, 2, 3
9. same

Chapter 3

1. $\vec{A}$ and $\vec{B}$
2. (a) — (c) yes (example: $5\hat{\imath}$ and $-2\hat{\jmath}$)
3. no, but $\vec{a}$ and $-\vec{b}$ are commutative: $\vec{a} + (-\vec{b}) = (-\vec{b}) + \vec{a}$
4. (a) yes; (b) yes; (c) no
5. (a) $\vec{a}$ and $\vec{b}$ are parallel; (b) $\vec{b} = 0$; (c) $\vec{a}$ and $\vec{b}$ are perpendicular
6. (a) — (d) negative
7. all but (e)
8. no (their orientations can differ)
9. (a) 0 (vectors are parallel); (b) 0 (vectors are antiparallel)
10. (a) $\vec{B}$ and $\vec{C}$, $\vec{D}$ and $\vec{E}$; (b) $\vec{D}$ and $\vec{E}$

Chapter 4

1. (a) $(7\,\text{m})\hat{\imath} + (1\,\text{m})\hat{\jmath} + (-2\,\text{m})\hat{k}$; (b) $(5\,\text{m})\hat{\imath} + (-3\,\text{m})\hat{\jmath} + (1\,\text{m})\hat{k}$; (c) $(-2\,\text{m})\hat{\imath}$
2. (a) 1 and 3: a_y is constant but a_x is not and thus $\vec{a}$ is not; 2: a_x is constant but a_y is not and thus $\vec{a}$ is not; 4: a_x and a_y are both constant and thus $\vec{a}$ is constant; (b) $-2\,\text{m/s}^2$, $-3\,\text{m/s}$
3. a, b, c
4. yes (the vertical component of $\vec{v}$ is downward)
5. (a) all tie; (b) 1 and 2 tie (the rocket is shot upward), then 3 and 4 tie (it is shot into the ground!)
6. $(2\,\text{m/s})\hat{\imath} - (4\,\text{m/s})\hat{\jmath}$
7. (a) 3, 2, 1; (b) 1, 2, 3; (c) all tie; (d) 6, 5, 4
8. (a) 0; (b) 350 km/h; (c) 350 km/h; (d) same nothing changed about the vertical motion
9. (a) less; (b) unanswerable; (c) equal; (d) unanswerable
10. (a) all tie; (b) all tie; (c) 3, 2, 1; (d) 3, 2, 1
11. (a) 2; (b) 3; (c) 1; (d) 2; (e) 3; (f) 1
12. 2, then 1 and 4 tie, then 3
13. (a) yes; (b) no; (c) yes

Chapter 5

1. (a) 5; (b) 7; (c) $(2\,\text{N})\hat{\imath}$; (d) $-(6\,\text{N})\hat{\jmath}$; (e) fourth; (f) fourth
2. (a) 2 and 3; (b) 2
3. (a) 2 and 4; (b) 2 and 4
4. increase
5. (a) 2, 3, 4; (b) 1, 3, 4; (c) 1, $+y$; 2, $+x$; 3, fourth quadrant; 4, third quadrant
6. 1, graphs a and e; 2, graphs b and d; 3, graphs b and f; 4, graphs c and f
7. (a) less; (b) greater
8. (a) increases from an initial value of mg; (b) decreases from mg to zero (after which the block moves up away from the floor)
9. (a) 20 kg; (b) 18 kg; (c) 10 kg; (d) all tie; (e) 3, 2, 1
10. (a) 17 kg; (b) 12 kg; (c) 10 kg; (d) all tie; (e) $\vec{F}$, $\vec{F}_{21}$, $\vec{F}_{32}$
11. (a) 4 or 5, choose 4; (b) 2; (c) 1; (d) 4 or 5, choose 5; (e) 3; (f) 6; (g) 3 and 6; 1, 2, and 5; (h) 3 and 6; (i) 1, 2, and 5
12. (a) increases; (b) increases; (c) decreases; (d) decreases; (e) a and b, 3; c and d, 2

Chapter 6

1. (a) F_1, F_2, F_3; (b) all tie

2. (a) rightward; (b) leftward; (c) decrease; (d) leftward; (e) rightward; (f) increase; (g) no
3. (a) same; (b) increases; (c) increases; (d) no
4. (a) upward; (b) horizontal, toward you; (c) no change; (d) increases; (e) increases
5. (a) decrease; (b) decrease; (c) increase; (d) increase; (e) increase
6. (a) decrease; (b) decrease; (c) decrease; (d) decrease; (e) decrease
7. (a) the block's mass m; (b) equal (they are a third-law pair); (c) that on the slab is in the direction of the applied force; that on the block is in the opposite direction; (d) the slab's mass M
8. (a) same; (b) less
9. 4, 3, then 1, 2, and 5 tie
10. (a) all tie; (b) all tie; (c) 2, 3, 1

Chapter 7

1. all tie
2. (a) positive; (b) negative; (c) negative
3. (a) positive; (b) zero; (c) negative; (d) negative; (e) zero; (f) positive
4. (a) 3 m; (b) 3 m; (c) 0 and 6 m; (d) negative direction of x
5. (a) A, B, C; (b) C, B, A; (c) C, B, A; (d) A, 2; B, 3; C, 1
6. b (positive work), a (zero work), c (negative work), d (more negative work)
7. all tie
8. (a) to (c) yes; (d) and (e) no
9. c, d, then a and b tie, then f, e
10. (a) A; (b) B
11. (a) $2F_1$; (b) $2W_1$
12. (a) 3 and 6; (b) 1, 4, then 2 and 5 tie (zero), then 6, then 3; (c) 1 and 4 tie, then 2 and 5 tie (zero), then 6, then 3
13. B, C, A

Chapter 8

1. (a) 12 J; (b) −2 J
2. c and d tie, then a and b tie
3. (a) all tie; (b) all tie
4. (a) and (b) all tie (same change in elevation)
5. (a) 4; (b) returns to its starting point and repeats the trip; (c) 1; (d) 1
6. (a) AB, CD, and then BC and DE tie (zero force); (b) 5 J; (c) 5 J; (d) 6 J; (e) FG; (f) DE
7. (a) fL; (b) 0.50; (c) 1.25; (d) 2.25; (e) b, center; c, right; d, left
8. (a) less; (b) equal
9. (a) increasing; (b) decreasing; (c) decreasing; (d) constant in AB and BC, decreasing in CD

Chapter 9

1. (a) to (d) at the origin
2. same (hang is a visual illusion, due primarily to skilled, rapid passing of the ball from hand to hand during the jump; so much occurs that the jump seems to last longer than normal; the illusion may also be enhanced if arms or legs are raised to flatten the body's path, as in Fig. 9–6.)
3. (a) at the center of the sled; (b) $L/4$, to the right; (c) not at all (no net external force); (d) $L/4$, to the left; (e) L; (f) $L/2$; (g) $L/2$
4. (a) less; (b) greater
5. (a) ac, cd, and bc; (b) bc; (c) bd and ad
6. (a) 2 N, rightward; (b) 2 N, rightward; (c) greater than 2 N, rightward
7. c, d, then a and b tie
8. (a) a, d, and f; (b) 2; (c) d, f, a
9. b, c, a

Chapter 10

1. all tie
2. 1B, $p_x = -2\,\text{kg}\cdot\text{m/s}$, $p_y = 4\,\text{kg}\cdot\text{m/s}$; 2C, $p_y = 1\,\text{kg}\cdot\text{m/s}$; 2D, $p_x = 11\,\text{kg}\cdot\text{m/s}$; 3E, $p_y = -2\,\text{kg}\cdot\text{m/s}$, $p_x = 4\,\text{kg}\cdot\text{m/s}$
3. b and c
4. (a) positive; (b) positive; (c) 2 and 3
5. (a) rightward; (b) rightward; (c) smaller
6. a, projectile momentum cannot increase, target cannot move in the negative x direction; b, projectile momentum cannot increase, total momentum is not conserved; e, target cannot move in the negative x direction, total momentum is not conserved; h, total momentum is not conserved; i, target cannot move in the negative x direction, total momentum is not conserved

7. (a) one stationary; (b) 2; (c) 5; (d) equal (pool player's result)
8. (a) lower (almost zero); (b) greater (baseball may smash into the ceiling)
9. (a) 2; (b) 1; (c) 3; (d) yes; (e) no
10. c

Chapter 11

1. (a) positive; (b) zero; (c) negative; (d) negative
2. (a) clockwise; (b) counterclockwise; (c) yes; (d) positive; (e) constant
3. finite angular displacements are not commutative
4. a and c
5. (a) c, a, then b and d tie; (b) b, then a and c tie, then d
6. (a) $+x$; (b) $+y$
7. 3, 1, 2
8. larger
9. 90°, then 70° and 110° tie
10. $\vec{F}_5$, $\vec{F}_4$, $\vec{F}_2$, $\vec{F}_1$, $\vec{F}_3$ (zero)
11. (a) decrease; (b) clockwise; (c) counterclockwise
12. (a) 3 and 6; (b) 1, 4, then 2 and 5 tie (zero), then 6, then 3; (c) 1 and 4 tie, then 2 and 5 tie (zero), then 6, then 3

Chapter 12

1. (a) same; (b) block; (c) block
2. (a) greater; (b) same
3. (a) L; (b) $1.5L$
4. (a) 0 or 180°; (b) 90°
5. b, then c and d tie, then a and e tie (zero)
6. (a) 0 ($\vec{r}$ and $\vec{F}$ are both radial); (b) same
7. a, then b and c tie, then e, d (zero)
8. (a) same; (b) greater (smaller rotational inertia)
9. (a) same; (b) increase; (c) decrease; (d) same, decrease, increase
10. (a) 4, 6, 7, 1, and then 2, 3, and 5 tie (zero); (b) 1, 4, and 7
11. (a) 1, 2, 3 (zero); (b) 1 and 2 tie, then 3; (c) 1 and 3 tie, then 2
12. (a) 5 and 6; (b) 1 and 4 tie, then the rest tie
13. (a) 3, 1, 2; (b) 3, 1, 2

Chapter 13

1. (a) yes; (b) yes; (c) yes; (d) no
2. (a) 1, 2, 3 (zero), 4, 5, 6; (b) 6, 5, 4, 3, 2, 1 (zero)
3. a and c (forces and torques balance)
4. b
5. $m_2 = 12\,\text{kg}$, $m_3 = 3\,\text{kg}$, $m_4 = 1\,\text{kg}$
6. (a) same; (b) smaller; (c) smaller; (d) same
7. (a) 15 N (the key is the pulley with the 10-N piñata); (b) 10 N
8. (a) $\sin\theta$; (b) same; (c) larger
9. A, then tie of B and C
10. 4, 1, then 2 and 3 tie

Chapter 14

1. (a) between, closer to less massive particle; (b) no; (c) no (other than infinity)
2. Gm^2/r^2, upward
3. $3GM^2/d^2$, leftward
4. b, tie of a and c, then d
5. (a) 1 and 2 tie, then 3 and 4 tie; (b) 1, 2, 3, 4
6. a tie of b, d, and f, then a, c, e
7. $U_i/4$
8. 1, tie of 2 and 4, then 3
9. (a) all tie; (b) all tie
10. b, a, c
11. (a) — (d) zero
12. (a) at the center of mass of the system, halfway between the stars; (b) 1, 3, 2

Chapter 15

1. e, then b and d tie, then a and c tie
2. b, then a and d tie (zero), then c
3. (a) 2; (b) 1, less; 3, equal; 4, greater
4. (a) all tie; (b) 1, 2, 3; (c) 3, 2, 1
5. (a) moves downward; (b) moves downward
6. 3, 4, 1, 2
7. all tie
8. (a) same; (b) same; (c) lower; (d) higher
9. (a) downward; (b) downward; (c) same
10. b, then a and c tie

Chapter 16

1. c
2. vice versa
3. (a) 2; (b) positive; (c) between 0 and $+x_m$

4. a and b
5. (a) toward $-x_m$; (b) toward $+x_m$; (c) between $-x_m$ and 0; (d) between $-x_m$ and 0; (e) decreasing; (f) increasing
6. (a) $-\pi$, $-180°$; (b) $-\pi/2$, $-90°$; (c) $+\pi/2$, $+90°$
7. (a) π rad; (b) π rad; (c) $\pi/2$ rad
8. (a) 1; (b) 3
9. (a) varies; (b) varies; (c) $x = \pm x_m$; (d) more likely
10. (a) greater; (b) same; (c) same; (d) greater; (e) greater
11. b (infinite period; does not oscillate), c, a
12. (a) greater; (b) same; (c) greater
13. one system : $k = 1500$ N/m, $m = 500$ kg; other system: $k = 1200$ N/m, $m = 400$ kg; the same ratio $k/m = 3$ gives resonance for both systems

---

Chapter 17

1. 7d
2. a, upward; b, upward; c, downward; d, downward; e, downward; f, downward; g, upward; h, upward
3. (a) $\pi/2$ rad and 0.25 wavelength; (b) π rad and 0.5 wavelength; (c) $3\pi/2$ rad and 0.75 wavelength; (d) 2π rad and 1.0 wavelength; (e) $3T/4$; (f) $T/2$
4. (a) 1, 4, 2, 3; (b) 1, 4, 2, 3
5. (a) 4; (b) 4; (c) 3
6. intermediate (closer to fully destructive interference)
7. a and d tie, then b and c tie
8. (a) 8; (b) antinode; (c) longer; (d) lower
9. d
10. (a) node; (b) antinode
11. (a) decrease; (b) disappears

Chapter 18

1. pulse along path 2
2. (a) same as first wave; (b) λ; (c) s_m; (d) π rad
3. (a) 2.0 wavelengths; (b) 1.5 wavelengths; (c) fully constructive, fully destructive
4. $\lambda/2$
5. (a) exactly out of phase; (b) exactly out of phase
6. (a) exactly out of phase; (b) same
7. (a) one; (b) nine
8. (a) two; (b) antinode
9. (a) increase; (b) decrease
10. 150 Hz and 450 Hz
11. all odd harmonics
12. d, fundamental
13. d, e, b, c, a
14. (a) 3, then 1 and 2 tie; (b) 1, then 2 and 3 tie; (c) 3, 2, 1

Chapter 19

1. 25 S°, 25 U°, 25 R°
2. c, then the rest tie
3. A and B tie, then C, D
4. B, then A and C tie
5. (a) both clockwise; (b) both clockwise
6. (a) cycle 2; (b) cycle 2
7. c, a, b
8. upward (with liquid water on the exterior ice and on the ice at the bottom, there can be no temperature differences horizontally and downward)
9. sphere, hemisphere, cube
10. (a) greater; (b) 1, 2, 3; (c) 1, 3, 2; (d) 1, 2, 3; (e) 2, 3, 1
11. (a) at freezing point; (b) no liquid freezes; (c) ice partly melts
12. (a) f, ice temperature will not rise to freezing point and then drop; (b) b and c, at freezing point; d, above; e, below; (c) b, liquid partly freezes, no ice melts; c, no liquid freezes, no ice melts; d, no liquid freezes, ice fully melts; e, liquid fully freezes, no ice melts

---

Chapter 20

1. increased but less than doubled
2. (a) 1, 2, 3; (b) all tie (zero); (c) 1, 2, 3; (d) 1, 2, 3; (e) all tie (zero); (f) 1, 2, 3
3. tie of a and c, then b, then d
4. 1180 J
5. 1–4
6. d, tie of a and b, then c
7. 20 J
8. constant-volume process
9. (a) 3; (b) 1; (c) 4; (d) 2; (e) yes
10. (a) same; (b) increases; (c) decreases; (d) increases

11. (a) 1, 2, 3, 4; (b) 1, 2, 3
12. −4 J

Chapter 21

1. unchanged
2. 9 and −8, 8 and −5, 5 and −3, 3 and −2
3. b, a, c, d
4. (a) all tie; (b) all tie; (c) all tie (zero)
5. equal
6. c, a, b
9. (a) same; (b) increase; (c) decrease
8. increase
9. (a) same; (b) increase; (c) decrease
10. more than the age of the universe
11. (a) 0; (b) 0.25; (c) 0.50

Chapter 22

1. No, only for charged particles, charged particle-like objects, and spherical shells
2. all tie
3. a and b
4. (a) between; (b) positively charged; (c) unstable
5. $2q^2/4\pi\epsilon_0 r^2$, up the page
6. a and d tie, then b and c tie
7. (a) same; (b) less than; (c) cancel; (d) add; (e) the adding components; (f) positive direction of y; (g) negative direction of y; (h) positive direction of x; (i) negative direction of x
8. (a) neutral; (b) negatively
9. (a) possibly; (b) definitely
10. Once enough electrons have moved to the far end, any other conduction electron is repelled by the electrons at that end as much as it is repelled by the negatively charged rod at the near end.
11. no (the person and the conductor share the charge)

Chapter 23

1. (a) toward positive x; (b) downward and to the right; (c) A
2. (a) to their left; (b) no
3. two points: one to the left of the particles, the other between the protons
4. $q/4\pi\epsilon_0 d^2$, leftward
5. (a) yes; (b) toward; (c) no (the field vectors are not along the same line; (d) cancel; (e) add; (f) adding components; (g) toward negative y
6. all tie
7. e, b, then a and c tie, then d (zero)
8. (a)rightward; (b) $+q_1$ and $-q_3$, increase; $+q_2$, decrease; n, same
9. (a) toward the bottom; (b) 2 and 4 toward the bottom, 3 toward the top
10. (a) positive; (b) same
11. (a) 4, 3, 1, 2; (b) 3, then 1 and 4 tie, then 2
12. Accumulated excess charge creates an electric field; proximity of a second body concentrates the excess charge and increases the field, causing electric breakdown in the air (sparking).

Chapter 24

1. (a) $8\,\mathrm{N\cdot m^2/C}$; (b) 0
2. (a) a^2; (b) πr^2; (c) $2\pi rh$
3. (a) all four; (b) neither (they are equal)
4. all tie
5. (a) S_3, S_2, S_1; (b) all tie; (c) S_3, S_2, S_1; (d) all tie (zero)
6. all tie
7. 2σ, σ, 3σ; or 3σ, σ, 2σ
8. (a) 2, 1, 3; (b) all tie ($+4q$)
9. (a) all tie ($E = 0$); (b) all tie
10. (a) a, b, c, d; (b) a and b tie, then c, d

Chapter 25

1. (a) higher; (b) positive; (c) negative; (d) all tie
2. (a) 1 and 2; (b) none; (c) no; (d) 1 and 2, yes; 3 and 4, no
3. $-4q/4\pi\epsilon_0 d$
4. b, then a, c, and d tie
5. (a)–(c) $Q/4\pi\epsilon_0 R$; (d) a, b, c
6. (a) 1, then 2 and 3 tie; (b) 3
7. (a) 2, 4, and then a tie of 1, 3, and 5 (where $E = 0$); (b) negative x direction; (c) positive x direction
8. (a) 3 and 4 tie, then 1 and 2 tie; (b) 1 and 2, increase; 3 and 4, decrease
9. (a)–(d) zero
10. (a) positive; (b) positive; (c) negative;

(d) all tie

Chapter 26

1. a, 2; b, 1; c, 3
2. (a) $V/3$; (b) $+CV/3$; (c) $+CV/3$ (not $+CV$)
3. *a*, series; *b*, parallel; *c*, parallel
4. (a) no; (b) yes; (c) all tie
5. (a) $C/3$; (b) $3C$; (c) parallel
6. parallel, C_1 alone, C_2 alone, series
7. (a) same; (b) same; (c) more; (d) more
8. (a) – (d) less
9. (a) 2; (b) 3; (c) 1
10. (a) less; (b) more; (c) equal; (d) more
11. (a) increases; (b) increases; (c) decreases; (d) decreases; (e) same, increases, increases, increases

Chapter 27

1. a, b, and c tie, then d (zero)
2. a, b, and c tie, then d
3. b, a, c
4. increase
5. tie of A, B, and C, then tie of A + B and B + C, then A + B + C
6. (a)–(d) top–bottom, front–back, left–right
7. (a)–(c) 1 and 2 tie, then 3
8. C, then A and B tie, then D
9. C, A, B
10. (a) conductors: 1 and 4; semiconductors: 2 and 3; (b) 2 and 3; (c) all four

Chapter 28

1. 3, 4, 1, 2
2. (a) series; (b) parallel; (c) parallel
3. (a) no; (b) yes; (c) all tie
4. (a) equal; (b) more
5. parallel, R_2, R_1, series
6. 2.0 A
7. (a) same; (b) same; (c) less; (d) more
8. 60 μC
9. (a) less; (b) less; (c) more
10. (a) all tie; (b) 1, 3, 2
11. c, b, a

Chapter 29

1. (a) no, $\vec{v}$ and $\vec{F}_B$ must be perpendicular; (b) yes; (c) no, $\vec{B}$ and $\vec{F}_B$ must be perpendicular
1. (a) all tie; (b) 1 and 2 (charge is negative)
2. tie of a, b, and c, and then d (zero)
3. (a) $\vec{F}_E$; (b) $\vec{F}_B$
4. 2, 5, 6, 9, 10
5. (a) negative; (b) equal; (c) equal; (d) half-circle
6. into page: a, d, e; out of page: b, c, f (the particle is negatively charged)
7. (a) $\vec{B}_1$; (b) $\vec{B}_1$ into page, $\vec{B}_2$ out of page; (c) less
8. 1i, 2e, 3c, 4a, 5g, 6j, 7d, 8b, 9h, 10f, 11k
9. (a) 1, 180°; 2, 270°; 3, 90°; 4, 0°; 5, 315°; 6, 225°; 7, 135°; 8, 45°; (b) 1 and 2 tie, then 3 and 4 tie; (c) 8, then 5 and 6 tie, then 7
10. (a) positive; (b) (1) and (2) tie, then (3), which is zero

Chapter 30

1. *c*, *d*, then *a* and *b* tie
2. (a) into; (b) greater
3. *c*, *a*, *b*
4. *b*, *d*, *a* (zero)
5. (a) 1, 3, 2; (b) less
6. *a*, tie of *b* and *d*, then *c*
7. *c* and *d* tie, then *b*, *a*
8. *b*, *a*, *d*, *c* (zero)
9. *d*, then tie of *a* and *e*, then *b*, *c*
10. (a) 2 opposite 4; (b) 2 and 4 opposite 6; (c) 1 and 5 opposite 3 and 6; (d) 1 and 5 opposite 2, 3, and 4

Chapter 31

1. (a) all tie (zero); (b) 2, then tie of 1 and 3 (zero)
2. out
3. (a) into; (b) counterclockwise; (c) larger
4. (a) leftward; (b) rightward
5. *c*, *a*, *b*
6. *d* and *c* tie, then *b*, *a*
7. *c*, *b*, *a*
8. *c*, *a*, b
9. (a) more; (b) same; (c) same; (d) same (zero)

10. a, 2; b, 4; c, 1; d, 3

Chapter 32

1. supplied
2. b
3. (a) all down; (b) 1 up, 2 down, 3 zero
4. (a) increase; (b) increase
5. (a) 1 up, 2 up, 3 down; (b) 1 down, 2 up, 3 zero
6. (a) 1 down, 2 down, 3 up; (b) 1 up, 2 down, 3 zero
7. (a) 1, up; 2, up; 3, down; (b) and (c) 2, then 1 and 3 tie
8. a, decreasing; b, decreasing
9. (a) rightward; (b) leftward; (c) into
10. 1/4
11. 1, a; 2, b; 3, c and d

Chapter 33

1. (a) $T/4$; (b) $T/4$; (c) $T/2$ (see Fig. 33–2); (d) $T/2$ (see Eq. 31–37)
2. with n zero or a positive integer, (a) $0 \pm n2\pi$; (c) $\pi/2 \pm n2\pi$; (e) $\pi \pm n2\pi$; (g) $3\pi/2 \pm n2\pi$
3. b, a, c
4. (a) less; (b) greater
5. (a) 3, 1, 2; (b) 2, tie of 1 and 3
6. (a) decrease; (b) same (U_B equals U_E, which has not been changed)
7. a, inductor; b, resistor; c, capacitor
8. (a) 1 and 4; (b) 2 and 3
9. (a) leads; (b) capacitive; (c) less
10. a less, b equal, c greater
11. (a) rightward, increase (X_L increases, closer to resonance); (b) rightward, increase (X_C decreases, closer to resonance); (c) rightward, increase (ω_d/ω increases, closer to resonance)
12. (a) positive; (b) decrease L (to decrease X_L and get closer to resonance); (c) decrease C (to increase X_C and get closer to resonance)

Chapter 34

1. (a) positive direction of z; (b) x
2. into
3. (a) same; (b) increase; (c) decrease
4. (a) and (b) $A = 1$, $n = 4$, $\theta = 30°$
5. c
6. b, 30°; c, 60°; d, 60°; e, 30°; f, 60°
7. a, b, c
8. d, b, a, c
9. none
10. (a) b; (b) blue; (c) c
11. b
12. 1.5

Chapter 35

1. c
2. (a) a; (b) c
3. (a) a and c; (b) three times; (c) you
4. (a) from infinity to the focal point; (b) decrease continually
5. convex
6. d (infinite), tie of a and b, then c
7. (a) decrease; (b) increase; (c) increase
8. mirror, equal; lens, greater
9. (a) all but variation 2; (b) for 1, 3, and 4: right, inverted; for 5 and 6: left, same
10. (a) less; (b) less

Chapter 36

1. a, c, b
2. (a) peak; (b) valley
3. (a) 300 nm; (b) exactly out of phase
4. (a) $2d$; (b) (odd number)$\lambda/2$; (c) $\lambda/4$
5. (a) intermediate closer to maximum, $m = 2$; (b) minimum, $m = 3$; (c) intermediate closer to maximum, $m = 2$; (d) maximum, $m = 1$
6. (a) increase; (b) 1λ
7. (a)–(c) decrease; (d) blue
8. a and c tie, then b and d tie (zero)
9. (a) maximum; (b) minimum; (c) alternates
10. d
11. (a) 0.5 wavelength; (b) 1 wavelength
12. (a) no; (b) $2(0) = 0$; (c) $2L$

Chapter 37

1. (a) contract; (b) contract
2. (a) the $m = 5$ minimum; (b) (approximately) the maximum between the $m = 4$ and $m = 5$ minima
3. with megaphone (larger opening, less diffraction)

4. (a) 1 and 3 tie, then 2 and 4 tie; (b) 1 and 2 tie , then 3 and 4 tie
5. four
6. (a) larger; (b) red
7. (a) less; (b) greater; (c) greater
8. (a) decrease; (b) same; (c) in place
9. (a) decrease; (b) decrease; (c) to the right
10. (a) A; (b) left; (c) left; (d) right
11. (a) increase; (b) first order

Chapter 38

1. all tie (pulse speed is c)
2. (a) C_1'; (b) C_1'
3. (a) C_1; (b) C_1
4. (a) Sam; (b) neither
5. (a) negative; (b) positive
6. (a) no; (b) yes; (c) yes
7. less
9. (a) 3, tie of 1 and 2, then 4; (b) 4, tie of 1 and 2, then 3; (c) 1, 4, 2, 3
9. b, a, c, d
10. (a) 3, then 1 and 2 tie; (b) 2, then 1 and 3 tie; (c) 2, 1, 3; (d) 2, 1, 3

Chapter 39

1. (a) microwave; (b) x ray; (c) x ray
2. (a) true; (b) false; (c) false; (d) true; (e) true; (f) false
3. potassium
4. only b
5. Positive charge builds up on the plate, inhibiting further electron emission
6. only e
7. none
8. The fractional wavelength change for visible light is too small
9. (a) greater; (b) less
10. (a) B; (b) – (d) A
11. no essential change
12. electron
13. (a) decreases by a factor of $1/\sqrt{2}$; (b) decreases by a factor of 1/2
14. electron, neutron, alpha particle
15. (a) decreasing; (b) increasing; (c) same; (d) same
16. proton
17. a
18. amplitude of reflected wave is less than that of incident wave
19. (a) zero; (b) yes
20. all tie

Chapter 40

1. (a) 1/4; (b) same factor
2. a, c, b
3. c
4. (a) 18; (b) 17
5. (a) $(\sqrt{1/L})\sin(\pi/2L)x$; (b) $(\sqrt{4/L})\sin(2\pi/L)x$; (c) $(\sqrt{2/L})\cos(\pi/L)x$
6. 12 eV ($4 \longrightarrow 2$ in A) matches $1 \longrightarrow 2$ in C; 9 eV ($5 \longrightarrow 4$ in A) matches $1 \longrightarrow 2$ in D; 24 eV ($5 \longrightarrow 1$ in A) matches $1 \longrightarrow 3$ in D; 15 eV ($4 \longrightarrow 1$ in A) matches $1 \longrightarrow 2$ in E
7. less
8. equal
9. (a) wider; (b) deeper
10. (a) 3; (b) 4
11. $n = 1$, $n = 2$, $n = 3$
12. (a) greater; (b) less; (c) less
13. b, c, and d
14. same
15. (a) first Lyman plus first Balmer; (b) Lyman series limit minus Paschen series limit
16. (a) $n = 3$; (b) $n = 1$; (c) $n = 5$

Chapter 41

1. 0, 2, and 3
2. same number (10)
3. 6p
4. 2, −1, 0, and 1
5. (a) 2, 8; (b) 5, 50
6. (a) bromine; (b) rubidium; (c) hydrogen
7. (a) n; (b) n and ℓ
8. all true
9. a, c, e, f
10. (a) rubidium; (b) krypton
11. (a) unchanged; (b) decrease; (c) decrease
12. (a) 2; (b) 3
13. a and b
14. In addition to the quantized energy, a helium atom has kinetic energy; its total energy can equal 20.66 eV

Chapter 42

1. 4
2. 8
3. b and c
4. c
5. (a) anywhere in the lattice; (b) in any silicon-silicon bond; (c) in a silicon ion core, at a lattice site
6. a and b
7. b and d
8. $4s^2$ and $4p^2$
9. $+4e$
10. (a) arsenic, antimony; (b) gallium, indium; (c) tin
11. none
12. a
13. (a) right to left; (b) back bias
14. zero
15. a, b, and c
16. b

Chapter 43

1. less
2. more protons than neutrons
3. ^{240}U
4. above
5. less
6. (a) ^{196}Pt; (b) no
7. (a) on the $N = Z$ line; (b) positrons; (c) about 120
8. (a) below; (b) below; (c) radioactive
9. no
10. yes
11. yes
12. A and C tie, then B
13. (a) increases; (b) same
14. no effect
15. 7h
16. ^{209}Po
17. d
18. (a) all except ^{198}Au; (b) ^{132}Sn and ^{208}Pb

Chapter 44

1. a
2. a
3. b
4. b, e, a, c, d
5. (a) ^{93}Sr; (b) ^{140}I; (c) ^{155}Nd
6. c
7. c
8. c, a, d, b
9. a
10. d
11. c
12. c

Chapter 45

1. d
2. into
3. the π^+ pion, whose track curves downward at the left
4. b, c, d
5. a, b, c, d
6. f
7. c, f
8. c
9. 1d, 2e, 3a, 4b, 5c
10. (a) lepton; (b) antiparticle; (c) fermion; (d) yes
11. 1b, 2c, 3d, 4e, 5a
12. b, f, c, d, a, g, e
13. (a) 0; (b) +1; (c) −1; (d) +1; (e) −1

SECTION SIX
ANSWERS TO PROBLEMS

The following are the answers to the end-of-chapter exercises and problems.

Chapter 1

1. (a) 10^9; (b) 10^{-4}; (c) 9.1×10^5
2. 403 L
3. (a) 160 rods; (b) 40 chains
4. (a) 23 points; (b) 1.9 picas
5. (a) 4.00×10^4 km; (b) 5.10×10^8 km^2; (c) 1.08×10^{12} km^3
6. (a) 14.5 roods; (b) 1.47×10^4 m^2
7. 1.9×10^{22} cm^3
8. (a) 1.0 m^3; (b) 6.0×10^{-4} m^3
9. 1.1×10^3 acre · feet
10. (a) 52.6 min; (b) 5.2%
11. (a) 0.98 ft/ns; (b) 0.30 mm/ps
12. (a) yes; (b) 8.6 s
13. C, D, A, B, E; the important criterion is the consistency of the daily variation, not its magnitude
14. (a) 495 s; (b) 141 s; (c) 198 s; (d) −245 s
15. 0.12 AU/min
16. 15°
17. 2.1 h
18. (a) 3.88×10^8; (b) 1557.806 448 872 75 s; (c) $\pm 3 \times 10^{-11}$ s
19. 9.0×10^{49}
20. (a) 1.430 m^2; (b) 72.84 km
21. (a) 10^3 kg; (b) 158 kg/s
22. 1.2×10^9 kg
23. (a) 1.18×10^{-29} m^3: (b) 0.282 nm
24. 0.260 kg
25. (a) 60.8 W; (b) 43.3 Z
26. (a) 22 pecks; (b) 5.5 bushels; (c) 200 L
27. 89 km
28. 1.2 m
29. $\approx 1 \times 10^{36}$

Chapter 2

1. 414 ms
2. 6.150 s
3. (a) +40 km/h; (b) 40 km/h
4. 1.29 s
5. (a) 73 km/h; (b) 68 km/h; (c) 70 km/h; (d) 0
6. (a) 1.74 m/s; (b) 2.14 m/s
7. (a) 0, −2, 0, 12 m/s; (b) +12 m; (c) +7 m/s
8. (a) 60 km
9. 1.4 m
10. (a) $2\,s < t < 4\,s$; (b) $0 < t < 3\,s$; (c) $3\,s < t < 7\,s$; (d) $t = 3\,s$
11. (a) −6 m/s; (b) negative x direction; (c) 6 m/s; (d) first smaller, then zero, and then larger; (e) yes ($t = 2$ s); (f) no
12. (a) 28.5 cm/s; (b) 18.0 cm/s; (c) 40.5, cm/s; (d) 28.1 cm/s; (e) 30.3 cm/s
13. 100 m
14. (e) situations (a), (b), and (d)
15. (a) velocity squared; (b) acceleration; (c) m^2/s^2, m/s^2
17. 20 m/s^2, in the direction opposite to its initial velocity
18. (a) 1.10 m/s; (b) 6.11 mm/s^2; (c) 1.47 m/s; (d) 6.11 mm/s^2
19. (a) 80 m/s; (b) 110 m/s; (c) 20 m/s^2
20. 5.9 m
21. (a) m/s^2, m/s^3; (b) 1.0 s; (c) 82 m; (d) −80 m; (e) 0, −12, −36, −72 m/s; (f) −6, −18, −30, −42 m/s^2
22. each, 0.28 m/s^2
23. 0.10 m
24. 0.556 s
25. (a) 1.6 m/s; (b) 18 m/s
26. 3.75 ms
27. (a) 3.1×10^6 s (1.2 months); (b) 4.6×10^{13} m
28. 2.8 m/s^2
29. 1.62×10^{15} m/s^2
30. $21g$
31. 2.65 s
32. 4 m/s^2, positive x direction
33. (a) 3.56 m/s^2; (b) 8.43 m/s
34. yes; red train, 0; green train, 10 m/s
35. (a) 5.00 m/s; (b) 1.67 m/s^2; (c) 7.50 m
36. (a) 82 m; (b) 19 m/s
37. (a) 0.74 s; (b) −6.2 m/s^2

38. (a) $0.994\,\text{m/s}^2$
39. (a) 10.6 m; (b) 41.5 s
40. 183 m/s; no
41. (a) 29.4 m; (b) 2.45 s
42. (a) 1.54 s; (b) 27.1 m/s
43. (a) 31 m/s; (b) 6.4 s
44. (a) 5.44 s; (b) 53.3 m/s; (c) 5.80 m
45. (a) 3.2 s; (b) 1.3 s
46. (a) $v = \sqrt{v_0^2 + 2gh}$; (b) $t = [\sqrt{v_0^2 + 2gh} - v_0]/g$; (c) same as (a); (d) $t = [\sqrt{v_0^2 + 2gh} + v_0]/g$, greater
47. (a) 3.70 m/s; (b) 1.74 m/s; (c) 0.154 m
48. 34 m
49. 4.0m/s
50. 22.0 m/s
51. $857\,\text{m/s}^2$, upward
52. (a) 101 m; (b) 13.0 s
53. $+1.26 \times 10^3\,\text{m/s}^2$, upward
54. (a) 350 ms; (b) 82 ms (each includes both ascent and descent through the 15 cm)
55. 22 cm and 89 cm below the nozzle
56. (a) $8.0\,\text{m/s}^2$; (b) 20 m/s
57. 1.5 s
58. $4H$
59. (a) 5.4 s; (b) 41 m/s
60. (a) 12.3 m/s
61. (a) 76 m; (b) 4.2 s
62. (a) 8.85 m/s; (b) 1.00 m
63. (a) 1.23 cm; (b) 4 times, 9 times, 16 times, 25 times
64. (a) 17 s; (b) 290 m
65. 2.34 m

Chapter 3

1. The displacements should be (a) parallel, (b) antiparallel, (c) perpendicular
2. Walpole (where the state prison is located)
3. (a) −2.5 m, −6.9 m
4. (a) 0.349 rad; (b) 0.873 rad; (c) 1.75 rad; (d) 18.9°; (e) 120°; (f) 441°
5. (a) 47.2 m; (b) 122°
6. (a) 13 m; (b) 7.5 m
7. (a) 168 cm; (b) 32.5° above the floor
8. (a) 27.8 m; (b) 13.4 m
9. (a) 6.42 m; (b) no; (c) yes; (d) yes; (e) a possible answer: $(4.30\,\text{m})\hat{\imath} + (3.70\,\text{m})\hat{\jmath} + (3.00\,\text{m})\hat{k}$; (f) 7.96 m
10. (a) 81 km; (b) 40° north of east
11. (a) 370 m; (b) 36° north of east; (c) 425 m; (d) the distance
12. (b) 3.2 km; (c) 41° south of west
13. (a) $(-9\,\text{m})\hat{\imath} + (10\,\text{m})\hat{\jmath}$; (b) 13 m; (c) +132°
14. (a) 12 m; (b) −5.8 m; (c) −2.8 m
15. (a) 4.2 m; (b) 40° east of north; (c) 8.0 m; (d) 24° north of west
16. (a) $(8.0\,\text{m})\hat{\imath} + (2.0\,\text{m}\,\hat{\jmath}$; (b) 8.2 m; (c) 14°; (d) $(2.0\,\text{m})\hat{\imath} - (6.0\,\text{m})\hat{\jmath}$; (e) 6.3 m; (f) −72°
17. (a) $(3.0\,\text{m})\hat{\imath} - (2.0\,\text{m})\hat{\jmath} + (5.0\,\text{m})\hat{k}$; (b) $(5.0\,\text{m})\hat{\imath} - (4.0\,\text{m})\hat{\jmath} - (3.0\,\text{m})\hat{k}$; (c) $(-5.0\,\text{m})\hat{\imath} + (4.0\,\text{m})\hat{\jmath} + (3.0\,\text{m})\hat{k}$
18. (a) 5.0 m; (b) −37°; (c) 10 m; (d) 53°; (e) 11 m; (f) 27°; (g) 11 m; (h) 80°; (i) 11 m; (j) 260°; (k) 180°
19. (a) 38 m; (b) 320°; (c) 130 m; (d) 1.2°; (e) 62 m; (f) 130°
20. (a) $(1.28\,\text{m})\hat{\imath} + (6.60\,\text{m})\hat{\jmath}$; (b) 6.72 m, at 79.1° and 1.38 rad
21. (a) 1.59 m; (b) 12.1 m; (c) 12.2 m; (d) 82.5°
22. (a) 26.6 m; (b) 209° or −151°
24. (a) $(10.0\,\text{m})\hat{\imath} + (1.63\,\text{m})\hat{\jmath}$; (b) 10.2 m; (c) 9.24° counterclockwise
26. (a) $(-3.18\,\text{m})\hat{\imath} + (4.72\,\text{m})\hat{\jmath}$; (b) 5.69 m; (c) +124°
27. (a) Put axes along cube edges with origin at one corner. Diagonals are $a\hat{\imath} + a\hat{\jmath} + a\hat{k}$, $a\hat{\imath} + a\hat{\jmath} - a\hat{k}$, $a\hat{\imath} - a\hat{\jmath} - a\hat{k}$, $a\hat{\imath} - a\hat{\jmath} + a\hat{k}$; (b) 54.7°; (c) $\sqrt{3}a$
28. (a) $(9.19\,\text{m})\hat{\imath} + (7.71\,\text{m})\hat{\jmath}$; (b) $(14.0\,\text{m})\hat{\imath} + (3.41\,\text{m})\hat{\jmath}$
29. (a) 30; (b) 52
31. 22°
34. $\vec{B} = -3.00\,\hat{\imath} - 3.00\,\hat{\jmath} - 4.00\,\hat{k}$
35. (b) $a^2 b \sin\phi$
36. 540
37. (a) 3.0 m; (b) 0; (c) 3.46 m; (d) 2.00 m; (e) −5.00 m; (f) 8.66 m; (g) −6.67; (h) 4.33
38. (a) 57°; (b) 2.2 m; (c) −4.5 m; (d) −2.2 m; (e) 4.5 m

Chapter 4

1. (a) $(-5.0\,\text{m})\hat{\imath}+(8.0\,\text{m})\hat{\jmath}$; (b) 9.4 m; (c) 122°; (e) $(8\,\text{m})\hat{\imath} - (8\,\text{m})\hat{\jmath}$; (f) 11 m; (g) −45°
2. (a) 6.2 m
3. (a) $(-7.0\,\text{m})\hat{\imath} + (12\,\text{m})\hat{\jmath}$; (b) xy plane
4. 1030 m, horizontally to the west

5. 7.59 km/h, 22.5° east of north
6. $(-0.70\,\text{m/s})\hat{\imath} + (1.4\,\text{m/s})\hat{\jmath} - (0.40\,\text{m/s})\hat{k}$
7. (a) $(3.00\,\text{m/s})\hat{\imath} - (8.00\,\text{m/s}^2)t\,\hat{\jmath}$; (b) $(3.00\,\text{m/s})\hat{\imath} - (16.0\,\text{m/s})\hat{\jmath}$; (c) 16.3 m/s; (d) −79.4°
8. (a) 63.1 km, 18.3° south of west; (b) 0.70 km/h, 18.3° south of east; (c) 1.56 km/h; (d) 1.20 km/h, 33.4° north of east
9. (a) $(8\,\text{m/s}^2)t\,\hat{\jmath} + (1\,\text{m/s})\hat{k}$; (b) $(8\,\text{m/s}^2)\hat{\jmath}$
10. (a) $(-1.5\,\text{m/s}^2)\hat{\imath} + (0.50\,\text{m/s}^2)\hat{k}$; (b) $1.6\,\text{m/s}^2$, in the xz plane, 162° from $+x$
11. (a) $(6.00\,\text{m})\hat{\imath} - (106\,\text{m})\hat{\jmath}$; (b) $(19.0\,\text{m/s})\hat{\imath} - (224\,\text{m/s})\hat{\jmath}$; (c) $(24.0\,\text{m/s}^2)\hat{\imath} - (336\,\text{m/s}^2)\hat{\jmath}$; (d) −85.2° to $+x$
12. $(-2.10\,\text{m/s}^2)\hat{\imath} + (2.81\,\text{m/s}^2)\hat{\jmath}$
13. (a) $(-1.5\,\text{m/s})\hat{\jmath}$; (b) $(4.5\,\text{m})\hat{\imath} - (2.25\,\text{m})\hat{\jmath}$
14. (a) $(-18\,\text{m/s}^2)\hat{\imath}$; (b) 0.75 s; (c) never; (d) 2.2 s
15. (a) 45 m; (b) 22 m/s
16. 60.0°
17. (a) 62 ms; (b) 480 m/s
18. (a) 0.495 s; (b) 3.07 m/s
19. (a) 0.205 s; (b) 0.205 s; (c) 20.5 cm; (d) 61.5 cm
20. (a) 18 cm; (b) 1.9 m
21. (a) 2.00 ns; (b) 2.00 ns; (c) 1.0×10^7 m/s; (d) 2.00×10^6 m/s
22. 25.9 cm
23. (a) 16.9 m; (b) 8.21 m; (c) 27.6 m; (d) 7.26 m; (e) 40.2 m; (f) 0
24. (a) 95 m; (b) 31 m
25. 4.8 cm
26. (a) 38 ft/s; (b) 32 ft/s; (c) 9.3 ft
28. (a) 12.0 m; (b) 19.2 m/s; (c) 4.80 m/s; (d) no
29. (a) 11 m; (b) 23 m; (c) 17 m/s; (d) 63° below horizontal
30. (a) 20 m/s; (b) 36 m/s; (c) 74 m
31. (a) 24 m/s; (b) 65° above the horizontal
32. (a) 1.60 m; (b) 6.86 m; (c) 2.86 m
33. (a) 10 s; (b) 897 m
34. 78.5°
35. the third
36. 5.8 m/s
37. (a) 202 m/s; (b) 806 m; (c) 161 m/s; (d) −171 m/s
38. (a) 20.3 m/s; (b) 21.7 m/s
39. (a) yes; (b) 2.56 m
40. (a) yes; (b) 20 cm; (c) no; (d) 0.86 cm
41. between the angles 31° and 63° above the horizontal
42. $4.0\,\text{m/s}^2$
43. (a) 7.49 km/s; (b) $8.00\,\text{m/s}^2$
44. (a) 0.94 m; (b) 19 m/s; (c) $2400\,\text{m/s}^2$; (d) 0.05 s
45. (a) 19 m/s; (b) 35 rev/min; (c) 1.7 s
46. (a) 7.32 m west of the center; (b) 7.32 m north of the center
47. (a) $0.034\,\text{m/s}^2$; (b) 84 min
48. (a) 7.3 km; (b) 80 km/h
49. (a) 12 s; (b) $4.1\,\text{m/s}^2$, down; (c) $4.1\,\text{m/s}^2$, up
50. (a) 1.3×10^5 m/s; (b) $7.9 \times 10^5\,\text{m/s}^2$; (c) increase
51. $160\,\text{m/s}^2$
52. (a) 4.2 m, 45°; (b) 5.5 m, 68°; (c) 6.0 m, 90°; (d) 4.2 m, 135°; (e) 0.85 m/s, 135°; (f) 0.94 m/s, 90°; (g) 0.94 m/s, 180°; (h) $0.30\,\text{m/s}^2$, 180°; (i) $0.30\,\text{m/s}^2$, 270°
53. (a) $13\,\text{m/s}^2$, eastward; (b) $13\,\text{m/s}^2$, eastward
54. (a) 5 km/h, upstream; (b) 1 km/h, downstream
55. 36 s, no
56. 130°
57. 60°
58. (a) $(80\,\text{km/h})\hat{\imath} - (60\,\text{km//h})\hat{\jmath}$; (b) $\vec{v}$ happens to be along the line of sight; (c) answers do not change
59. 32 m/s
60. (a) $(-32\,\text{km/h})\hat{\imath} - (46\,\text{km/h})\hat{\jmath}$; (b) $[(2.5\,\text{km}) - (32\,\text{km/h})t]\hat{\imath} + [(4.0\,\text{km}) - (46\,\text{km/h})t]\hat{\jmath}$; (c) 0.084 h; (d) ≈ 200 m
61. (a) 38 knots, 1.5° east of north; (b) 4.2 h; (c) 1.5° west of south
62. 93° from the car's direction of motion
63. (a) 37° west of north; (b) 62.6 s
64. (a) A: 10.1 km, 0.556 km; B: 12.1 km, 1.51 km; C: 14.3 km, 2.68 km; D: 16.4 km, 3.99 km; E: 18.5 km, 5.53 km; (b) the rocks form a curtain that curves upward and away from you

---

Chapter 5

1. (a) $F_x = 1.88$ N; (b) $F_y = 0.684$ N; (c) $(1.88\text{ N})\hat{\imath} + (0.684\text{ N})\hat{\jmath}$
2. (a) 0; (b) $(4\text{ m/s}^2),\hat{\jmath}$; (c) $(3\text{ m/s}^2)\hat{\imath}$
3. 2.9 m/s^2
4. $(-2\text{ N})\hat{\imath} + (6\text{ N})\hat{\jmath}$
5. $(3\text{ N})\hat{\imath} - (11\text{ N})\hat{\jmath} + (4\text{ N})\hat{k}$
6. (a) $(0.86\text{ m/s}^2)\hat{\imath} - (0.16\text{ m/s}^2)\hat{\jmath}$; (b) 0.88 m/s^2; (c) $-11°$ from $+x$
7. (a) $(-32\text{ N})\hat{\imath} - (21\text{ N})\hat{\jmath}$; (b) 38 N; (c) 213° from $+x$
8. (a) $\vec{F}_2$ and $\vec{F}_3$ are in the $-x$ direction, $\vec{a} = 0$; (b) $\vec{F}_2$ and $\vec{F}_3$ are in the $-x$ direction, $\vec{a}$ is on the x axis, $a = 0.83\text{ m/s}^2$; (c) $\vec{F}_2$ and $\vec{F}_3$ are 34° above and below the $-x$ direction; $\vec{a} = 0$
9. (a) 108 N; (b) 108 N; (c) 108 N
10. 2.0 N, down
11. (a) 11 N; (b) 2.2 kg; (c) 0; (d) 2.2 kg
12. (a) 740 N; (b) 290 N; (c) 0; (d) 75 kg at each location
13. 16 N
14. (a) 328 N, down; (b) 0; (c) 0; (d) 284 N, up
15. (a) 42 N; (b) 72 N; (c) 4.9 m/s^2
16. (a) 590 N, up; (b) 340 N, up; (c) 590 N, down
17. (a) 0.02 m/s^2; (b) 8×10^4 km; (c) 2×10^3 m/s
18. 310 N
19. 1.2×10^5 N
20. (a) 5500 N; (b) 2.7 s; (c) 4.0; (d) 2.0
21. 1.5 mm
22. 6.8×10^3 N
23. (a) $(285\text{ N})\hat{\imath} + (705\text{ N})\hat{\jmath}$; (b) $(285\text{ N})\hat{\imath} - (115\text{ N})\hat{\jmath}$; (c) 307 N; (d) $-22°$ from $+x$; (e) 3.67 m/s^2; (f) $-22°$ from $+x$
24. (a) 68 N; (b) 73 N
25. (a) 0.62 m/s^2; (b) 0.13 m/s^2; (c) 2.6 m
26. (a) $\cos\theta$; (b) $\sqrt{\cos\theta}$
27. (a) 494 N, up; (b) 494 N, down
28. (a) $(5\text{ m/s})\hat{\imath} + (4.3\text{ m/s})\hat{\jmath}$; (b) $(15\text{ m})\hat{\imath} + (6.4\text{ m})\hat{\jmath}$
29. (a) 2.2×10^{-3} N; (b) 3.7×10^{-3} N
30. (a) 68 N, up the slope; (b) 28 N, up the slope; (c) 12 N, down the slope
31. (a) 1.1 N
32. (a) 4.6×10^3 N; (b) 5.8×10^3 N
33. 1.8×10^4 N
34. 23 kg
35. (a) 620 N; (b) 580 N
36. (a) 0.970 m/s^2; (b) 11.6 N; (c) 34.9 N
37. (a) 3260 N; (b) 2.7×10^3 kg; (c) 1.2 m/s^2
38. (a) 0.74 m/s^2; (b) 7.3 m/s^2
39. (a) 180 N; (b) 640 N
40. 5.1 m/s
41. (a) 1.23 N; (b) 2.46 N; (c) 3.69 N; (d) 4.92 N; (e) 6.15 N; (f) 0.25 N
42. 839 kN
43. (a) 0.735 m/s^2; (b) downward; (c) 20.8 N
44. (a) 3.1 N; (b) 14.7 N
45. (a) 1.18 m; (b) 0.674 s; (c) 3.50 m/s
46. (a) 120 m/s^2; (b) $12g$; (c) 1.4×10^8 N; (d) 4.2 y
47. (a) 4.9 m/s^2; (b) 2.0 m/s^2; (c) upward; (d) 120 N
48. (a) 6800 N; (b) 21° to the barge's line of motion
49. (a) 2.18 m/s^2; (b) 116 N; (c) 21.0 m/s^2
50. (a) 466 N; (b) 527 N; (c) 931 N; (d) 1050 N; (e) 931 N; (f) 1050 N; (g) 1860 N; (h) 2110 N
51. (b) $F/(m+M)$; (c) $MF/(m+M)$; (d) $F(m+2M)/2(m+M)$
52. (a) 566 N; (b) 1130 N
53. $2Ma/(a+g)$
54. 18 kN
55. (a) 31.3 kN; (b) 24.3 kN
56. (a) 7.3 kg; (b) 89 N

Chapter 6

1. (a) 200 N; (b) 120 N
2. 2°
3. 0.61
4. 160 N
5. (a) 190 N; (b) 0.56 m/s^2
6. 20°
7. (a) 0.13 N; (b) 0.12
8. (b) 240 N; (c) 0.60
9. (a) no; (b) $(-12\text{ N})\hat{\imath} + (5\text{ N})\hat{\jmath}$
10. (a) 6.0 N, to the left; (b) 3.6 N, to the left; (c) 3.1 N, to the left
12. (a) 110 N; (b) 130 N; (c) no; (d) 46 N; (e) 17 N
13. (a) 300 N; (b) 1.3 m/s^2
14. 0.53
15. (a) 66 N; (b) 2.3 m/s^2
16. (a) 11 N; (b) 0.14 m/s^2, rightward
17. (b) 3.0×10^7 N

18. (a) 8.6 N; (b) 46 N; (c) 39 N
19. 100 N
20. (a) 17 N, up the incline; (b) 20 N, up the incline; (c) 15 N, up the incline
21. (a) 0; (b) 3.9 m/s^2 down the incline; (c) 1.0 m/s^2 down the incline
22. 3.3 kg
23. (a) 3.5 m/s^2; (b) 0.21 N; (c) blocks move independently
24. 8.5 N
25. 490 N
26. (a) 1.05 N, in tension; (b) 3.62 m/s^2 down the plane; (c) answers are the same except that the rod is under compression
27. (a) 6.1 m/s^2, leftward; (b) 0.98 m/s^2, leftward
28. (a) 3.0×10^5 N; (b) 1.2°
29. $g(\sin\theta - \sqrt{2}\mu_k \cos\theta)$
30. (a) 19°; (b) 3300 N
31. 9.9 s
32. (a) 320 km/h; (b) 650 km/h, no
33. 6200 N
34. 3.75
35. 2.3
36. 9.7g
37. about 48 km/h
38. (a) 3.7 kN, up; (b) 1.3 kN, down
39. 21 m
40. (a) 3.7 kN, up; (b) 2.3 kN, down
41. (a) $\sqrt{Mgr/m}$
42. (a) 275 N; (b) 877 N
43. (a) light; (b) 778 N; (c) 223 N
44. 12°
45. 2.2 km
46. (a) 210 N; (b) 44 m/s
47. (b) 8.74 N; (c) 37.9 N, radially inward; (d) 6.45 m/s
48. (a) 70 km/h; (b) 140 km/h; (c) yes

Chapter 7

1. 1.2×10^6 m/s
2. (a) 5×10^{14} J; (b) 0.1 megaton TNT; (c) 8 bombs
3. (a) 3610 J; (b) 1900 J; (c) 1.1×10^{10} J
4. (a) 2.4 m/s; (b) 4.8 m/s
5. (a) 2.9×10^7 m/s; (b) 2.1×10^{-13} J
6. 5000 J
7. (a) 590 J; (b) 0; (c) 0; (d) 590 J
8. 0.32 J
9. (a) 170 N; (b) 340 m; (c) -5.8×10^4 J; (d) 340 N, (e) 170 m; (f) -5.8×10^4 J
10. 530 J
11. (a) 1.50 J; (b) increases
12. 20 J
13. 15.3 J
14. (a) 36 kJ; (b) 200 J
15. (a) 98 N; (b) 4.0 cm; (c) 3.9 J; (d) −3.9 J
16. (a) 270 N; (b) −400 J; (c) 400 J; (d) 0; (e) 0
17. (a) 1.2×10^4 J; (b) -1.1×10^4 J; (c) 1100 J; (d) 5.4 m/s
18. (a) 8840 J; (b) 7840 J; (c) 6840 J
19. (a) $-3Mgd/4$; (b) Mgd; (c) $Mgd/4$; (d) $\sqrt{gd/2}$
20. 1250 J
21. (a) −0.043 J; (b) −0.13 J
22. (a) 0.29 J; (b) −1.8 J; (c) 3.5 m/s; (d) 23 cm
23. (a) 6.6 m/s; (b) 4.7 m
24. 25 J
25. 800 J
26. (a) 12 J; (b) 4.0 m; (c) 18 J
27. 0, by both methods
28. (a) 2.3 J; (b) 2.6 J
29. −6 J
30. 270 kW
31. 490 W
32. (a) 28 W; (b) (6 m/s)$\hat{j}$
33. (a) 0.83 J; (b) 2.5 J; (c) 4.2 J; (d) 5.0 W
34. (a) 900 J; (b) 113 W; (c) 225 W
35. 740 W
36. (a) 0; (b) −350 W
37. 68 kW
38. (a) 1.7 W; (b) 0; (c) −1.7 W
39. (a) 1.8×10^5 ft · lb; (b) 0.55 hp
40. (a) 100 J; (b) 67 W; (c) 33 W

Chapter 8

1. 89 N/cm
2. (a) 167 J; (b) −167 J; (c) 196 J; (d) 29 J (e) 167 J; (f) −167 J; (g) 296 J; (h) 129 J
3. (a) 4.31 mJ; (b) −4.31 mJ; (c) 4.31 mJ; (d) −4.31 mJ; (e) all increase
4. (a) 0; (b) $mgh/2$; (c) mgh; (d) $mgh/2$; (e) mgh; (f) increase
5. (a) mgL; (b) $-mgL$; (c) 0; (d) $-mgL$; (e) mgL; (f) 0; (g) same
6. (a) $4mgR$; (b) $3mgR$; (c) $5mgR$; (d) mgR; (e) $2mgR$; (f) all the same

7. (a) 184 J; (b) −184 J; (c) −184 J
8. (a) $mgL(1-\cos\theta)$; (b) $-mgL(1-\cos\theta)$; (c) $mgL(1-\cos\theta)$; (d) all increase
9. (a) 2.08 m/s; (b) 2.08 m/s; (c) increase
10. (a) 12.9 m/s; (b) 12.9 m/s; (c) increase
11. (a) $\sqrt{2gL}$; (b) $2\sqrt{gL}$; (c) $\sqrt{2gL}$; (d) all the same
12. (a) v_0; (b) $(v_0^2+gh)^{1/2}$; (c) $(v_0^2+2gh)^{1/2}$; (d) $v_0^2/2g+h$; (e) all the same
13. (a) 260 m; (b) same; (c) decrease
14. (a) 2.29 m/s; (b) same
15. (a) 21.0 m/s; (b) 21.0 m/s; (c) 21 m/s
16. (a) 784 N/m; (b) 62.7 J; (c) 62.7 J; (d) 80.0 cm
17. (a) 0.98 J; (b) −0.98 J; (c) 3.1 N/cm;
18. (a) $\sqrt{v_0^2+2gL(1-\cos\theta_0)}$; (b) $\sqrt{2gL\cos\theta_0}$; (c) $\sqrt{gL(3+2\cos\theta_0)}$; (d) both decrease
19. (a) 39.2 J; (b) 39.2 J; (c) 4.00 m
20. (a) $8mg$ leftward; (b) mg downward; (c) $2.5R$
21. (a) 35 cm (b) 1.7 m/s
22. −320 J
23. (a) 4.8 m/s; (b) 2.4 m/s
24. (a) 4.4 m; (b) same
25. 10 cm
26. (a) no; (b) 930 N
27. 1.25 cm
28. (a) 7.2 J; (b) −7.2 J; (c) 86 cm; (d) 26 cm
30. (a) 1.4 m/s; (b) 1.9 m/s; (c) 28°
31. (a) $2\sqrt{gL}$: (b) $5mg$; (c) 71°
32. (a) 2.40 m/s; (b) 4.19 m/s
33. $mgL/32$
34. (a) $U_g=0$, $U_e=12.5$ J, $K=0$; (b) $U_g=2.0$ J, $U_e=8.0$ J, $K=2.5$ J; (c) $U_g=4.0$ J, $U_e=4.5$ J, $K=4.0$ J; (d) $U_g=6.0$ J, $U_e=2.0$ J, $K=4.5$ J; (e) $U_g=8.0$ J, $U_e=0.50$ J, $K=4.0$ J; (f) $U_g=10$ J, $U_e=0$; $K=2.5$ J; (g) $U_g=12$ J, $U_e=0$; $K=0.50$ J; (h) 31.3 cm
36. (a) 4.8 N, positive x direction; (b) between $x\approx 1.5$ m and $x\approx 13.5$ m; (c) 3.5 m/s
37. (a) $1.12(A/B)^{1/6}$; (b) repulsive; (c) attractive
38. (a) −3.73 J; (c) 1.29 m; (d) 9.12 m; (e) 2.16 J; (f) 4.0 m; (g) $F(x)=(4-x)e^{-x/4}$ N; (h) 4.0 m
39. (a) 5.6 J; (b) 3.5 J
40. 25 J
41. (a) 30.1 J; (b) 30.1 J; (c) 0.22;
42. (a) 560 J; (b) 560 J
43. (a) −2900 J; (b) 390 J; (c) 210 N
44. (a) −3800 J; (b) 31 kN
45. 11 kJ
46. 0.53 J
47. 20 ft · lb
48. (a) 2700 MJ; (b) 2700 MW; (c) 240 M$
49. (a) 1.5 MJ; (b) 0.51 MJ; (c) 1.0 MJ; (d) 63 m/s
50. 0.15
51. (a) 67 J; (b) 67 J; (c) 46 cm
52. 75 J
53. (a) 31.0 J; (b) 5.35 m/s; (c) conservative
54. 4.3 m
55. (a) 44 m/s; (b) 0.036
56. (a) 150 J; (b) 5.5 m/s
57. (a) −0.90 J; (b) 0.46 J; (c) 1.0 m/s
58. (a) 5.5 m/s; (b) 5.4 m; (c) same
59. 1.2 m
60. (a) 13 cm.; (b) 2.7 m; (c) both increase
62. (a) 10 m; (b) 50 N; (c) 4.1 m; (d) 120 N
63. in the center of the flat part
64. (a) 7.4 m/s; (b) 90 cm; (c) 2.8 m; (d) 15 m
65. (a) 216 J; (b) 1180 N; (c) 432 J; (d) motor also supplies thermal energy to crate and belt
66. (a) 3.0 mm; (b) 1.1 J; (d) yes; (e) 37 J; (f) no
67. (b) $\rho(L-x)/2$; (c) $v=v_0[2(\rho L+m_f)/(\rho L+2m_f-\rho x)]^{1/2}$; (e) 35 m/s

---

Chapter 9

1. (a) 4600 km; (b) $0.73R_e$
2. 6.46×10^{-11} m, along the line joining the atoms
3. (a) 1.1 m; (b); 1.3 m; (c) shifts toward topmost particle
4. $0.2L$ down from the heavy rod, along symmetry axis
5. −0.25 m; (b) 0
6. in the iron, at midheight and midwidth, 2.7 cm from midlength
7. 6.8×10^{-12} m toward the nitrogen atom, along axis of symmetry
8. 20 cm; (b) 20 cm; (c) 16 cm

9. (a) $H/2$; (b) $H/2$; (c) descends to lowest point and then ascends to $H/2$; (d) $\frac{HM}{m}\left(\sqrt{1+\frac{m}{M}}-1\right)$
10. 6.2 m
11. 72 km/h
12. (a) down, $mv/(m+M)$; (b) balloon again stationary
13. (a) 28 cm; (b) 2.3 m/s
14. (a) 22 m; (b) 9.3 m/s
15. 53 m
16. $(-4.0\text{ m})\hat{\imath}+(4.0\text{ m})\hat{\jmath}$
17. (a) half way between the containers; (b) 26 mm toward the heavier container; (c) down; (d) $-1.6\times10^{-2}\text{ m/s}^2$
18. 58 kg
19. 4.2 m
20. (a) 52.0 km/h; (b) 28.8 km/h
21. 24 km/h
22. 4.9 kg · m/s
23. (a) 7.5×10^4 J; (b) 3.8×10^4 kg·m/s; (c) 38° south of east
24. (a) 30°; (b) $(-0.572\text{ kg}\cdot\text{m/s})\hat{\jmath}$
25. (a) $(-4.0\times10^4\text{ kg}\cdot\text{m/s})\hat{\imath}$; (b) west; (c) 0
26. (a) 5.0 kg · m/s; (b) 10 kg · m/s
27. 3.0 mm/s, away from the stone
28. 0.57 m/s, toward center of mass
29. increases by 4.4 m/s
30. $-1.4\text{ m/s})\hat{\imath}$
31. 4400 km/h
32. $wv_{\text{rel}}/(W+w)$
33. (a) 7290 m/s; (b) 8200 m/s; (c) 1.271×10^{10} J; (d) 1.275×10^{10} J
34. 3.5 m/s
35. (a) 1.4×10^{-22} kg · m/s; (b) 150°; (c) 120°; (d) 1.6×10^{-19} J
36. (a) 20 J; (b) 40 J
37. (a) 1014 m/s, 9.48° clockwise from the positive x direction; (b) 3.23 MJ
38. $mv^2/6$
39. 14 m/s, 135° from the other pieces
40. one chunk stops, the other moves ahead with a speed of 4.0 m/s
41. 108 m/s m/s
42. (a) 8.0×10^4 N; (b) 27 kg/s
43. (a) 1.57×10^6 N; (b) 1.35×10^5 kg; (c) 2.08 km/s
44. (a) 2.72; (b) 7.39
45. 2.2×10^{-3}
46. 28.8 N
47. (a) 46 N; (b) none
48. (a) 50 kg/s; (b) 160 kg/s
49. (a) 0.2 to 0.3 MJ; (b) same amount
50. (a) 7.8 MJ; (b) 6.2
51. (a) 8.8 m/s; (b) 2600 J; (c) 1.6 kW
52. 5.5×10^6 N
53. 24 W
54. 100 m
55. (a) 860 N; (b) 2.4 m/s
56. (a) 3.0×10^5 J; (b) 10 kW; (c) 20 kW
57. (a) 2.1×10^6 kg; (b) $\sqrt{100+1.5t}$ m/s; (c) $(1.5\times10^6)/\sqrt{100+1.5t}$ N; (d) 6.7 km
58. 69 hp
59. 1.5 cm downward (the bubbles rise but the layers descend)

Chapter 10

1. 2.5 m/s
2. 6.2×10^4 N
3. 3000 N
4. (a) 25 to 30 kN; (b) 1000 to 1200 kg · m/s
5. 67 m/s, in opposite direction
6. (a) 1.1 m; (b) 4.8×10^3 kg · m/s
7. (a) 42 N · s; (b) 2100 N
8. 5 N
9. (a) $(7.4\times10^3\text{ N}\cdot\text{s})\hat{\imath}-(7.4\times10^3\text{ N}\cdot\text{s})\hat{\jmath}$; (b) $(-7.4\times10^3\text{ N}\cdot\text{s})\hat{\imath}$; (c) 2.3×10^3 N; (d) 2.1×10^4 N; (e) −45°
10. 2.9×10^3 N
11. 10 m/s
12. (a) 0.48 g; (b) 7200 N
13. (a) 1.0 kg · m/s; (b) 250 J; (c) 10 N; (d) 1700 N (e) answer for (c) includes time between pellet collisions
14. 990 N
15. 41.7 cm/s
16. (a) 3.7 m/s; (b) 1.3 N · s; (c) 180 N
17. (a) 1.8 N·s, upward in the figure; (b) 180 N, downward in the figure
18. (a) 1.95×10^5 kg · m/s for each direction of thrust; (b) backward: −50.9 MJ, forward: +66.1 MJ, sideways: +7.61 MJ
19. (a) 9.0 kg · m/s; (b) 3000 N; (c) 4500 N; (d) 20 m/s
20. (a) 1.81 m/s; (b) 5.0 m/s
21. 3.0 m/s
22. 310 m/s

23. $\approx 2\,\text{mm/y}$
24. (a) 2.7 m/s; (b) 1400 m/s
25. (a) 4.6 m/s; (b) 3.9 m/s; (c) 7.5 m/s
26. (a) 721 m/s; (b) 937 m/s
27. (a) $mR(\sqrt{2gh} + gt)$; (b) 5.06 kg
28. (a) +2.0 m/s; (b) −1.3 J; (c) +40 J; (d) got energy from some source, such as a small explosion
29. 1.18×10^4 kg
30. 7.3 cm
31. (a) $mv_i/(m + M)$; (b) $M/(m + M)$
32. 120 J
33. 25 cm
34. 0.33 m
35. (a) 1.9 m/s, to the right; (b) yes; (c) no, total kinetic energy would have increased
36. 0.22%
37. (a) 99 g; (b) 1.9 m/s; (c) 0.93 m/s
38. 38 km/s
39. 7.8%
40. (a) 2.47 m/s; (b) 1.23 m/s
41. (a) 1.2 kg; (b) 2.5 m/s
42. (a) 0.60 cm; (b) 4.9 cm; (c) 9.0 cm; (d) 0
43. (a) 100 g; (b) 1.0 m/s
44. $m_1/3$
45. (a) 1/3; (b) $4h$
46. (a) $(10\,\text{m/s})\,\hat{\imath} + (15\,\text{m/s})\,\hat{\jmath}$; (b) 500 J lost
47. (a) 4.15×10^5 m/s; (b) 4.84×10^5 m/s
48. (a) 250 m/s; (b) 433 m/s
49. (a) 41°; (b) 4.76 m/s; (c) no
50. (a) 117° from the final direction of B
51. 120°
52. (a) 1.9 m/s, 30° to initial direction, on opposite side from first ball; (b) no
53. (a) 6.9 m/s, 30° to positive x direction; (b) 6.9 m/s, −30° to positive x direction; (c) 2.0 m/s, in negative x direction
54. 103°
56. (a) 4.50×10^{-3} kg · m/s; (b) 0.529 kg · m/s; (c) stroke
57. (a) $5mg$; (b) $7mg$; (c) 5 m

Chapter 11

1. (a) $a + 3bt^2 - 4ct^3$; (b) $6bt - 12ct^2$
2. (a) 0.105 rad/s; (b) 1.75×10^{-3} rad/s; (c) 1.45×10^{-4} rad/s
3. (a) 5.5×10^{15} s; (b) 26
4. (a) 4.0 rad/s; (b) 28 rad/s; (c) 12 rad/s^2; (d) 6.0 rad/s^2; (e) 18 rad/s^2
5. (a) 2 rad; (b) 0; (c) 130 rad/s; (d) 32 rad/s^2; (e) no
6. (a) 4.0 m/s: (b) no
7. 11 rad/s
8. (a) 9000 rev/min^2; (b) 420 rev
9. (a) −67 rev/min^2; (b) 8.3 rev
10. (a) 30 s; (b) 1800 rad
11. 200 rev/min
12. (a) 2.0 rad/s^2; (b) 5.0 rad/s; (c) 10 rad/s; (d) 75 rad
13. 8.0 s
14. (a) 13.5 s; (b) 27.0 rad/s
15. (a) 44 rad; (b) 5.5 s, 32 s; (c) −2.1 s, 40 s
16. (a) 1.0 rev/s^2; (b) 4.8 s; (c) 9.6 s; (d) 48 rev
17. (a) 340 s; (b) -4.5×10^{-3} rad/s^2; (c) 98 s
18. (a) 12 rad/s; (b) 3.0 s
19. 1.8 m/s^2, toward the center
20. (a) 3.5 rad/s; (b) 52 cm/s; (c) 26 cm/s
21. 0.13 rad/s
22. (a) 20.9 rad/s; (b) 12.5 m/s; (c) 800 rev/min^2; (d) 600 rev
23. (a) 3.0 rad/s; (b) 30 m/s; (c) 6.0 m/s^2; (d) 90 m/s^2
24. (a) 2.50×10^{-3} rad/s; (b) 20.2 m/s^2; (c) 0
25. (a) 3.8×10^3 rad/s; (b) 190 m/s
26. (a) −1.1 rev/min^2; (b) 9900 rev; (c) −0.99 mm/s^2; (d) 31 m/s^2
27. (a) 7.3×10^{-5} rad/s; (b) 350 m/s; (c) 7.3×10^{-5} rad/s; (d) 460 m/s
28. (a) 40.2 cm/s^2; (b) 2.36×10^3 m/s^2; (c) 83.2 m
29. 16 s
30. (a) 6.4 cm/s^2; (b) 2.6 cm/s^2
31. (a) -2.3×10^{-9} rad/s^2; (b) 2600 y; (c) 24 ms
32. (a) 73 cm/s^2; (b) 0.075; (c) 0.11
33. 12.3 kg · m^2
34. 6.75×10^{12} rad/s
35. (a) 1100 J; (b) 9700 J
36. (a) 221 kg · m^2; (b) 1.10×10^4 J
37. (a) $5md^2 + \frac{8}{3}Md^2$; (b) $(\frac{5}{2}m + \frac{4}{3}M)d^2\omega^2$
38. (a) 6490 kg · m^2; (b) 4.36 MJ
39. 0.097 kg · m^2
40. (a) 2.0 kg · m^2; (b) 6.0 kg · m^2; (c) 2.0 kg · m^2
41. $\frac{1}{3}M(a^2 + b^2)$
42. (a) 1300 g · cm^2; (b) 550 g · cm^2; (c) 1850 g · cm^2; (d) $A + B$
44. (a) 49 MJ; (b) 100 min

45. 4.6 N · m
46. (a) 8.4 N · m; (b) 17 N · m; (c) 0
47. (a) $r_1F_1 \sin\theta_1 - r_2F_2 \sin\theta_2$; (b) −3.8 N · m
48. 12 N · m
49. (a) 28.2 rad/s^2; (b) 338 N · m
50. 1.28 kg · m^2
51. (a) 155 kg · m^2; (b) 64.4 kg
52. 9.7 rad/s^2, counterclockwise
53. 130 N
54. 4.6 rad/s^2
55. (a) 6.00 cm/s^2; (b) 4.87 N; (c) 4.54 N; (d) 1.20 rad/s^2; (e) 0.0138 kg · m^2
56. (a) 420 rad/s^2; (b) 500 rad/s
57. (a) 1.73 m/s^2; (b) 6.92 m/s^2
58. (a) 1.4 m/s; (b) 1.4 m/s
59. 396 N · m
60. (a) 19.8 kJ; (b) 1.32 kW
61. (a) $mL^2\omega^2/6$; (b) $L^2\omega^2/6g$
62. (a) 8.2×10^{28} N · m; (b) 2.6×10^{29} J; (c) 3.0×10^{21} kW
63. 5.42 m/s
64. (a) 0.15 kg · m^2; (b) 11 rad/s
65. $\sqrt{9g/4L}$
66. $\sqrt{\dfrac{2gh}{1+(2M/3m)+(I/mr^2)}}$
67. (a) $[(3g/H)(1-\cos\theta)]^1/2$; (b) $3g(1-\cos\theta)$; (c) $(3/2)g\sin\theta$; (d) 41.8°
68. (a) 5.1 h; (b) 8.1 h
69. 17
70. (a) 0.32 rad/s; (b) 100 km/h

Chapter 12

1. (a) 59.3 rad/s; (b) 9.31 rad/s^2; (c) 70.7 m
2. (a) 0; (b) 0; (c) 22 m/s, positive x direction; (d) 1500 m/s^2; (e) 22 m/s, negative x direction; (f) 1500 m/s^2; (g) 22 m/s, in $+x$ direction; (h) 0; (i) 44 m/s; (j) 1500 m/s^2, in $+x$ direction; (k) 0; (l) 1500 m/s^2
3. −3.15 J
4. 1.00
5. 1/50
6. (a) $\frac{1}{2}mR^2$; (b) a solid circular cylinder
7. (a) 8.0°; (b) more
8. (a) 4.0 N, to the left; (b) 0.60 kg · m^2
9. 4.8 m
10. (a) $mg(R-r)$; (b) 2/7; (c) $(17/7)mg$
11. (a) 63 rad/s; (b) 4.0 m
12. (a) $2.7R$; (b) $(50/7)mg$, to the left
13. (a) 8.0 J; (b) 3.0 m/s; (c) 6.9 J; (d) 1.8 m/s
14. (a) -0.11ω; (b) −2.1 m/s^2; (c) −47 rad/s^2; (d) 1.2 s; (e) 8.6 m; (f) 6.1 m/s
15. (a) 13 cm/s^2; (b) 4.4 s; (c) 55 cm/s; (d) 1.8×10^{-2} J; (e) 1.4 J; (f) 27 rev/s
16. (a) 0.89 s; (b) 9.4 J; (c) 1.4 m/s; (d) 0.12 J; (e) 440 rad/s; (f) 9.2 J
18. (a) 24 N · m, in positive y direction; (b) 24 N · m, in negative y direction; (c) 12 N · m, in positive y direction; (d) 12 N · m, in negative y direction
19. (a) 10 N · m, parallel to yz plane, at 53° to $+y$; (b) 22 N · m, $-x$
20. (a) $(-1.5\text{ N·m})\hat{\imath}-(4.0\text{ N·m})\hat{\jmath}-(1.0\text{ N·m})\hat{k}$; (b) $(-1.5\text{ N·m})\hat{\imath}-(4.0\text{ N·m})\hat{\jmath}-(1.0\text{ N·m})\hat{k}$
21. (a) $(50\text{ N·m})\hat{k}$; (b) 90°
22. (a) $(6.0\text{ N·m})\hat{\imath}-(3.0\text{ N·m})\hat{\jmath}-(6.0\text{ N·m})\hat{k})$; (b) $(26\text{ N·m})\hat{\imath}+(3.0\text{ N·m})\hat{\jmath}-(18\text{ N·m})\hat{k}$; (c) $(32\text{ N·m})\hat{\imath}-(24\text{ N·m})\hat{k})$; (d) 0
23. 9.8 kg · m^2/s
24. (a) 12 kg · m^2/s, out of page; (b) 3.0 N · m, out of page
25. (a) 0; (b) $(8.0\text{ N·m})\hat{\imath}+(8.0\text{ N·m})\hat{k}$
26. (a) $(600\text{ kg·m}^2/\text{s})\hat{k}$; (b) $(720\text{ kg·m}^2/\text{s})\hat{k}$
27. (a) mvd; (b) no; (c) 0, yes
28. (a) $(-32\text{ kg·m}^2/\text{s})\hat{k}$; (b) $(-32\text{ kg·m}^2/\text{s})\hat{k}$; (c) $(12\text{ N·m})\hat{k}$; (d) 0
29. (a) $(-170\text{ kg·m}^2/\text{s})\hat{k}$; (b) $(+56\text{ N·m})\hat{k}$; (c) $(+56\text{ kg·m}^2/\text{s})\hat{k}$
30. 4.5 N · m, parallel to xy plane at −63° from $+x$
31. (a) 0; (b) $(8t)$ N · m, in $-z$ direction; (c) $(2/\sqrt{t})$ N · m, in $-z$ direction; (d) $(8/t^3)$ N · m, in $+z$ direction
32. (a) $(-24t^2\text{ kg·m}^2/\text{s})\hat{k}$; (b) $(-48t\text{ N·m})\hat{k}$; (c) $(+12t^2\text{ kg·m}^2/\text{s})\hat{k}$, $(+24t\text{ N·m})\hat{k}$
33. (a) −1.47 N · m; (b) 20.4 rad; (c) −29.9 J; (d) 19.9 W
34. (a) 0.53 kg · m^2/s; (b) 4200 rev/min
35. (a) $14md^2$; (b) $4md^2\omega$; (c) $14md^2\omega$
37. $\omega_0R_1R_2I_1/(I_1R_2^2+I_2R_1^2)$
38. (a) 1.6 kg · m^2; (b) 4.0 kg · m^2/s
39. (a) 3.6 rev/s; (b) 3.0; (c) work done by man in moving bricks inward
40. 500 rev
41. (a) 267 rev/min; (b) 2/3

42. (a) 750rev/min; (b) 450 rev/min, in the original direction of the second disk
43. (a) 149kg · m^2; (b) 158kg · m^2/s; (c) 0.746rad/s
44. 3
45. $\frac{m}{M+m}\left(\frac{v}{R}\right)$
46. (a) they revolve in a circle of radius 1.5 m at an angular speed of 0.93 rad/s; (b) 98 J; (c) 8.4 rad/s; (d) 880 J; (e) energy is transferred from internal energy of the skaters to kinetic energy
47. (a) $(mRv - I\omega_0)/(I+mR^2)$; (b) no, energy transferred to internal energy of cockroach
48. (a) $mvR/(I+MR^2)$; (b) $mvR^2/(I+MR^2)$
49. 3.4 rad/s
50. 1300 m/s
51. (a) 0.148 rad/s; (b) 0.0123; (c) 181°
52. (a) $+0.14\omega_0$; (b) 1.14; (c) energy is transferred from internal energy of the cockroach to kinetic energy
53. the day would be longer by about 0.8 s
54. 2.6 rad/s
55. (a) 18 rad/s; (b) 0.92
56. 0.070 rad/s
57. (a) 0.24 kg · m^2; (b) 1800 m/s
58. 1.5 rad/s
59. $\theta = \cos^{-1}\left[1 - \frac{6m^2h}{d(2m+M)(3m+M)}\right]$

Chapter 13

1. (a) 2; (b) 7
2. (a) 2.5 m; (b) 7.3°
3. (a) $(-27\,\text{N})\hat{\imath} + (2\,\text{N})\hat{\jmath}$; (b) 176° counterclockwise from the positive x direction
4. 120°
5. 7920 N
6. (a) 840 N; (b) 530 N
7. (a) $(mg/L)\sqrt{L^2+r^2}$; (b) mgr/L
8. (a) 2770 N; (b) 3890 N
9. (a) 1160 N, down; (b) 1740 N, up; (c) left; (d) right
10. 8.3 kN
11. 74 g
12. 0.536 m
13. (a) 280 N; (b) 880 N, 71° above the horizontal
14. (a) $3W$, up; (b) $4W$, down
15. (a) 8010 N; (b) 3.65 kN; (c) 5.66 kN
16. (a) 49 N; (b) 28 N; (c) 57 N; (d) 29°
17. 71.7 N
18. (a) 1900 N, up; (b) 2100 N, down
19. (a) 5.0 N; (b) 30 N; (c) 1.3 m
20. (a) 408 N; (b) 245 N, to the right; (c) 163 N, up
21. $mg\frac{\sqrt{2rh-h^2}}{r-h}$
22. (a) 340 N; (b) 0.88 m; (c) increases, decreases
23. (a) 192 N; (b) 96.1 N; (c) 55.5 N
24. (a) $L/2$; (b) $L/4$; (c) $L/6$; (d) $L/8$; (e) $25L/24$
25. (a) 6630 N; (b) 5740 N; (c) 5960 N
26. (a) 130 N, up; (b) 80 N, away from door at top, toward door at bottom
27. 2.20 m
28. (a) $Wx/(L\sin\theta)$; (b) $Wx/(L\tan\theta)$; (c) $W(1-x/L)$
29. 0.34
30. (a) 1.50 m; (b) 433 N; (c) 250 N
31. (a) 211 N; (b) 534 N; (c) 320 N
32. (a) −797 N, 265 N; (b) 797 N, 265 N; (c) 797 N, 931 N; (d) −797 N, −265 N
33. (a) 445 N; (b) 0.50; (c) 315 N
34. (a) $a_1 = L/2$, $a_2 = 5L/8$, $h = 9L/8$; (b) $b_1 = 2L/3$, $b_2 = L/2$, $h = 7L/6$
35. (a) slides at 31°; (b) tips at 34°
36. (a) 7.5×10^{10} N/m^2; (b) 2.9×10^{8} N/m^2
37. (a) 6.5×10^{6} N/m^2; (b) 1.1×10^{-5} m
38. (a) 4/5; (b) 1/5; (c) 1/4
39. (a) 867 N; (b) 143 N; (c) 0.165
40. (a) 1.4×10^{9} N; (b) 75
41. (a) 51°; (b) $0.64Mg$
42. (a) 6.78 m^3; (b) 1.22×10^{5} N; (c) $\sigma_0 + (\sigma_m - \sigma_0)r/r_m = (40\,000 + 13.19r)$ N/m^2, with r in meters; (d) $2\pi r\,dr$; (e) $[(40\,000 + 13.19r)\,\text{N/m}^2]2\pi r\,dr$, with r in meters; (f) 1.04×10^{4} N; (g) −0.13

Chapter 14

1. 19 m
2. (a) 1×10^{-8} N; (b) 1×10^{-6} N; (c) 5×10^{-7} N; (d) no, the force due to the planet is 50 to 100 times greater
3. 29 pN
4. 2.16
5. 1/2

6. 3.4×10^5 km
7. 2.60×10^5 km
8. 2.1×10^{-8} N
9. 0.017 N, toward the 300-kg sphere
10. (a) $M = m$; (b) 0
11. 3.2×10^{-7} N
12. (a) 3.7×10^{-5} N, positive y direction
13. $\dfrac{GmM}{d^2}\left[1 - \dfrac{1}{8(1 - R/2d)^2}\right]$
14. 0.068 N
15. 2.6×10^6 m
16. (a) 17 N; (b) 2.5
17. (b) 1.9 h
18. (a) $7.6\,\mathrm{m/s^2}$; (b) $4.2\,\mathrm{m/s^2}$
20. (a) $a_g = (3.03 \times 10^{43}\,\mathrm{kg \cdot m/s^2})/M_h$; (b) decrease; (c) $9.82\,\mathrm{m/s^2}$; (d) $7.30 \times 10^{-15}\,\mathrm{m/s^2}$; (e) no
21. 4.7×10^{24} kg
22. (a) $G(M_1 + M_2)m/a^2$; (b) GM_1m/b^2; (c) 0
23. (a) $(3.0 \times 10^{-7}\,\mathrm{N/kg})m$; (b) $(3.3 \times 10^{-7}\,\mathrm{N/kg})m$; (c) $(6.7 \times 10^{-7}\,\mathrm{N/kg \cdot m})mr$
24. $R/3$ and $\sqrt{3}R$
25. (a) $9.83\,\mathrm{m/s^2}$; (b) $9.84\,\mathrm{m/s^2}$; (c) $9.79\,\mathrm{m/s^2}$
26. (a) -4.4×10^{-11} J; (b) -2.9×10^{-11} J; (c) 2.9×10^{-11} J
27. (a) -1.3×10^{-4} J; (b) less; (c) positive; (d) negative
28. 1/2
29. (a) 0.74; (b) $3.7\,\mathrm{m/s^2}$; (c) 5.0 km/s
30. (a) 0.0451; (b) 28.5
31. (a) 5.0×10^{-11} J; (b) -5.0×10^{-11} J
32. (a) 2.2×10^7 J; (b) 6.9×10^7 J
34. 2.4×10^4 m/s
35. (a) 1700 m/s; (b) 250 km; (c) 1400 m/s
36. (a) 3.8×10^7 J; (b) 1030 km
37. (a) 82 km/s; (b) 1.8×10^4 km/s
38. (a) -1.67×10^{-8} J; (b) 0.56×10^{-8} J
39. 2.5×10^4 km
40. 1.87 y
41. 6.5×10^{23} kg
42. 5.93×10^{24} kg
43. 5×10^{10}
44. 0.35 lunar months
45. (a) 7.82 km/s; (b) 87.5 min
46. (a) 5.01×10^9 m; (b) 7.20 solar radii
47. (a) 6640 km; (b) 0.0136
48. 3.58×10^4 km
49. (a) 1.9×10^{13} m; (b) $3.5R_P$
50. (a) 6×10^{16} kg; (b) $4 \times 10^3\,\mathrm{kg/m^3}$
52. 5.8×10^6 m
53. 0.71 y
54. $2\pi r^{3/2}/\sqrt{G(M + m/4)}$
55. $\sqrt{GM/L}$
56. (a) $-GM_Em/r$; (b) $-2GM_Em/r$; (c) it falls radially to Earth
57. (a) 2.8 y; (b) 1.0×10^{-4}
58. (a) 1/2; (b) 1/2; (c) B, by 1.1×10^8 J
60. (a) 54 km/s; (b) 960 m/s; (c) $R_p/R_a = v_a/v_p$
61. (a) no; (b) same; (c) yes
62. (a) 4.6×10^5 J; (b) 260
63. (a) 7.5 km/s; (b) 97 min; (c) 410 km; (d) 7.7 km/s; (e) 92 min; (f) 3.2×10^{-3} N; (g) no; (h) yes, if the satellite-Earth system is considered isolated
64. 1.1 s

Chapter 15

1. 1.1×10^5 Pa or 1.1 atm
2. 18 N
3. 2.9×10^4 N
4. (a) 190 kPa; (b) 15.9/10.6
5. 0.074
6. 38 kPa
7. 26 kN
8. 1.90×10^4 Pa
9. 5.4×10^4 Pa
10. 130 km
11. (a) 5.3×10^6 N; (b) 2.8×10^5 N; (c) 7.4×10^4 N; (d) no
12. (a) 6.06×10^9 N; (b) 20 atm
13. 7.2×10^5 N
14. 2.0
15. $\frac{1}{4}\rho g A(h_2 - h_1)^2$
16. 44 km
17. 1.7 km
18. (a) 5.0×10^6 N; (b) 5.6×10^6 N
19. (a) $\rho g W D^2/2$; (b) $\rho g W D^3/6$; (c) $D/3$
20. -3.9×10^{-3} atm
21. (a) 7.9 km; (b) 16 km
22. (a) fA/a; (b) 103 N
23. 4.4 mm
24. (a) 35.6 kN; (b) yes, decreases by $0.330\,\mathrm{m^3}$
25. (a) $2.04 \times 10^{-2}\,\mathrm{m^3}$; (b) 1570 N
26. (a) 37.5 kN; (b) 39.6 kN; (c) 2.23 kN; (d) 2.18 kN

27. (a) $670\,kg/m^3$; (b) $740\,kg/m^3$
28. 390 kg
29. (a) 1.2 kg; (b) $1300\,kg/m^3$
30. $1.5\,g/cm^3$
31. 57.3 cm
32. (a) 1.8 kg; (b) 2.0 kg
33. $0.126\,m^3$
34. $0.12\left(\frac{1}{\rho}-\frac{1}{8}\right)$ %, with ρ in g/cm^3
35. (a) $45\,m^2$; (b) car should be over center of slab if slab is to be level
36. five
37. (a) 9.4 N; (b) 1.6 N
38. (a) $1.80\,m^3$; (b) $4.75\,m^3$
39. 1.9 m/s
40. 4.0 m
41. 66 W
42. (a) 56 L/min; (b) 1.0
43. (a) 2.5 m/s; (b) 2.6×10^5 Pa
44. (a) 2.40 m/s; (b) 245 Pa
45. (a) 3.9 m/s; (b) 88 kPa
46. 1.7 MPa
47. (a) $1.6 \times 10^{-3}\,m^3/s$; (b) 0.90 m
49. 116 m/s
50. (a) 2; (b) $R_1/R_2 = 1/2$; (c) drain it until $h_2 = h_1/4$
51. (a) $6.4\,m^3$; (b) 5.4 m/s; (c) 9.8×10^4 Pa
52. (a) 3.1 m/s; (b) 9.5 m/s
53. (a) 74 N; (b) $150\,m^3$
54. (b) $H - h$; (c) $H/2$
55. (b) $2.0 \times 10^{-2}\,m^3/s$
56. (a) 4.1 m/s; (b) 21 m/s; (c) $8.0 \times 10^{-3}\,m^3/s$
57. (b) 63.3 m/s
58. 110 m/s
59. (a) 180 kN; (b) 81 kN; (c) 20 kN; (d) 0; (e) 78 kPa; (f) no
60. (a) $0.13v_s$; (b) $0.96v_s$ (almost supersonic; that is, almost faster than the speed of sound)
61. (a) 0.050; (b) 0.41; (c) no; (d) lay back on the surface, slowly pull your legs free, then roll over to the shore
62. (a) 1.5 m/s; (b) $d = R/2\pi v_1 r$; (c) decreases; (d) 42 μm; (e) 3.1 cm/s; (f) $10\,kJ/m^3$; (g) $0.49\,kJ/m^3$; (h) 18 Pa; (i) no (mechanical energy is not conserved by the jump)

Chapter 16

1. (a) 0.50 s (b) 2.0 Hz; (c) 18 cm
2. (a) 0.75 s; (b) 1.3 Hz; (c) 8.4 rad/s
3. (a) 0.50 s; (b) 2.00 Hz; (c) 12.6 rad/s; (d) 79.0 N/m; (e) 4.40 m/s; (f) 27.6 N
4. $37.8\,m/s^2$
5. $f > 500$ Hz
6. (a) 1.23 kN/m; (b) 76.0 N
7. (a) 6.28×10^5 rad/s; (b) 1.59 mm
8. (a) 10 N; (b) 120 N/m
9. (a) 1.0 mm; (b) 0.75 m/s; (c) $570\,m/s^2$
10. (a) 2800 rad/s; (b) 2.1 m/s; (c) $5.7\,km/s^2$
11. (a) 1.29×10^5 N/m; (b) 2.68 Hz
12. (a) 3.0 m; (b) −49 m/s; (c) $-270\,m/s^2$; (d) 20 rad; (e) 1.5 Hz; (f) 0.67 s
13. 7.2 m/s
14. (b) 12.47 kg; (c) 54.43 kg
15. 2.08 h
16. 22 cm
17. 3.1 cm
18. (a) 25 cm; (b) 2.2 Hz
19. (a) 5.58 Hz; (b) 0.325 kg; (c) 0.400 m
20. (a) 0.500 m; (b) −0.251 m; (c) 3.06 m/s
21. (a) 2.2 Hz; (b) 56 cm/s; (c) 0.10 kg; (d) 20.0 cm below y_i
22. $2\pi/3$ rad
23. (a) $0.183A$; (b) same direction
26. (a) $1.6 \times 10^4\,m/s^2$; (b) 2.5 m/s; (c) $7.9 \times 10^3\,m/s^2$; (d) 2.2 m/s
28. (a) 0.525 m; (b) 0.686 s
29. (a) $(n+1)k/n$; (b) $(n+1)k$; (c) $\sqrt{(n+1)/n}f$; (d) $\sqrt{n+1}f$
30. (a) 1.1 Hz; (b) 5.0 cm
31. 37 mJ
32. (a) 200 N/m; (b) 1.39 kg; (c) 1.91 Hz
33. (a) 2.25 Hz; (b) 125 J; (c) 250 J; (d) 86.6 cm
34. (a) 7.25×10^6 N/m; (b) 49,400
35. (a) 130 N/m; (b) 0.62 s; (c) 1.6 Hz; (d) 5.0 cm; (e) 0.51 m/s
36. (a) $mv/(m+M)$; (b) $mv/\sqrt{k(m+M)}$
37. (a) 3/4; (b) 1/4; (c) $x_m/\sqrt{2}$
38. (a) $-(80\,N)\cos[(2000\,rad/s)t - \pi/3\,rad]$; (b) 3.1 ms; (c) 4.0 m/s; (d) 0.080 J
39. (a) 16.7 cm; (b) 1.23%
40. (a) $0.735\,kg\cdot m^2$; (b) $0.024\,N\cdot m$; (c) 0.181 rad/s
41. (a) 39.5 rad/s; (b) 34.2 rad/s; (c) $124\,rad/s^2$

42. (a) 8.3 s; (b) no
43. 99 cm
44. 8.77 s
45. 5.6 cm
46. $2\pi\sqrt{\dfrac{R^2 + 2d^2}{2gd}}$
47. (a) $2\pi\sqrt{\dfrac{L^2 + 12d^2}{12gd}}$; (b) increases for $d < L/\sqrt{12}$, decreases for $d > L/\sqrt{12}$; (c) increases; (d) no change
48. (a) 0.869 s; (b) $r = R/2 = 6.25$ cm
49. (a) 0.205 kg · m^2; (b) 47.7 cm; (c) 1.50 s
50. (a) 1.64 s; (b) equal
52. (a) $2\pi\sqrt{\dfrac{L^2 + 12x^2}{12gx}}$; (b) 0.289
53. $2\pi\sqrt{m/3k}$
54. $\dfrac{1}{2\pi}\left(\dfrac{\sqrt{g^2 + v^4/R^2}}{L}\right)^{1/2}$
55. (a) 0.35 Hz; (b) 0.39 Hz; (c) 0
56. 14.0°
57. (b) smaller
58. (a) $(r/R)\sqrt{k/m}$; (b) $\sqrt{k/m}$; (c) no oscillation
59. 0.39
60. 6.0%
61. (a) 14.3 s; (b) 5.27
62. (a) 490 N/cm; (b) 1100 kg/s
63. (a) $F_m/b\omega_d$; (b) F_m/b
64. 5.0 cm

Chapter 17

1. (a) 3.49 m^{-1}; (b) 31.5 m/s
2. (a) 7.5×10^{14} Hz to 4.3×10^{14} Hz; (b) 1.0 to 200 m; (c) 6.0×10^{16} to 3.0×10^{19} Hz
3. (a) 0.68 s; (b) 1.47 Hz; (c) 2.06 m/s
4. $(0.010\,\text{m}) \sin \pi \left[(3.33\,\text{m}^{-1})x + (1100\,\text{s}^{-1})t\right]$
6. (a) 6.0 cm; (b) 100 cm; (c) 2.0 Hz; (d) 200 cm/s; (e) $-x$ direction; (f) 75 cm/s; (g) -2.0 cm
7. (a) $(2.0\,\text{cm}) \sin(0.63x - 2500t)$, where x is in centimeters and t is in seconds; (b) 50 m/s; (c) 40 m/s
8. (b) 2.0 cm/s; (c) $(4.0\,\text{cm}) \sin(\pi x/10 - \pi t/5 + \pi)$, where x is in cm and t is in s; (d) -2.5 cm/s
9. (a) 11.7 cm; (b) π rad
10. 3.2
11. 129 m/s
12. $\sqrt{2}$
13. (a) 15 m/s; (b) 0.036 N
14. (a) 30 m/s; (b) 17 g/m
15. $(0.12\,\text{mm}) \sin\left[(141\,\text{m}^{-1})x + (628\,\text{s}^{-1})t\right]$
16. 300 m/s
17. (a) $2\pi y_m/\lambda$; (b) no
18. (a) 0.64 Hz; (b) 63 cm; (c) $(5\,\text{cm}) \sin\left[(0.1\,\text{cm}^{-1})x - (4.0\,\text{s}^{-1})t\right]$; (d) 0.064 N
19. (a) 5.0 cm ; (b) 40 cm; (c) 12 m/s; (d) 0.033 s; (e) 9.4 m/s; (f) $(5.0\,\text{cm}) \sin(16x + 190t + 0.93)$, where x is in meters and t is in seconds
20. (a) 28.6 m/s; (b) 22.1 m/s; (c) 188 g; (d) 313 g
21. 2.63 m from the end of the wire from which the later pulse originates
22. (a) $\sqrt{k(\Delta\ell)(\ell + \Delta\ell)/m}$
24. 198 Hz
25. (a) 3.77 m/s; (b) 12.3 N; (c) zero; (d) 46.3 W; (e) zero; (f) zero; (g) ± 0.50 cm
26. (a) 82.8°; (b) 1.45 rad; (c) 0.23 wavelength
27. $1.4y_m$
28. (a) 0.31 m; (b) 1.64 rad; (c) 2.2 mm
29. 5.0 cm
30. 84°
31. (a) $0.83y_1$; (b) 37°
32. (a) $2f_3$; (b) λ_3
33. (a) 140 m/s; (b) 60 cm; (c) 240 Hz
34. 10 cm
35. (a) 82.0 m/s; (b) 16.8 m; (c) 4.88 Hz
36. (a) 66.1 m/s; (b) 26.4 Hz
37. 7.91 Hz, 15.8 Hz, 23.7 Hz
38. $f_{1A} = f_{4B}$; $f_{2A} = f_{8B}$
39. (a) 105 Hz; (b) 158 m/s
40. (b) the energy is entirely kinetic energy of the transversely moving sections of the flat string
41. (a) 0.25 cm; (b) 120 cm/s; (c) 3.0 cm; (d) zero
42. (a) 0.50 m; (b) 0, 0.25 s, 0.50 s
43. (a) 50 Hz; (b) $(0.50\,\text{cm}) \sin\left[(\pi\,\text{m}^{-1})x \pm (100\pi\,\text{s}^{-1})t\right]$
44. 36 N
45. (a) 1.3 m;

(b) $(0.002\,\text{m})\sin(9.4x)\cos(3800t)$, where x is in meters and t is in seconds
46. (a) 4.0 m; (b) 24 m/s; (c) 1.4 kg; (d) 0.11 s
47. (a) 2,0 Hz; (b) 200 cm; (c) 400 cm/s; (d) 50 cm, 150 cm, 250 cm, etc.; (e) 0, 100 cm, 200 cm, etc.
48. (a) 0, 0.20 m, 0.40 m; (b) 0.050 s; (c) 8.0 m/s; (d) 0.020 m; (e) 0, 0.025 s, 0.050 s
50. (a) +0.04 m; (b) 0; (c) 0; (d) −0.126 m/s
51. (a) 323 Hz; (b) eight
52. (a) 8.0 cm; (b) 1.0 cm

Chapter 18

1. divide the time by 3
2. the radio listener by about 0.85 s
3. (a) 79 m, 41 m; (b) 89 m
4. 170 m
5. 1900 km
6. (a) $L(V - v)/Vv$; (b) 364 m
7. 40.7 m
8. 17 m and 1.7 cm, respectively
9. (a) 0.0762 mm; (b) 0.333 mm
10. (a) 6.0 m/s; (b) $y = (0.30\,\text{cm})\sin(\pi x/12 + 50\pi t)$, with x in cm and t in s
11. (a) 1.50 Pa; (b) 158 Hz; (c) 2.22 m; (d) 350 m/s
12. (a) eight; (b) eight
13. (a) $343(1+2m)$ Hz, with m being an integer from 0 to 28; (b) $686m$ Hz, with m being an integer from 1 to 29
14. 4.12 rad
15. (a) 143 Hz, 429 Hz, 715 Hz; (b) 286 Hz, 572 Hz, 858 Hz
16. 17.5 cm
17. 15.0 mW
18. (a) $0.080\,\text{W/m}^2$; (b) $0.013\,\text{W/m}^2$
19. 36.8 nm
20. 1.26
21. (a) 1000; (b) 32
22. (a) $8.84\,\text{nW/m}^2$; (b) 39.5 dB
23. (a) 59.7; (b) 2.81×10^{-4}
24. $s_m \propto r^{-1/2}$
25. (b) $5.76 \times 10^{-17}\,\text{J/m}^3$
26. (a) 5000; (b) 71; (c) 71
27. (b) length^2
28. (a) $5.97 \times 10^{-5}\,\text{W/m}^2$; (b) 4.48 nW
29. (a) 5200 Hz; (b) $\text{amplitude}_{\text{SAD}}/\text{amplitude}_{\text{SBD}} = 2$
30. (a) 833 Hz; (b) 0.418 m
31. (a) 57.2 cm; (b) 42.9 cm
32. water filled to a height of 7/8, 5/8, 3/8, 1/8 meter
33. (a) 405 m/s; (b) 596 N; (c) 44.0 cm; (d) 37.3 cm
34. (a) 5.0 cm from one end; (b) 1.2; (c) 1.2
35. (a) 1129 Hz, 1506 Hz, and 1882 Hz
36. (a) $L(1 - 1/r)$; (b) 13 cm; (c) 5/6
37. 12.4 m
38. (a) 71.5 Hz; (b) 64.8 N
39. (a) node; (c) 22 s
40. (a) 0.20 m, 0.60 m, 1.0 m; (b) 0.60 m; (c) 143 Hz
41. 45.3 N
42. 2.25 ms
43. 387 Hz
44. (a) ten; (b) four
45. 0.02
46. zero
47. 17.5 kHz
48. 4.61 m/s
49. (a) 526 Hz; (b) 555 Hz
50. 0.195 MHz
51. (a) 1.02 kHz; (b) 1.04 kHz
52. (a) 1584 Hz; (b) 0.208 m; (c) 2160 Hz; (d) 0.152 m
53. 155 Hz
54. 41 kHz
55. (a) 485.8 Hz; (b) 500.0 Hz; (c) 486.2 Hz; (d) 500.0 Hz
56. (a) 2.0 kHz; (b) 2.0 kHz
57. (a) 598 Hz; (b) 608 Hz; (c) 589 Hz
58. 30°
59. (a) 42°; (b) 11 s
60. 33.0 km

Chapter 19

1. 0.05 kPa, nitrogen
2. 1.366
3. 348 K
4. (a) 320°F; (b) −12.3°F
5. (a) −40°; (b) 575°; (c) Celsius and Kelvin cannot give the same reading
6. (a) −96°F; (b) 56.7°C
7. (a) Dimensions are reciprocal time
8. 1/2

9. $-91.9°X$
10. 1.1 cm
11. 960 μm
12. (a) 9.996 cm; (b) 68°C
13. 2.731 cm
14. 49.87 cm^3
15. 29 cm^3
17. 0.26 cm^3
18. $23 \times 10^{-6}/C°$
19. 360°C
20. (a) −0.69 %; (b) aluminum
22. (a) 0.36 %; (b) 0.18 %; (c) 0.54 %; (d) 0.00 %; (e) $1.8 \times 10^{-5}/C°$
23. 0.68 s/h, fast
24. 0.217 K/s
25. 7.5 cm
26. 94.6 L
27. (a) 523 J/kg · K; (b) 26.2 J/mol · K; (c) 0.600 mol
28. 109 g
29. 42.7 kJ
30. 1.30 MJ
31. 1.9 times as great
32. 250 g
33. (a) 33.9 Btu; (b) 172 F°
34. (a) 52 MJ; (b) 0°C
35. 160 s
36. (a) 20,300 cal; (b) 1110 cal; (c) 873°C
37. 2.8 days
38. 3.0 min
39. 742 kJ
40. 73 kW
41. 82 cal
42. 33 m^2
43. 33 g
44. (a) 5.3°C, no ice remaining; (b) 0°C, 60 g of ice left
45. (a) 0°C; (b) 2.5°C
46. 13.5 C°
47. 8.72 g
48. (a) −200 J; (b) −293 J; (c) −93 J
49. A: 120 J, B: 75 J, C: 30 J
50. (a) A → B: + + +, B → C: + 0 +, C → A: − − −; (b) −20 J
51. −30 J
52. −5.0 J
53. (a) 6.0 cal; (b) −43 cal; (c) 40 cal; (d) 18 cal, 18 cal
54. 766°C
55. (a) 0.13 m; (b) 2.3 km
56. (a) 230 J/s; (b) heat flows out about 15 times as fast
57. 1660 J/s
58. (a) 800 W; (b) 2×10^4 J
59. (a) 16 J/s; (b) 0.048 g/s
60. arrangement (b)
61. 0.50 min
62. (a) 1230; (b) 2270; (c) 1040 W
63. (a) 17 kW/m^2; (b) 18 W/m^2
64. −4.2°C
65. 0.40 cm/h
66. 1.1 m
67. (a) 90 W; (b) 230 W; (c) 330 W
68. (a) $P_i = \sigma\epsilon T^4\left[a + 2h(\pi a)^{1/2}\right]$; (b) $P_h = \sigma\epsilon T^4\left[Na + 2h(\pi Na)^{1/2}\right]$; (d) 5; (e) 10; (f) 26; (g) 150; (h) 1900; (i) 0.13

Chapter 20

1. 0.933 kg
2. (a) 0.0127; (b) 7.65×10^{21}
3. 6560
4. number of molecules in the ink $\approx 3 \times 10^{16}$; number of people $\approx 5 \times 10^{20}$; statement is wrong, by a factor of about 20,000
5. (a) 5.47×10^{-8} mol; (b) 3.29×10^{16}
6. 25
7. (a) 0.0388 mol; (b) 220°C
8. 186 kPa
9. (a) 106; (b) 0.892 m^3
10. 653 J
11. $A(T_2 - T_1) - B(T_2^2 - T_1^2)$
12. 1/5
13. 5600 J
14. (a) 1.5 mol; (b) 1800 K; (c) 600 K; (d) 5.0 kJ
15. 100 cm^3
16. 22.8 m
17. 2.0×10^5 Pa
18. 2.50 km/s
19. 180 m/s
20. 442 m/s
21. 9.53×10^6 m/s
22. (a) 511 m/s; (b) −200°C; (c) 899°C
23. 1.9 kPa
24. (a) 494 m/s; (b) 28 g/mol, N_2
25. 3.3×10^{-20} J
26. (a) 5.65×10^{-21} J; (b) 7.72×10^{-21} J; (c) 3400 J; (d) 4650 J

27. (a) 6.75×10^{-20} J; (b) 10.7
30. 0.32 nm
31. (a) 6×10^9 km
32. 3.7 GHz
33. 15 cm
34. (a) 1.7; (b) 5.0×10^{-5} cm; (c) 7.9×10^{-6} cm
35. (a) 3.27×10^{10}; (b) 172 m
36. (a) 3.2 cm/s; (b) 3.4 cm/s; (c) 4.0 cm/s
37. (a) 6.5 km/s; (b) 7.1 km/s
38. (a) 420 m/s, 458 m/s
39. (a) 1.0×10^4 K; (b) 1.6×10^5 K; (c) 440 K, 7000 K; (d) hydrogen, no; oxygen, yes
40. 1.50
41. (a) 7.0 km/s; (b) 2.0×10^{-8} cm; (c) 3.5×10^{10} collisions/s
42. 4.7
43. (a) $2N/3v_0$; (b) $N/3$; (c) $1.22v_0$; (d) $1.31v_0$
44. 3400 J
45. $RT \ln(V_f/V_i)$
46. (a) 15.9 J; (b) 34.4 J/mol · K; (c) 26.1 J/mol · K
47. $(n_1C_1 + n_2C_2 + n_3C_3)/(n_1 + n_2 + n_3)$
48. (a) −5.0 kJ; (b) 2.0 kJ; (c) 5.0 kJ
49. (a) 6.6×10^{-26} kg; (b) 40 g/mol
50. 50 J
51. 8000 J
52. (a) 0.375 mol; (b) 1090 J; (c) 0.714
53. (a) 6980 J; (b) 4990 J; (c) 1990 J; (d) 2990 J
54. (a) 2.5 atm, 340 K; (b) 0.40 L
55. (a) 14 atm; (b) 620 K
56. 1500 $\text{N} \cdot \text{m}^{2.2}$
59. 1.40
60. (a) $p_0/3$; (b) polyatomic (ideal); (c) $K_f/K_i = 1.44$
61. (a) In joules, in the order Q, ΔE_{int}, W:
 $1 \to 2$: 3740, 3740, 0;
 $2 \to 3$: 0, −1810, 1810;
 $3 \to 1$: −3220, −1930, −1290;
 Cycle: 520, 0, 520;
 (b) $V_2 = 0.0246\ \text{m}^3$, $p_2 = 2.00$ atm, $V_3 = 0.0373\ \text{m}^3$, $p_3 = 1.00$ atm

---

Chapter 21

1. 14.4 J/K
2. 1.86×10^4 J
3. (a) 9220 J; (b) 23.0 J/K; (c) 0
4. 2.75 mol
5. (a) 5.79×10^4 J; (b) 173 J/K
6. (a) AE; (b) AC; (c) AF; (d) none
7. (a) 14.6 J/K; (b) 30.2 J/K
8. (a) 4500 J; (b) −5000 J; (c) 9500 J
9. (a) 57.0° C; (b) −22.1 J/K; (c) +24.9 J/K; (d) +2.8 J/K
10. (a) −710 mJ/K; (b) +710 mJ/K; (c) +723 mJ/K; (d) −723 mJ/K; (e) +13 mJ/K; (f) 0
12. (a) $p_1/3$, $p_1/3^{1.4}$, $T_1/3^{0.4}$;
 (b) in the order W, Q, ΔE_{int}, ΔS
 $1 \to 2$: $1.10RT_1$, $1.10RT_1$, 0, $1.10R$
 $2 \to 3$: 0, $-0.889RT_1$, $-0.889RT_1$, $-1.10R$
 $3 \to 1$: $-0.889RT_1$, 0, $0.889RT_1$, 0
13. (a) 320 K; (b) 0; (c) +1.72 J/K
14. (b) (I) constant T, $Q = pV \ln 2$; constant V, $Q = 4.5pV$; (II) constant T, $Q = -pV \ln 2$; constant p, $Q = 7.5pV$;
 (c) (I) constant T, $W = pV \ln 2$; constant V, $W = 0$; (II) constant T, $W = -pV \ln 2$; constant p, $W = 3pV$;
 (d) $4.5pV$ for either case;
 (e) $4R \ln 2$ for either case
15. +0.75 J/K
16. +0.64 J/K
17. (a) −943 J/K; (b) +943 J/K; (c) yes
18. (c) $Q_I = nRT_i \ln(V_f/V_i)$, $Q_{II} = \frac{3}{2}nR(T_f - T_x)$, no;
 (d) $\Delta S = \frac{3}{2}nR \ln(T_f/T_x)$, yes;
 (e) $T_x = 315$ K, $Q_I = 2880$ J, $Q_{II} = 2300$ J, $\Delta S = 5.76$ J/K
19. (a) $3p_0V_0$; (b) $\Delta E_{\text{int}} = 6RT_0$, $\Delta S = (3/2)R \ln 2$; (c) both are zero
20. (a) 1.84 kPa; (b) 441 K; (c) 3160 J; (d) 1.94 J/K
21. (a) 31%; (b) 16 kJ
22. 97 K
23. (a) 23.6%; (b) 1.49×10^4 J
24. 99.99995%
25. 266 K and 341 K
26. (a) 4.67 kJ/s; (b) 4.17 kJ/s
27. (a) 1470 J; (b) 554 J; (c) 918 J; (d) 62.4%
29. (a) 2270 J; (b) 14,800 J; (c) 15.4%; (d) 75.0%, greater
31. (a) 78%; (b) 81 kg/s
32. (a) monatomic; (b) 75%
33. (a) $T_2 = 3T_1$, $T_3 = 3T_1/4^{\gamma-1}$, $T_4 = T_1/4^{\gamma-1}$, $p_2 = 3p_1$, $p_3 = 3p_1/4^{\gamma}$, $p_4 = p_1/4^{\gamma}$; (b) $1 - 4^{1-\gamma}$

34. (a) 3.00; (b) 800 J
35. 21 J
36. 13 J
37. 440 W
38. (a) 0.071 J; (b) 0.50 J; (c) 2.0 J; (d) 5.0 J
39. 0.25 hp
40. 1.08 MJ
41. $[1-(T_2/T_1)]/[1-(T_4/T_3)]$
44. (a) 1.26×10^{14}; (b) 1.13×10^{15}; (c) 11.1%; (d) 1.01×10^{29}, 1.27×10^{30}, 8.0%; (e) 9.25×10^{58}, 1.61×10^{60}, 5.7%; (f) there are more microstates that can be occupied
45. (a) $W = N!/(n_1!\,n_2!\,n_3!)$; (b) $[(N/2)!\,(N/2)!]/[(N/3)!\,(N/3)!\,(N/3)!]$; (c) 4.2×10^{16}

Chapter 22

1. 1.38 mitem1.0.50 C
2. 2.81 N on each
3. (a) 4.9×10^{-7} kg; (b) 7.1×10^{-11} C
4. $3F/8$
5. (a) 0.17 N; (b) −0.046 N
6. (a) $q_1 = 9q_2$; (b) $q_1 = -25q_2$
7. either −1.00 μC and +3.00 μC or +1.00 μC and −3.00 μC
8. $q_1 = -4q_2$
9. (a) charge $-4q/9$ must be located on the line joining the two positive charges, a distance $L/3$ from charge $+q$
10. 14 cm from q_1, 24 cm from q_2
11. (a) 5.7×10^{13} C, no; (b) 6.0×10^5 kg
12. (a) 34.5 N, −10° from the positive direction of x; (b) $x = -8.4$ cm, $y = +2.7$ cm
13. $q = Q/2$
14. (a) $Q = -2\sqrt{2}q$; (b) no
15. (b) $\pm 2.4 \times 10^{-8}$ C
16. 3.1 cm
17. (a) $\frac{L}{2}\left(1 + \frac{1}{4\pi\epsilon_0}\frac{qQ}{Wh^2}\right)$; (b) $\sqrt{3qQ/4\pi\epsilon_0 W}$
18. 2.89×10^{-9} N
19. -1.32×10^{13} C
20. 0.19 MC
21. (a) 3.2×10^{-19} C; (b) two
22. (a) 8.99×10^{-19} N; (b) 625
23. 6.3×10^{11}
24. 5.1 m below first electron
25. 122 mA
26. 1.3×10^7 C
27. (a) 0; (b) 1.9×10^{-9} N
28. 1.7×10^8 N
29. (a) ^{9}B; (b) ^{13}N; (c) ^{12}C
30. (a) $F = (Q^2/4\pi\epsilon_0 d^2)\alpha(1-\alpha)$; (c) 0.5; (d) 0.15 and 0.85

Chapter 23

1. (a) 6.4×10^{-18} N; (b) 20 N/C
4. 0.111 nC
5. 56 pC
6. 6.4×10^5 N/C, toward the negative charge
7. 3.07×10^{21} N/C, radially outward
9. 50 cm from q_1 and 100 cm from q_2
10. (a) $x = 2.7d$
11. 0
12. $E = q/\pi\epsilon_0 a^2$, along bisector, away from triangle
13. 1.02×10^5 N/C, upward
15. 6.88×10^{-28} C·m
16. $(1/4\pi\epsilon_0)(p/r^3)$, antiparallel to $\vec{p}$
18. 0.51
20. $(1/4\pi\epsilon_0)(4q/\pi R^2)$, negative direction of y
21. $q/\pi^2\epsilon_0 r^2$, vertically downward
22. $R/\sqrt{2}$
23. (a) $-q/L$; (b) $q/4\pi\epsilon_0 a(L+a)$
26. 6.3×10^3 N/C
27. $R/\sqrt{3}$
28. 1.02×10^{-2} N/C, westward
29. 3.51×10^{15} m/s^2
30. 2.03×10^{-7} N/C, up
31. 6.6×10^{-15} N
32. (a) 4.8×10^{-13} N; (b) 4.8×10^{-13} N
33. (a) 1.5×10^3 N/C; (b) 2.4×10^{-16} N, up; (c) 1.6×10^{-26} N; (d) 1.5×10^{10}
34. (a) −0.029 C; (b) repulsive forces would explode the sphere
35. (a) 1.92×10^{12} m/s^2; (b) 1.96×10^5 m/s
36. (a) 7.12 cm; (b) 28.5 ns; (c) 11.2%
37. $-5e$
38. 1.64×10^{-19} C (≈ 3% high)
39. (a) 2.7×10^6 m/s; (b) 1000 N/C
40. (a) $(-2.1 \times 10^{13}\text{ m/s}^2)\,\hat{j}$; (b) $(1.5 \times 10^5\text{ m/s})\,\hat{i} - (2.8 \times 10^6\text{ m/s}\,\hat{j}$
41. 27 μm

42. (a) 0.245 N, 11.3° clockwise from the $+x$ axis; (b) $x = 108\,\text{m}$, $y = -21.6\,\text{m}$
43. (a) yes; (b) upper plate, 2.73 cm
44. (a) $9.30 \times 10^{-15}\,\text{C}\cdot\text{m}$; (b) $2.05 \times 10^{-11}\,\text{J}$
45. (a) 0; (b) $8.5 \times 10^{-22}\,\text{N}\cdot\text{m}$; (c) 0
46. $2pE\cos\theta_0$
47. $(1/2\pi)\sqrt{pE/I}$
48. (a) the field induces an electric dipole in a grain, which then moves toward a region of stronger electric field by moving toward the bee and then toward the stigma; if it were positively charged, it would not move to the bee; if it were negatively charged, it would move to the bee but not then to the stigma; (b) 1000 N/c; (c) no, because the grains would then fall off or be repelled off; (d) negative;

Chapter 24

1. (a) 693 kg/s; (b) 693 kg/s; (c) 347 kg/s; (d) 347 kg/s; (e) 575 kg/s
2. $-0.015\,\text{N}\cdot\text{m}^2/\text{C}$
3. (a) 0; (b) $-3.92\,\text{N}\cdot\text{m}^2/\text{C}$; (c) 0; (d) 0 for each field
4. (a) enclose $2q$ and $-2q$, or enclose all four charges; (b) enclose $2q$ and q; (c) not possible
5. $2.0 \times 10^5\,\text{N}\cdot\text{m}^2/\text{C}$
6. $-\pi a^2 E$
7. (a) $8.23\,\text{N}\cdot\text{m}^2/\text{C}$; (b) $8.23\,\text{N}\cdot\text{m}^2/\text{C}$; (c) 72.8 pC in each case
8. (a) $-1.3 \times 10^{-8}\,\text{C/m}^2$; (b) $8.2 \times 10^{10}\,em^3$
9. $3.54\,\mu\text{C}$
10. $-4.3\,\text{nC}$
11. 0 through each of the three faces meeting at q, $q/24\epsilon_0$ through each of the other faces
12. $2.0\,\mu\text{C/m}^2$
13. (a) $37\,\mu\text{C}$; (b) $4.1 \times 10^6\,\text{N}\cdot\text{m}^2/\text{C}$
14. (a) $4.5 \times 10^{-7}\,\text{C/m}^2$; (b) $5.1 \times 10^4\,\text{N/C}$
15. (a) $-3.0 \times 10^{-6}\,\text{C}$; (b) $+1.3 \times 10^{-5}\,\text{C}$
16. (a) $0.32\,\mu\text{C}$; (b) $0.14\,\mu\text{C}$
17. $5.0\,\mu\text{C/m}$
18. (a) $E = \lambda/2\pi\epsilon_0 r$; (b) 0
19. (a) $E = q/2\pi\epsilon_0 LR$, radially inward; (b) $-q$ on both inner and outer surfaces; (c) $E = q/2\pi\epsilon_0 Lr$, radially outward
20. $3.8 \times 10^{-8}\,\text{C/m}^2$
21. (a) $2.3 \times 10^6\,\text{N/C}$, radially out; (b) $4.5 \times 10^5\,\text{N/C}$, radially in
22. (a) 1.9 N/C; (b) 3.6 N/C
23. 3.6 nC
24. (a) 0.24 kN/C; (b) $-6.4\,\text{nC/m}^2$; (c) $+3.2\,\text{nC/m}^2$
25. (b) $\rho R^2/2\epsilon_0 r$
26. (a) $E = \sigma/\epsilon_0$, to the left; (b) $E = 0$; (c) $E = \sigma/\epsilon_0$, to the right
27. (a) $5.3 \times 10^7\,\text{N/C}$; (b) 60 N/C
28. $E = \dfrac{z\sigma}{2\epsilon_0 (z^2 + R^2)^{1/2}}$
29. $5.0\,\text{nC/m}^2$
30. (a) 0; (b) 0; (c) $7.9 \times 10^{-11}\,\text{N/C}$, leftward
31. 0.44 mm
32. $4.9 \times 10^{-10}\,\text{C}$
33. (a) $\rho x/\epsilon_0$; (b) $\rho d/2\epsilon_0$
34. (a) $-750\,\text{N}\cdot\text{m}^2/\text{C}$; (b) $-6.64\,\text{nC}$
35. $-7.5\,\text{nC}$
36. (a) $2.50 \times 10^4\,\text{N/C}$; (b) $1.35 \times 10^4\,\text{N/C}$
39. $-1.04\,\text{nC}$
40. (a) $-q$; (b) $+q$; (c) $E = q/4\pi\epsilon_0 r^2$, radially outward; (d) $E = 0$; (e) $E = q/4\pi\epsilon_0 r^2$, radially outward; (f) 0; (g) $E = q/4\pi\epsilon_0 r^2$, radially outward; (h) yes, charge is induced; (i) no; (j) yes; (k) no; (l) no
42. (a) $-e/\pi a_0^3$; (b) $5e[\exp(-2)]/4\pi\epsilon_0 a_0^2$, radially outward
43. (a) $E = (q/4\pi\epsilon_0 a^3)r$; (b) $E = q/4\pi\epsilon_0 r^2$; (c) 0; (d) 0; (e) inner, $-q$; outer, 0
45. $q/2\pi a^2$
47. $6K\epsilon_0 r^3$
48. (a) $E = |\rho|r/2\epsilon_0$; (b) increases; (c) inward; (d) $3 \times 10^6\,\text{N/C}$, at inside surface of pipe; (e) yes, along inside surface of pipe

Chapter 25

1. (a) $3.0 \times 10^5\,\text{C}$; (b) $3.6 \times 10^6\,\text{J}$
2. 1.2 GeV
3. (a) $3.0\times 10^{10}\,\text{J}$; (b) 7.7 km/s; (c) $9.0\times 10^4\,\text{kg}$
4. (a) 2.46 V; (b) 2.46 V; (c) 0
5. 8.8 mm
6. (a) $2.4 \times 10^4\,\text{V/m}$; (b) 2.9 kV
7. (a) 136 MV/m; (b) 8.82 kV/m
8. (a) $-qr^2/(8\pi\epsilon_0 R^3)$; (b) $q/(8\pi\epsilon_0 R)$; (c) center

9. (b) because $V = 0$ point is chosen differently; (c) $q/(8\pi\epsilon_0 R)$; (d) potential differences are independent of the choice for the $V = 0$ point
10. (b) $q_0\sigma z/2\epsilon_0$
11. (a) $Q/4\pi\epsilon_0 r$; (b) $\dfrac{\rho}{3\epsilon_0}\left(\dfrac{3}{2}r_2^2 - \dfrac{1}{2}r^2 - \dfrac{r_1^3}{r}\right)$,

    $\rho = \dfrac{Q}{\dfrac{4\pi}{3}(r_2^3 - r_1^3)}$;

    (c) $\dfrac{\rho}{2\epsilon_0}(r_2^2 - r_1^2)$, with ρ as in (b);

    (d) yes
12. $-1.1\,\text{nC}$
13. (a) $-4.5\,\text{kV}$; (b) $-4.5\,\text{kV}$
15. $x = d/4$ and $x = -d/2$
16. none
17. (a) 0.54 mm; (b) 790 V
18. (a) 3.3 nC; (b) $12\,\text{nC/m}^2$
19. 6.4×10^8 V
20. $0.94q/4\pi\epsilon_0 d$
21. $2.5q/4\pi\epsilon_0 d$
22. $16.3\,\mu\text{V}$
24. (a) $\dfrac{2\lambda}{4\pi\epsilon_0}\ln\left[\dfrac{L/2 + (L^2/4 + d^2)^{1/2}}{d}\right]$; (b) 0
25. (a) $-5Q/4\pi\epsilon_0 R$; (b) $-5Q/4\pi\epsilon_0(z^2 + R^2)^{1/2}$
26. $-Q/4\pi\epsilon_0 R$
27. $(\sigma/8\epsilon_0)\left[(z^2 + R^2)^{1/2} - z\right]$
28. $(Q/4\pi\epsilon_0 L)\ln(1 + L/d)$
29. $(c/4\pi\epsilon_0)[L - d\ln(1 + L/d)]$
30. 670 V/m
31. 17 V/m at 135° counterclockwise from $+x$
32. 39 V/m, in $-x$ direction
34. (a) $c/4\pi\epsilon_0\left[\sqrt{L^2 + y^2} - y\right]$;

    (b) $\dfrac{c}{4\pi\epsilon_0}\left[1 - \dfrac{y}{\sqrt{L^2 + y^2}}\right]$
35. (a) $\dfrac{Q}{4\pi\epsilon_0 d(d + L)}$, leftward; (b) 0
36. (a) 1.15×10^{-19} J; (b) decreases
37. $-0.21q^2/\epsilon_0 a$
38. -1.2×10^{-6} J
39. (a) $+6.0 \times 10^4$ V; (b) -7.8×10^5 V; (c) 2.5 J; (d) increase; (e) same; (f) same
40. 0
41. $W = \dfrac{qQ}{8\pi\epsilon_0}\left(\dfrac{1}{r_1} - \dfrac{1}{r_2}\right)$
42. (a) 27.2 V; (b) -27.2 eV; (c) 13.6 eV; (d) 13.6 eV
43. 2.5 km/s
44. 1.8×10^{-10} J
45. (a) 0.225 J; (b) A, $45.0\,\text{m/s}^2$; B, $22.5\,\text{m/s}^2$; (c) A, 7.75 m/s, B, 3.87 m/s
46. $\sqrt{2eV/m}$
47. 0.32 km/s
48. 1.48×10^7 m/s
49. 1.6×10^{-9} m
50. 400 V
51. 2.5×10^{-8} C
52. (a) $V_1 = V_2$; (b) $q_1 = q/3$, $q_2 = 2q/3$; (c) 2
53. (a) -180 V; (b) 2700 V, -8900 V
54. (a) 12,000 N/C; (b) 1800 V; (c) 5.8 cm
55. (a) -0.12 V; (b) 1.8×10^{-8} N/C, radially inward
56. $r < R_1$: $E = 0$, $V = \dfrac{1}{4\pi\epsilon_0}\left(\dfrac{q_1}{R_1} + \dfrac{q_2}{R_2}\right)$;

    $R_1 < r < R_2$: $E = q_1/4\pi\epsilon_0 r^2$,

    $V = \dfrac{1}{4\pi\epsilon_0}\left(\dfrac{q_1}{r} + \dfrac{q_2}{R_2}\right)$;

    $r > R_2$: $E = (q_1 + q_2)/4\pi\epsilon_0 r^2$,

    $V = (q_1 + q_2)/4\pi\epsilon_0 r$
57. (a) $V = \rho(R^2 - r^2)/4\epsilon_0$; (b) 78 kV

Chapter 26

1. 7.5 pC
2. (a) 3.5 pF; (b) 3.5 pF; (c) 57 V
3. 3.0 mC
5. (a) 140 pF; (b) 17 nC
6. 8.85×10^{-12} m
7. $5.04\pi\epsilon_0 R$
8. (a) 84.5 pF; (b) $191\,\text{cm}^2$
10. $7.33\,\mu\text{F}$
11. 9090
12. 315 mC
13. $3.16\,\mu\text{F}$
14. (a) 7.9×10^{-4} C; (b) 79 V
16. (a) $100\,\mu\text{C}$; (b) $20\,\mu\text{C}$
17. 43 pF
18. (a) $3\,\mu\text{F}$; (b) $60\,\mu\text{C}$; (c) 10 V, $30\,\mu\text{C}$; (d) 10 V, $20\,\mu\text{C}$; (e) 5.0 V, $20\,\mu\text{C}$
19. (a) 50 V; (b) 5.0×10^{-5} C; (c) 1.5×10^{-4} C
20. (a) $q_1 = 9.0\,\mu\text{C}$, $q_2 = 16\,\mu\text{C}$, $q_3 = 9.0\,\mu\text{C}$, $q_4 = 16\,\mu\text{C}$; (b) $q_1 = 8.4\,\mu\text{C}$, $q_2 = 17\,\mu\text{C}$, $q_3 = 11\,\mu\text{C}$, $q_4 = 14\,\mu\text{C}$
21. $q_1 = \dfrac{C_1C_2 + C_1C_3}{C_1C_2 + C_1C_3 + C_2C_3}C_1V_0$,

$$q_2 = q_3 = \frac{C_2C_3}{C_1C_2 + C_1C_3 + C_2C_3} C_1V_0$$

22. 99.6 nJ
23. 72 F
24. (a) 35 pF; (b) 21 nC; (c) 6.3 μJ; (d) 0.60 MV/m; (e) 1.6 J/m^3
25. 0.27 J
26. 10.4 cents
27. (a) 2.0 J
28. (a) $q_1 = 0.21$ mC, $q_2 = 0.11$ mC, $q_3 = 0.32$ mC; (b) $V_1 = V_2 = 21$ V, $V_3 = 79$ V; (c) $U_1 = 2.2$ mJ, $U_2 = 1.1$ mJ, $U_3 = 13$ mJ
29. (a) $2V$; (b) $U_i = \epsilon_0 AV^2/2d$, $U_f = 2U_i$; (c) $\epsilon_0 AV^2/2d$
30. (a) $q_1 = q_2 = 0.33$ mC, $q_3 = 0.40$ mC; (b) $V_1 = 33$ V, $V_2 = 67$ V, $V_3 = 100$ V; (c) $U_1 = 5.6$ mJ, $U_2 = 11$ mJ, $U_3 = 20$ mJ
32. 0.11 J/m^3
34. 4.0
35. Pyrex
36. (a) 6.2 cm; (b) 280 pF
37. 81 pF/m
38. (a) 0.73 nF; (b) 28 kV
39. 0.63 m^2
42. $\dfrac{\epsilon_0 A}{4d}\left(\kappa_1 + \dfrac{2\kappa_2\kappa_3}{\kappa_2+\kappa_3}\right)$
43. (a) 10 kV/m; (b) 5.0 nC; (c) 4.1 nC
44. (a) 13.4 pF; (b) 1.15 nC; (c) 1.13×10^4 N/C; (d) 4.33×10^3 N/C
45. (a) $C = 4\pi\epsilon_0\kappa\left(\dfrac{ab}{b-a}\right)$; (b) $q = 4\pi\epsilon_0\kappa V\left(\dfrac{ab}{b-a}\right)$; (c) $q' = q\left(1 - \dfrac{1}{\kappa}\right)$
46. (a) 7.1; (b) 0.77 μC
48. (a) 4.9 mJ; (b) no

Chapter 27

1. (a) 1200 C; (b) 7.5×10^{21}
2. 6.7 μC/m^2
3. 5.6 ms
4. (a) 2.4×10^{-5} A/m^2; (b) 1.8×10^{-15} m/s
5. (a) 6.4 A/m^2, north; (b) no, cross-sectional area
6. 14–gauge
7. 0.38 mm
8. (a) 0.654 μA/m^2; (b) 83.4 MA
9. (a) 2×10^{12}; (b) 5000; (c) 10 MV
10. (a) $J_0A/3$; (b) $2J_0A/3$
11. 13 min
12. 2.0×10^6 $(\Omega \cdot \text{m})^{-1}$
13. 2.0×10^{-8} $\Omega \cdot$ m
14. 0.536 Ω
15. 100 V
16. (a) 1.53 kA; (b) 54.1 MA/m^2; (c) 10.6×10^{-8} $\Omega \cdot$ m, platinum
17. 2.4 Ω
18. (a) 250 °C; (b) yes
19. 54 Ω
20. $2R$
21. 3.0
22. (a) 6.00 mA; (b) 1.59×10^{-8} V; (c) 21.2 nΩ
23. 8.2×10^{-4} $\Omega \cdot$ m
24. (a) 38.3 mA; (b) 109 A/m^2; (c) 1.28 cm/s; (d) 227 V/m
25. 2000 K
26. (a) 1.73 cm/s; (b) 3.24 pA/m^2
27. (a) 0.43%, 0.0017%, 0.0034%
28. (a) 0.40 Ω
29. (a) $R = \rho L/\pi ab$
31. 560 W
32. 14 kC
33. (a) 1.0 kW; (b) 25 ¢
34. 11.1 Ω
35. 0.135 W
36. (a) 28.8 Ω; (b) 2.60×10^{19} s^{-1}
37. (a) 10.9 A; (b) 10.6 Ω; (c) 4.5 MJ
38. (a) 5.85 m; (b) 10.4 m
39. 660 W
55. (a) $4.46 for a 31-day month; (b) 144 Ω; (c) 0.833 A
41. (a) 3.1×10^{11}; (b) 25 μA; (c) 1300 W, 25 MW
42. (a) 1.3×10^5 A/m^2; (b) 94 mV
43. (a) 17 mV/m; (b) 243 J
44. (a) $i = \rho\pi R^2 v$; (b) 17 μA; (c) no, current is perpendicular to the radial potential difference; (d) 1.3 W; (e) 260 mJ; (f) exit of the pipe into the silo
45. (a) $J = I/2\pi r^2$; (b) $E = \rho I/2\pi r^2$; (c) $\Delta V = \dfrac{\rho I}{2\pi}\left(\dfrac{1}{r} - \dfrac{1}{b}\right)$; (d) 0.16 A/$m^2$; (e) 16 V/m; (f) 0.16 MV

Chapter 28

1. (a) \$320; (b) 4.8 cents
2. 11 kJ
3. 14 h 24 min
4. (a) counterclockwise; (b) battery 1; (c) B
5. (a) 0.50 A; (b) $P_1 = 1.0$ W, $P_2 = 2.0$ W; (c) $P_1 = 6.0$ W supplied, $P_2 = 3.0$ W absorbed
6. (a) 80 J; (b) 67 J; (c) 13 J converted to thermal energy within battery
7. (a) 14 V; (b) 100 W; (c) 600 W; (d) 10 V, 100 W
8. (c) third plot gives rate of energy dissipation by R
9. (a) 50 V; (b) 48 V; (c) B is connected to the negative terminal
10. −10 V
11. 2.5 V
12. (a) 990 Ω; (b) 9.9×10^{-4} W
13. 8.0 Ω
14. the cable
15. (a) $r_1 - r_2$; (b) battery with r_1
16. (a) 1000 Ω; (b) 300 mV; (c) 2.3×10^{-3}
18. 4.0 Ω and 12 Ω
19. 5.56 A
20. 4.50 Ω
21. $i_1 = 50$ mA, $i_2 = 60$ mA, $V_{ab} = 9.0$ V
22. 0.00, 2.00, 2.40, 2.86, 3.00, 3.60, 3.75, 3.94 A
23. (a) bulb 2; (b) bulb 1
24. $V_d - V_c = +0.25$ V, by all paths
25. 3d
26. (a) 2.50 Ω; (b) 3.13 Ω
27. nine
28. (a) 120 Ω; (b) $i_1 = 51$ mA, $i_2 = i_3 = 19$ mA, $i_4 = 13$ mA
29. (a) $R = r/2$; (b) $P_{max} = \mathcal{E}^2/2r$
30. (a) series: $2\mathcal{E}/(2r + R)$, parallel: $2\mathcal{E}/(r + 2R)$; (b) series; (c) parallel
31. (a) 0.346 W; (b) 0.050 W; (c) 0.709 W; (d) 1.26 W; (e) −0.158 W
32. (a) 19.5 Ω; (b) 0; (c) ∞; (d) 82.3 W, 57.6 W
33. (a) battery 1: 0.67 A down, battery 2: 0.33 A up, battery 3: 0.33 A up; (b) 3.3 V
34. 1.43 Ω
35. (a) Cu: 1.11 A, Al: 0.893 A; (b) 126 m
36. (a) 13.5 kΩ; (b) 1500 Ω; (c) 167 Ω; (d) 1480 Ω
37. 0.45 A
38. (a) 12.5 V; (b) 50 A
39. −3.0%
42. (a) circuit a: 70.9 mA, 4.70 V; circuit b: 55.2 mA, 4.86 V; (b) circuit a: 66.3 Ω; circuit b: 88.0 Ω
44. (a) 0.41τ; (b) 1.1τ
45. 4.6
46. (a) 2.52 s; (b) 21.6 μC; (c) 3.40 s
47. (a) 2.41 μs; (b) 161 pF
48. 0.72 MΩ
49. (a) 0.955 μC/s; (b) 1.08 μW; (c) 2.74 μW; (d) 3.82 μW
51. (a) 2.17 s; (b) 39.6 mV
52. 2.35 MΩ
53. (a) 1.0×10^{-3} C; (b) 1.0×10^{-3} A; (c) $V_C = 1.0 \times 10^3\, e^{-t}$ V, $V_R = 1.0 \times 10^3\, e^{-t}$ V; (d) $P = e^{-2t}$ W
54. 24.8 Ω to 14.9 kΩ
55. (a) at $t = 0$, $i_1 = 1.1$ mA, $i_2 = i_3 = 0.55$ mA; at $t = \infty$, $i_1 = i_2 = 0.82$ mA, $i_3 = 0$; (c) at $t = 0$, $V_2 = 400$ V; at $t = \infty$, $V_2 = 600$ V; (d) after several time constants ($\tau = 7.1$ s) have elapsed
56. (a) 8 V; (b) 5 mA; (c) no
57. (a) 13 ms to 13 s; (b) 225 mJ; (c) 9.4 ms to 9.4 s; (d) low; (e) a ground connection on a pole can be extended to the conducting portion of the car before the first crew member touches the car
58. (a) 3000 V; (b) 10 s; (c) 11 GΩ

Chapter 29

1. (a) 6.2×10^{-18} N; (b) 9.5×10^8 m/s^2; (c) remains equal to 550 m/s
2. (a) 9.56×10^{-14} N, 0; (b) 0.267°
3. (a) 400 km/s; (b) 835 eV
4. (a) $(6.2 \times 10^{-14}\ \text{N})\,\hat{k}$; (b) $(-6.2 \times 10^{-14}\ \text{N})\,\hat{k}$
5. (a) east; (b) 6.28×10^{14} m/s^2; (c) 2.98 mm
6. (a) 1.4×10^{-18} N; (b) 1.6×10^{-19} N; (c) 1.0×10^{-18} N
7. (a) 3.4×10^{-4} T, horizontal and to the left as viewed along $\vec{v}_0$; (b) yes, if its velocity is the same as the electron's velocity
8. (a) 3.75 km/s
9. 0.27 mT
10. $(-11.4\ \text{V/m})\,\hat{\imath} - (6.00\ \text{V/m})\,\hat{\jmath} + (4.80\ \text{V/m})\,\hat{k}$

11. 680 kV/m
12. 7.4 μV
13. (b) 2.84×10^{-3}
14. 38.2 cm/s
15. 21 μT
16. (a) 1.11×10^7 m/s; (b) 0.316 mm
17. (a) 2.05×10^7 m/s; (b) 467 μT; (c) 13.1 MHz; (d) 76.3 ns
18. 127 u
19. (a) 0.978 MHz; (b) 96.4 cm
20. (a) 2.60×10^6 m/s; (b) 0.109 μs; (c) 0.140 MeV; (d) 70 kV
22. 0.53 m
23. (a) 1.0 MeV; (b) 0.5 MeV
24. $R_d = \sqrt{2}R_p$; $R_\alpha = R_p$.
25. (a) 495 mT; (b) 22.7 mA; (c) 8.17 MJ
26. $\vec{v} = v_{0x}\,\hat{\imath} + v_{0y}\cos(\omega t)\,\hat{\jmath} - v_{0y}\sin(\omega t)\,\hat{k}$, where $\omega = eB/m$
27. (a) 0.36 ns; (b) 0.17 mm; (c) 1.5 mm
28. (a) 0.253 T; (b) 130 ns
29. (a) $-q$; (b) $\pi m/qB$
30. (a) 18 MHz; (b) 17 MeV
31. 240 m
32. (a) 8.5 MeV; (b) 0.80 T; (c) 34 MeV; (d) 24 MHz; (e) 34 MeV, 1.6 T, 34 MeV, 12 MHz
33. 28.2 N, horizontally west
34. 20.1 N
35. 467 mA, from left to right
36. $(-2.5 \times 10^{-3}\ \text{N})\,\hat{\jmath} + (0.75 \times 10^{-3}\ \text{N})\,\hat{k}$
37. 0.10 T, at 31° from the vertical
38. (a) 3.3×10^8 A; (b) 1.0×10^{17} W; (c) totally unrealistic
39. 4.3×10^{-3} N · m, negative y
40. (a) 0, 0.138 N, 0.138 N
43. $2\pi aiB\sin\theta$, normal to the plane of the loop (up)
45. (a) 540 Ω, connected in series with the galvanometer; (b) 2.52 Ω, connected in parallel
46. $qvaB/2$
47. 2.45 A
48. 2.08 GA
49. (a) 12.7 A; (b) 0.0805 N · m
50. (a) 0.184 A · m²; (b) 1.45 N · m
51. (a) 0.30 J/T; (b) 0.024 N · m
52. (a) 20 min; (b) 5.9×10^{-2} N · m
53. (a) 2.86 A · m²; (b) 1.10 A · m²
54. 0.335 A·m², 297° counterclockwise from the positive y direction, in the yz plane
55. (a) $(8.0 \times 10^{-4}\ \text{N·m})(-1.2\,\hat{\imath} - 0.90\,\hat{\jmath} + 1.0\,\hat{k})$; (b) -6.0×10^{-4} J
56. (a) $(6.00 \times 10^{-4}\ \text{A})y\,dy\,\hat{k}$; (b) $(18.8\ \mu\text{N})\,\hat{k}$
57. $-(0.10\ \text{V/m})\,\hat{k}$
58. (a) 77°; (b) 77°
59. −2.0 T

Chapter 30

1. (a) 3.3 μT; (b) yes
2. 12 nT
3. (a) 16 A; (b) west to east
4. along a line parallel to the wire and 4.0 mm from it
5. (a) $\mu_0 qvi/2\pi d$, antiparallel to i; (b) same magnitude, parallel to i
6. 0
7. 2 rad
8. $\dfrac{\mu_0 i}{4}\left(\dfrac{1}{R_1} - \dfrac{1}{R_2}\right)$, into page
9. $\dfrac{\mu_0 i\theta}{4\pi}\left(\dfrac{1}{b} - \dfrac{1}{a}\right)$, out of page
10. (a) 0; (b) $\mu_0 i/4R$, into the page; (c) same as (b)
18. $\sqrt{2}\mu_0 i/8\pi a$, into page
19. $(\mu_0 i/2\pi w)\ln(1 + w/d)$, up
20. 200 μT, into page
21. (a) it is impossible to have other than $B = 0$ midway between them; (b) 30 A
22. at all points between the wires, on a line parallel to them, at a distance $d/4$ from the wire carrying current i
23. 4.3 A, out of page
24. From the left: $(46.9\ \mu\text{N/m})\,\hat{\jmath}$, $(18.8\ \mu\text{N/m})\,\hat{\jmath}$, 0, $(-18.8\ \mu\text{N/m})\,\hat{\jmath}$, $(-46.9\ \mu\text{N/m})\,\hat{\jmath})$
25. 80 μT, up the page
26. $0.338\mu_0 i^2/a$, toward the center of the square
27. $0.791\mu_0 i^2/\pi a$, 162° counterclockwise from the horizontal
28. (b) 2.3 km/s
29. 3.2 mN, toward the wire
30. $+5\mu_0 i$
31. (a) $(-2.0\ \text{A})\mu_0$; (b) 0
34. 1: $(-2.0\ \text{A})\mu_0$; 2: $(-13\ \text{A})\mu_0$
35. $\mu_0 J_0 r^2/3a$
36. (a) 0.13 μT; (b) 0.14 μT
38. $3i/8$, into page

40. 5.71 mT
41. 0.30 mT
42. 108 m
43. (a) 533 μT; (b) 400 μT
46. 0.272 A
47. (a) 4.77 cm; (b) 35.5 μT
48. (a) 4; (b) 1/2
49. 0.47 A $\cdot$ m^2
50. $8\mu_0 Ni/5\sqrt{5}R$
51. (a) 2.4 A $\cdot$ m^2; (b) 46 cm
52. (b) ia^2
54. (b) $(0.060\,\text{A}\cdot\text{m}^2)\hat{\jmath}$; (c) $(9.6 \times 10^{-11}\,\text{T})\hat{\jmath}$, $(-4.8 \times 10^{-11}\,\text{T})\hat{\jmath}$
56. (a) $\dfrac{\mu_0 i}{4}\left(\dfrac{1}{a}+\dfrac{1}{b}\right)$, into page;
    (b) $\frac{1}{2}i\pi(a^2+b^2)$, into page
57. (a) 79 μT; (b) 1.1×10^{-6} N $\cdot$ m
58. (a) $(\mu_0 i/2R)(1+1/\pi)$, directly out of page;
    (b) $(\mu_0 i/2\pi R)(\sqrt{1+\pi^2}$, out of the page at 18° to the page

---

Chapter 31

1. 1.5 mV
2. $-\mu_0 n A i_0 \omega \cos \omega t$
3. (a) 31 mV; (b) right to left
4. (a) -11 mV; (b) 0; (c) 11 mV
5. (a) $1.1 \times 10^{-3}\,\Omega$; (b) 1.4 T/s
6. (b) 58 mA
7. 30 mA
8. 0.452 V
9. (a) $\mu_0 i R^2 \pi r^2/2x^3$; (b) $3\mu_0 i\pi R^2 r^2 v/2x^4$;
   (c) in the same direction as the current in the large loop
10. 0
11. (b) no
12. (a) 1.26×10^{-4} T, 0, -1.26×10^{-4} T;
    (b) 5.04×10^{-8} V
13. 29.5 mC
14. 15.5 μC
15. (a) 21.7 V; (b) counterclockwise
16. (a) 24 μV; (b) from c to b
17. (b) design it so that $Nab = (5/2\pi)\,\text{m}^2$
18. (a) f; (b) $\pi^2 a^2 f B$
19. 5.50 kV
20. 0
21. 80 μV, clockwise
22. (a) 0.598 μV; (b) counterclockwise
23. (a) 13 μWb/m; (b) 17%; (c) 0
24. (a) $\dfrac{\mu_0 i a}{2\pi}\left(\dfrac{2r+b}{2r-b}\right)$;
    (b) $2\mu_0 iabv/\pi R(4r^2-b^2)$
25. 3.66 μW
26. $A^2B^2/R\Delta t$
27. (a) 48.1 mV; (b) 2.67 mA; (c) 0.128 mW
28. $v_t = mgR/B^2L^2$
29. (a) 600 mV, up the page; (b) 1.5 A, clockwise;
    (c) 0.90 W; (d) 0.18 N; (e) same as (c)
30. (a) 85.2 T $\cdot$ m^2; (b) 56.8 V; (c) 1
31. (a) 240 μV; (b) 0.600 mA; (c) 0.144 μW;
    (d) 2.88×10^{-8} N; (e) same as (c)
32. 1, -1.07 mV; 2, -2.40 mV; 3, 1.33 mV
33. (a) 71.5 μV/m; (b) 143 μV/m
34. 0.15 V/m
36. (a) 2.45 mWb; (b) 0.645 mH
37. 0.10 μWb
38. (a) $\mu_0 i/W$; (b) $\pi\mu_0 R^2/W$
40. (a) decreasing; (b) 0.68 mH
41. let the current change at 5.0 A/s
42. (a) 16 kV; (b) 3.1 kV; (c) 23 kV
43. (b) so that the changing magnetic field of one does not induce current in the other;
    (c) $L_{\text{eq}} = \sum_{j=1}^{N} L_j$
44. (b) so that the changing magnetic field of one does not induce current in the other;
    (c) $\dfrac{1}{L_{\text{eq}}} = \sum_{j=1}^{N}\dfrac{1}{L_j}$
45. $6.91\tau_L$
46. 12.3 s
47. 46 Ω
48. (a) $\mathcal{E}$; (b) $0.135\,\mathcal{E}$; (c) $0.693\tau_L$
49. (a) 8.45 ns; (b) 7.37 mA
50. $(42+20t)$ V
51. 12.0 A/s
52. (a) 0.29 mH; (b) 0.29 ms
53. (a) $i_1 = i_2 = 3.33$ A; (b) $i_1 = 4.55$ A, $i_2 = 2.73$ A; (c) $i_1 = 0$, $i_2 = 1.82$ A (reversed);
    (d) $i_1 = i_2 = 0$
54. I. (a) 2.0 A; (b) 0; (c) 2.0 A; (d) 0; (e) 10 V; (f) 2.0 A/s
    II. (a) 2.0 A; (b) 1.0 A; (c) 3.0 A; (d) 10 V; (e) 0; (f) 0
55. (a) $i(1-e^{-Rt/L})$
56. $1.23\tau_L$

57. 25.6 ms
58. (a) 240 W; (b) 150 W; (c) 390 W
59. (a) 97.9 H; (b) 0.196 mJ
60. (a) 18.7 J; (b) 5.10 J; (c) 13.6 J
62. 5.58 A
63. (a) 34.2 J/m^3; (b) 49.4 mJ
64. 3×10^{36} J
65. 1.5×10^8 V/m
66. (a) 1.3 mT; (b) 0.63 J/m^3
67. (a) 1.0 J/m^3; (b) 4.8×10^{-15} J/m^3
68. (a) 1.5 μWb, 100 mV; (b) 90 nWb, 12 mV
69. (a) 1.67 mH; (b) 6.00 mWb
70. 13 H
71. (b) have the turns of the two solenoids wrapped in opposite directions
72. magnetic field exists only within the cross section of the solenoid
73. magnetic field exists only within the cross section of solenoid 1
75. (a) $\frac{\mu_0 N l}{2\pi} \ln\left(1 + \frac{b}{a}\right)$; (b) 13 μH

Chapter 32

1. (b) sign is minus; (c) no, there is compensating positive flux through open end near magnet
2. +3 Wb
3. 47.4 μWb, inward
4. 13 MWb, outward
5. 55 μT
7. (a) 31.0 μT, 0°; (b) 55.9 μT, 73.9°; (c) 62.0 μT, 90°
8. (a) 1660 km; (b) 383 μT; (c) 61.1 μT, 84.2°
9. (a) -9.3×10^{-24} J/T; (b) 1.9×10^{-23} J/T
10. 4.6×10^{-24} J
11. (a) 0; (b) 0; (c) 0; (d) $\pm 3.2 \times 10^{-25}$ J; (e) -3.2×10^{-34} J · s, 2.8×10^{-23} J/T, $+9.7 \times 10^{-25}$ J, $\pm 3.2 \times 10^{-25}$ J
12. (b) $\vec{\mu}$ directed away from magnet; i clockwise; (c) away from magnet
13. $\Delta\mu = e^2 r^2 B/4m$
14. 0.48 K
15. 20.8 mJ/T
16. (b) $\vec{\mu}$ directed toward magnet, i counterclockwise; (c) toward magnet
17. yes
18. (a) 150 T; (b) 600 T; (c) no
19. (b) K_i/B, opposite to the field; (c) 310 A/m
20. 25 km
21. (a) 3.0 μT; (b) 5.6×10^{-10} eV
22. (a) 8.9 A · m^2; (b) 13 N · m
23. 5.15×10^{-24} A · m^2
25. (a) 180 km; (b) 2.3×10^{-5}
26. $r = 27.5$ mm and $r = 110$ mm
27. 2.4×10^{13} V/m · s
28. (a) 1.9 pT
30. 7.5×10^5 V/s
32. 7.2×10^{12} V/m · s
33. (a) 0.63 μT; (b) 2.3×10^{12} V/m · s
34. (a) 2.1×10^{-8} A, downward; (b) clockwise
35. (a) 710 mA; (b) 0; (c) 1.1 A
36. 7.2×10^{12} V/m · s
37. (a) 2.0 A; (b) 2.3×10^{11} V/m · s; (c) 0.50 A; (d) 0.63 μT · m
38. (a) 1.33 A; (b) $0.25R$ and $4.00R$

Chapter 33

1. 9.14 nF
2. 45.2 mA
3. (a) 1.17 μJ; (b) 5.58 mA
4. (a) 6.00 μs; (b) 167 kHz; (c) 3.00 μs
5. with n a positive integer: (a) $t = n(5.00\ \mu s)$; (b) $t = (2n - 1)(2.50\ \mu s)$; (c) $t = (2n - 1)(1.25\ \mu s)$
6. (a) 89 rad/s; (b) 70 ms; (c) 25 μF
7. (a) 1.25 kg; (b) 372 N/m; (c) 1.75×10^{-4} m; (d) 3.02 mm/s
8. 38 μH
9. 7.0×10^{-4} s
11. (a) 3.0 nC; (b) 1.7 mA; (c) 4.5 nJ
12. (a) 3.60 mH; (b) 1.33 kHz; (c) 0.188 ms
13. (a) 275 Hz; (b) 364 mA
14. 600, 710, 1100, 1300 Hz
15. (a) 6.0:1; (b) 36 pF, 0.22 mH
16. (a) $Q/2$; (b) $0.866I$
17. (a) 1.98 μJ; (b) 5.56 μC; (c) 12.6 mA; (d) −46.9°; (e) +46.9°
19. (a) 0.180 mC; (b) $T/8$; (c) 66.7 W
20. ω
21. (a) 356 μs; (b) 2.50 mH; (c) 3.20 mJ
22. (a) 0; (b) $2i(t)$
23. Let T_2 (= 0.596 s) be the period of the inductor plus the 900 μF capacitor and let T_1 (= 0.199 s) be the period of the inductor plus the 100 μF capacitor. Close S_2, wait $T_2/4$; quickly close S_1, then open S_2; wait $T_1/4$ and then open S_1.

25. 8.66 mΩ
26. 5.85 μC; 5.52 μC; 1.93 μC
27. $(L/R)\ln 2$
28. (a) $\pi/2$ rad; (b) $q = (I/\omega)\, e^{-Rt/2L} \sin \omega' t$
30. (a) 0.283 A; (b) 2.26 A
31. (a) 0.0955 A; (b) 0.0119 A
32. (a) 0.600 A; (b) 0.600 A
33. (a) 0.65 kHz; (b) 24 Ω
34. (a) 5.22 mA; (b) 0; (c) 4.51 mA
35. (a) 6.73 ms; (b) 11.2 ms; (c) inductor; (d) 138 mH
36. (a) 39.1 mA; (b) 0; (c) 33.8 mA
37. (a) $X_C = 0$, $X_L = 86.7\,\Omega$, $Z = 182\,\Omega$, $I = 198$ mA, $\phi = 28.5°$
38. (a) $X_C = 177\,\Omega$, $X_L = 0$, $Z = 239\,\Omega$, $I = 151$ mA, $\phi = -47.9°$
39. (a) $X_C = 37.9\,\Omega$, $X_L = 86.7\,\Omega$, $Z = 167\,\Omega$, $I = 216$ mA, $\phi = 17.1°$
40. (a) 2.36 mH; (b) they move away from 1.40 kHz
41. 1000 V
42. (a) 36.0 V; (b) 27.3 V; (c) 17.0 V; (d) −8.34 V
43. 89 Ω
44. (a) 16.6 Ω; (b) 422 Ω; (c) 0.521 A; (d) increases; (e) decreases; (f) increases
45. (a) 224 rad/s; (b) 6.00 A; (c) 228 rad/s, 219 rad/s; (d) 0.040
46. 1.13 kHz, 1.45 kHz, 1.78 kHz, 2.30 kHz
48. (a) 796 Hz; (b) no change; (c) decreased; (d) increased
49. 1.84 A
50. 100 V
51. 141 V
52. (a) taking; (b) supplying
53. 0, 9.00 W, 3.14 W, 1.82 W
55. (a) 12.1 Ω; (b) 1.19 kW
56. (a) 41.4 W; (b) −17.1 W; (c) 44.1 W; (d) 14.4 W
57. (a) 0.743; (b) leads; (c) capacitive; (d) no; (e) yes, no, yes; (f) 33.4 W
59. (a) 117 μF; (b) 0; (c) 90.0 W, 0; (d) 0°, 90°; (e) 1, 0
60. (a) 76.4 mH; (b) 17.8 Ω
61. (a) 2.59 A; (b) 38.8 V, 159 V, 224 V, 64.2 V, 75.0 V; (c) 100 W for R, 0 for L and C.
62. 1000 V
63. (a) 2.4 V; (b) 3.2 mA, 0.16 A
64. step up: 5.00, 4.00, 1.25; step down: 0.800, 0.250, 0.200
65. 10

Chapter 34

1. (a) 0.50 ms; (b) 8.4 min; (c) 2.4 h; (d) 5500 B.C.
2. (a) 4.7×10^{-3} Hz; (b) 3 min 32 s
3. (a) 515 nm, 610 nm; (b) 555 nm, 5.41×10^{14} Hz, 1.85×10^{-15} s
4. 7.49 GHz
5. it would steadily increase; (b) the summed discrepancies between the apparent time of eclipse and those observed from x; the radius of Earth's orbit
6. 4.7 m
7. 5.0×10^{-21} H
8. 1.07 pT
9. $B_x = 0$, $B_y = -6.7 \times 10^{-9} \cos[\pi \times 10^{15}(t - x/c)]$, $B_z = 0$ in SI units
11. 0.10 MJ
12. 4.8×10^{-29} W/m^2
13. 8.88×10^4 m^2
14. 1.2 MW/m^2
15. (a) 16.7 nT; (b) 33.1 mW/m^2
16. 1.03 kV/m; 3.43 μT
17. (a) 6.7 nT; (b) 5.3 mW/m^2; (c) 6.7 W
18. (a) 1.4×10^{-22} W; (b) 1.1×10^{15} W
19. (a) 87 mV/m; (b) 0.30 nT; (c) 13 kW
20. 3.3×10^{-8} Pa
21. 1.0×10^7 Pa
22. (a) 6.0×10^8 N; (b) $F_{grav} = 3.6 \times 10^{22}$ N
23. 5.9×10^{-8} Pa
24. (a) 3.97 GW/m^2; (b) 13.2 Pa; (c) 1.67×10^{-11} N; (d) 3.14×10^3 m/s^2
25. (a) 100 MHz; (b) 1.0 μT along the z axis; (c) 2.1 m^{-1}, 6.3×10^8 rad/s; (d) 120 W/m^2; (e) 8.0×10^{-7} N, 4.0×10^{-7} Pa
26. 491 nm
29. 1.9 mm/s
30. 0.96 km^2
31. (b) 580 nm
32. (a) $-y$ direction; (b) $E_z = -cB\sin(ky + \omega t)$, $E_x = E_y = 0$; (c) plane polarized with $\vec{E}$ along the z axis
33. (a) 1.9 V/m; (b) 1.7×10^{-11} Pa
34. 4.5×10^{-2} %
35. 3.1%

36. 20° or 70°
37. $4.4\,W/m^2$
38. $19\,W/m^2$
39. 2/3
40. (a) 0.16; (b) 0.84
41. (a) 2 sheets; (b) 5 sheets
42. 180°
43. 1.48
44. 1.26
45. (a) yes; (b) 1.3
46. (a) 56.7°; (b) 35.2°
47. 1.07 m
52. 34°
53. 1.22
54. 182 cm
55. (a) 49°; (b) 29°
56. (a) yes; (b) no; (c) between 42.9° and 43.3°
57. (a) cover the center of each face with an opaque disk of radius 4.5 mm; (b) about 0.63
58. (a) 35.6°; (b) 53.1°
59. (a) $\sqrt{1+\sin^2\theta}$; (b) $\sqrt{2}$; (c) light emerges at the right; (d) no light emerges at the right
60. (a) about 53°; (b) yes
61. 49.0°
62. 55.5°; 55.8°
63. (a) 15 m/s; (b) 8.7 m/s; (c) higher; (d) 72°
64. (a) 3.15 m; (b) 10 m; (c) 2.4 m at the west end of the tomb, 1.2 m at the east end (the floor of the east end is 1.2 m above that of the west end, perhaps as an alter)
65. 1.0
66. 30 cm

Chapter 35

1. 40 cm
2. 9.10 m
3. (a) 3
4. (a) 7; (b) 5; (c) 1 to 3; (d) depends on the position of O and your perspective
6. 1.5 m
7. new illumination is 10/9 of the old
8. 351 cm
9. 10.5 cm
10. (a) +, +40, −20, +2.0, no, no; (b) plane, ∞, ∞, −10, no; (c) concave, +40, +60, −2.0, yes, yes; (d) concave, +20, +40, +30, yes, yes; (e) convex, −20, +20, +0.50, no, no; (f) convex, −, −40, −18, +180, no, no; (g) −20, −, −, +5, +0.8, no, no; (h) concave, +8, +16, +12, −, yes
12. (b) 0.56 cm/s; (c) 11 m/s; (d) 6.7 cm/s
13. (a) 2.00; (b) none
14. (a) −18, no; (b) −33, no; (c) +71, yes; (d) any n_2 possible, no; (e) +30, no; (f) +10, no; (g) −26, no; (h) 1.0, yes
16. 42 mm
17. $i = -12$ cm
18. 1.85 mm
25. (b) separate the lenses by a distance $f_2 - |f_1|$, where f_2 is the focal length of the converging lens
19. 45 mm, 90 mm
20. (a) +40 cm; (b) at infinity
22. 5.0 mm
23. 22 cm
24. An X means that the quantity cannot be found from the given data: (a) +, X, X, +20, X, −1.0, yes, yes; (b) converging, X, X, −10, X, +2.0, no, no; (c) converging, +, X, X, −10, X, no, no; (d) diverging, −, X, X, −3.3, X, no, no; (e) converging, +30, −15, +1.5, no, no; (f) diverging, −30, −7.5, +0.75, no, no; (g) diverging, −120, −9.2, +0.92, no, no; (h) diverging, −10, X, X, −5, X, +, no; (i) converging, +3.3, X, X, +5, X, yes, yes
26. (a) 36 cm beyond the converging lens; (b) 1.2 cm; (c) real; (d) inverted
27. same orientation, virtual, 30 cm to the left of the second lens; $m = 1$
28. (a) coincides in location with the original object and is enlarged 5.0 times; (c) virtual; (d) yes
30. (a) converging; (b) 26.7 cm; (c) 8.89 cm
32. 2.1 mm
33. (a) 13.0 cm; (b) 5.23 cm; (c) −3.25; (d) 3.13; (e) −10.2
34. (a) $m_\theta = 1 + (25\text{ cm})/f$; (b) $m_\theta = (25\text{ cm})/f$; (c) 3.5, 2.5
35. (a) 2.35 cm; (b) decrease
36. −125
37. (a) 5.3 cm; (b) 3.0 mm
39. (b) 8.4 mm; (c) 3.0 mm

Chapter 36

1. (a) 5.09×10^{14} Hz; (b) 388 nm; (c) 1.97×10^8 m/s
2. 4.55×10^7 m/s
3. 1.56
4. 2.1×10^8 m/s
5. 22°, refraction reduces θ
6. (a) pulse 2; (b) $0.03L/c$
7. (a) 3.60 μm; (b) intermediate, closer to fully constructive interference
8. (a) 1.70 (or 0.70); (b) 1.70 (or 0.70); (c) 1.30 (or 0.30); (d) brightnesses are identical, close to fully destructive interference
9. (a) 0.833; (b) intermediate, closer to fully constructive interference
10. (a) 1.55 μm; (b) 4.65 μm
11. (a) 0.216 rad; (b) 12.4°
12. $(2m+1)\pi$
13. 2.25 mm
14. (a) 0.010 rad; (b) 5.0 mm
15. 648 nm
16. 0.15°
17. 16
18. 0
19. 0.072 mm
20. 8.75λ
21. 6.64 μm
22. (a) 0.253 mm; (b) the pattern shifts so that the 2.5λ minimum replaces the central maximum
23. 2.65
24. $y = 17\sin(\omega t + 13°)$
25. $y = 27\sin(\omega t + 8.5°)$
27. (a) 1.17 m, 3.00 m, 7.50 m; (b) no
29. $I = \frac{1}{9}I_m[1+8\cos^2(\pi d\sin\theta/\lambda)]$, I_m = intensity of central maximum
30. (a) 155 nm; (b) 310 nm
31. fully constructively
32. $L = (m+\frac{1}{2})\lambda/2$, for $m = 0, 1, 2, \ldots$
33. 0.117 μm, 0.352 μm
34. $\lambda/5$
35. 70.0 nm
36. none
37. 120 nm
38. (a) and (c)
39. (a) 552 nm; (b) 442 nm
40. 673 nm
42. 338 nm
43. 140
44. (a) dark; (b) blue end
45. 1.89 μm
46. 840 nm
47. 2.4 μm
48. 1.00025
49. $\sqrt{(m+\frac{1}{2})\lambda R}$, for $m = 0, 1, 2, \ldots$
50. (a) 34; (b) 46
51. 1.00 m
53. $x = (D/2a)(m+\frac{1}{2})\lambda$, for $m = 0, 1, 2, \ldots$.
54. 5.2 μm
55. 588 nm
56. 0.354 mm
57. 1.00030
58. $I = I_m\cos^2(2\pi x/\lambda)$, where I_m is the maximum intensity
59. (a) 0; (b) fully constructive; (c) increase (d)

Phase Difference	Position x (μm)	Type
0	$\approx \infty$	fc
0.50λ	7.88	fd
1.00λ	3.75	fc
1.50λ	2.29	fd
2.00λ	1.50	fc
2.50λ	0.975	fd

60. (f) 0.4

Chapter 37

1. 60.4 μm
2. (a) 0.430°; (b) 0.118 mm
3. (a) $\lambda_a = 2\lambda_b$; (b) coincidences occur when $m_b = 2m_a$
4. (a) 2.5 mm; (b) 2.2×10^{-4} rad
5. (a) 70 cm; (b) 1.0 mm
6. 41.2 m from the central axis
7. 1.77 mm
8. 160°
10. (a) 0.18°; (b) 0.46 rad; (c) 0.93
11. (d) 53°, 10°, 5.1°
13. (b) 0, 4.493 rad, etc.; (c) −0.50, 0.93, etc.
14. 30 μm
15. (a) 1.3×10^{-4} rad; (b) 10 km
16. (a) 1.3×10^{-4} rad; (b) 21 m
17. 50 m
18. 30 m
19. (a) 1.1×10^4 km; (b) 11 km
20. 53 m
21. 27 cm

22. 4.7 cm
23. (a) 0.347°; (b) 0.97°
24. (a) red; (b) 130 μm
25. (a) 8.7×10^{-7} rad; (b) 8.4×10^{7} km; (c) 0.025 mm
26. about 10^{-13}
27. five
28. three
29. (a) 4; (b) every fourth bright fringe is missing
30. $\lambda D/d$
31. (a) nine; (b) 0.255
32. (a) 5.05 μm; (b) 20.2 μm
33. (a) 3.33 μm; (b) 0, ±10.2°, ±20.7°, ±32.0°, ±45.0°, ±62.2°
34. all wavelengths shorter than 635 nm
35. three
36. 2 μm
37. (a) 6.0 μm; (b) 1.5 μm; (c) m = 0, 1, 2, 3, 5, 6, 7, 9
38. (a) three; (b) 0.051°
39. 1100
40. 523 nm
44. 470 nm to 560 nm
46. 491
47. 3650
48. (a) 56 pm; (b) none
50. (a) 1.0×10^{4} nm; (b) 3.3 mm
52. (a) $\tan\theta$; (b) 0.89
53. 0.26 nm
54. 26 pm, 39 pm
55. 39.8 pm
56. (a) 170 pm; (b) 130 pm
58. 0.570 nm
59. (a) $a_0/\sqrt{2}$, $a_0/\sqrt{5}$, $a_0/\sqrt{10}$, $a_0/\sqrt{13}$, $a_0/\sqrt{17}$
60. λ = 130 pm for m = 3, λ = 97.2 pm for $m = 4$
61. 30.6°, 15.3° (clockwise); 3.08°, 37.8° (counterclockwise)
62. (a) decrease; (b) 11°; (c) 0.23°
63. (a) 50 m; (b) no, the width of 10 m is too narrow to resolve; (c) not during daylight, but light pollution during the night would be a sure sign
64. 15 μm to 220 μm

Chapter 38

1. (a) 6.7×10^{-10} s; (b) 2.2×10^{-18} m
2. (a) 3×10^{-18}; (b) 8.2×10^{-8}; (c) 1.1×10^{-6}; (d) 3.7×10^{-5}; (e) 0.10
3. 0.99c
4. (a) 0.140; (b) 0.9950; (c) 0.999 950; (d) 0.999 999 50
5. 0.445 ps
6. (a) 0.999 999 50c
7. 1.32 m
8. 1.53 cm
9. 0.63 m
10. (a) 0.866c; (b) 2.00
11. (a) 87.4 m; (b) 394 ns
12. (b) 0.999 999 15c
13. (a) 26 y; (b) 52 y; (c) 3.7 y
14. (a) $x' = 0$, $t' = 2.29$ s; (b) $x' = 6.55 \times 10^{8}$ m, $t' = 3.16$ s
15. $x' = 138$ km, $t' = -374$ μs
16. $t_1' = 0$, $t_2' = -2.5$ μs
17. (a) 25.8 μs; (b) small flash
18. (a) S' must move toward S, along their common axis, at a speed of 0.480c; (b) big flash; (c) 4.39 μs
19. (a) 1.25; (b) 0.800 μs
20. 2.40 μs
21. 0.81c
22. (a) 0.84c, in the positive direction of x; (b) 0.21c, in the positive direction of x; the classical predictions are 1.1c and 0.15c
23. (a) 0.35c; (b) 0.62c
24. 0.588c, moving away
25. 1.2 μs
26. (a) 1.25 yr; (b) 1.6 yr; (c) 4.0 yr
27. 22.9 MHz
28. (a) 7000 km/s; (b) away
29. 1×10^{6} m/s, receding
30. 0.13c
31. yellow (550 nm)
32. (a) 79 keV; (b) 3.11 MeV; (c) 10.9 MeV
33. (a) 0.0625, 1.00196; (b) 0.941, 2.96; (c) 0.999 999 87, 1960
34. (a) 0.9988, 20.6; (b) 0.145, 1.01; (c) 0.073, 1.0027
35. 0.999 987c
36. 8.12 MeV
37. 18 smu/y
38. (a) 1.0 keV; (b) 1.1 MeV
39. (a) 0.707c; (b) 1.41; (c) 0.414mc^2

40. (a) $0.943c$; (b) $0.866c$
41. $\sqrt{8}mc$
42. (c) $207m_e$, the particle is a muon
43. 1.01×10^7 km, or about 250 Earth circumferences
44. (a) $0.948c$; (b) 226 MeV; (c) 314 MeV/c
45. 110 km
46. (a) 0.776 mm; (b) 16.0 mm; (c) 0.335 ns, no
47. 4.00 u, probably a helium nucleus
48. 660 km
49. 330 mT
50. (a) 534; (b) 0.999 998 25; (c) 2.23 T
51. (a) 2.08 MeV; (b) −1.18 MeV
52. (a) 1.93 m; (b) $x_{g2} = 6.00$ m, $t_{g2} = 1.36 \times 10^{-8}$ s; (c) 1.36×10^{-8} s; (d) 0.379 m; (e) $x_{c2} = 30.5$ m, $t_{c2} = -1.01 \times 10^{-7}$ s; (f) no; (g) event 2; (i) both Carman and Garageman are correct
53. (a) $vt \sin\theta$; (b) $t[1 - (v/c)\cos\theta]$; (c) $3.24c$

Chapter 39

1. 4.14 eV · fs
2. 2.1 μm, infrared
4. 2.11 eV
5. 1.0×10^{45} photons/s
6. 1.7×10^{21} photons/m^2 · s
7. 5.9 μeV
8. 8.6×10^5 m/s
9. 2.047 eV
10. 3.6×10^{-17} W
11. 4.7×10^{26} photons
12. 3.3×10^{18} photons/s
13. (a) infrared lamp; (b) 1.4×10^{21} photons/s
14. (a) 3.61 kW; (b) 1.00×10^{22} photons/s; (c) 60.2 s
15. (a) 2.96×10^{20} photons/s; (b) 48,600 km; (c) 5.89×10^{18} photons/m^2 · s
16. (a) no; (b) 544 nm, green
17. barium and lithium
18. (a) cesium only; (b) both elements
19. 170 nm
20. 10 eV
21. 676 km/s
22. (a) 1.3 V; (b) 680 km/s
23. (a) 2.00 eV; (b) 0; (c) 2.00 eV; (d) 295 nm
24. 1.07 eV
25. 233 nm
26. (a) 6.60×10^{-34} J · s; (b) 2.27 eV; (c) 545 nm
27. (a) 382 nm; (b) 1.82 eV
28. (a) 6.7×10^{-34} J · s; (b) 2.4 eV
29. 9.68×10^{-20} A
30. (a) 3.1 keV; (b) 14 keV
31. (a) 2.7 pm; (b) 6.05 pm
32. (a) 2.73×10^{-22} kg · m/s = 0.511 MeV/c; (b) 2.43 pm; (c) 1.24×10^{20} Hz
33. (a) 8.57×10^{18} Hz; (b) 35.4 keV; (c) 1.89×10^{-23} kg · m/s = 35.4 keV/c
34. (a) +4.8 pm; (b) −41 keV; (c) 41 keV; (d) same as incident x rays
36. (a) 2.43 pm; (b) 4.86 pm; (c) 0.255 MeV
37. (a) 2.43 pm; (b) 1.32 fm; (c) 0.511 MeV; (d) 938 MeV
38. 2.65 fm
39. 300 %
40. (a) 8.1×10^{-9} %; (b) 4.9×10^{-4} %; (c) 8.8 %; (d) 66 %; (e) the shorter the wavelength, the easier to measure the Compton shift
43. (a) 41.8 keV; (b) 8.2 keV
44. (a) 2.43 pm; (b) 4.11×10^{-6}; (c) -8.66×10^{-6} eV; (d) 2.43 pm, 9.76×10^{-2}, −4.45 keV
45. 1.12 keV
47. 44°
50. 1.7×10^{-35} m
51. 7.75 pm
52. (a) 38.7 pm; (b) 1.24 nm; (c) 904 fm
53. 4.3 μeV
54. (a) 3.96×10^6 m/s; (b) 81.9 kV
55. (a) 38.8 meV; (b) 146 pm
56. (a) 3.3×10^{-24} kg · m/s for each; (b) 38 eV for the electron, 6.2 keV for the proton
57. (a) photon: 1.24 μm, electron: 1.22 nm; (b) 1.24 fm for each
58. (a) 73 pm; (b) 3.4 nm; (c) yes, their average de Broglie wavelength is smaller than their average separation
59. (a) 1.9×10^{-21} kg · m/s; (b) 346 fm
60. (a) photon: 1.24 keV, electron: 1.50 eV; (b) 1.24 GeV for each
61. 0.025 fm, about 200 times smaller than a nuclear radius
62. (a) 5.2 fm; (b) no, the de Broglie wavelength is much less than the distance of closest approach
63. neutron
64. (a) 15 keV; (b) 120 keV

65. 9.70 kV (relativistic calculation), 9.76 kV (classical calculation)
73. (d) $x = n(\lambda/2)$, where $n = 0, 1, 2, 3, \ldots$
74. (a) no; (b) plane wavefronts of infinite extent, perpendicular to the x axis
75. 0.19 m
76. 2.1×10^{-24} kg · m/s
78. (a) 124 keV; (b) 41 keV
79. (a) proton: 9.02×10^{-6}, deuteron: 7.33×10^{-8}; (b) 3.0 MeV for each; (c) 3.0 MeV for each
80. 5.1 eV
81. (a) −20 %; (b) −10 %; (c) +15 %
82. (a) 10^{104} years; (b) 2×10^{-19} s (the smaller mass of the electron makes an enormous difference)
83. $T = 10^{-x}$, where $x = 7.2 \times 10^{39}$ (T is very small)

Chapter 40

1. (a) 37.7 eV; (b) 0.0206 eV
2. multiply it by $\sqrt{2}$
3. 1900 MeV
4. 850 pm
5. 0.020 eV
6. 0.65 eV
7. 90.3 eV
8. (a) $n = 12$ and 13; (b) impossible
10. (a) $n = 10$ and 11; (b) impossible
11. 68.7 nm, 25.8 nm, 13.7 nm, and 8.59 nm
12. (a) 72.2 eV; (b) 68.7 nm ($2 \longrightarrow 1$); 41.2 nm ($3 \longrightarrow 2$); 29.4 nm ($4 \longrightarrow 3$); 25.8 nm ($3 \longrightarrow 1$); 17.2 nm ($4 \longrightarrow 2$); 13.7 nm ($4 \longrightarrow 1$); (c) 13.7 nm alone; 29.4 nm and then 25.8 nm; 29.4 nm, then 41.2 nm, and then 68.7 nm; 17.2 nm and then 68.7 nm
13. (a) 1.3×10^{-19} eV; (b) about 1.2×10^{19}; (c) 0.95 J = 5.9×10^{18} eV; (d) yes
14. (a) decrease; (b) increase
15. (b) no; (c) no; (d) yes
16. (a) 0.091; (b) 0.091; (c) 0.818
17. (a) 0.050 %; (b) 0.10 %; (c) 0.0095 %
18. (b) meter^{-3}
19. 59 eV
20. 280 eV
21. (b) $k = (2\pi/h)\,[2m\,(U_0 - E)]^{1/2}$
22. (b) $k = (2\pi/h)(2mE)^{1/2}$
24. 0.734 eV
25. 3.08 eV
26. (a) 1.25; (b) 2.00; (c) 5.00; (d) 1.00
27. 0.75, 1.00; 1.25, 1.75, 2.00, 2.25, 3.00, 3.75
28. (a) 3.00; (b) 9.00; (c) 2.00; (d) three; (e) six
29. 1.00, 2.00, 3.00, 5.00, 6.00, 8.00, 9.00
31. 2.6 eV
32. 1.17 eV
33. 4.0
34. (a) −3.4 eV; (b) 3.4 eV
35. (a) 12 eV; (b) 6.5×10^{-27} kg·m/s; (c) 103 nm
36. (a) 658 nm; (b) 366 nm
38. (a) 291 nm^{-3}; (b) 10.2 nm^{-1}
39. (a) 0; (b) 10.2 nm^{-1}; (c) 5.54 nm^{-1}
40. (a) 12.7 eV; (b) 12.7 eV ($4 \longrightarrow 1$), 2.55 eV ($4 \longrightarrow 2$), 0.66 eV ($4 \longrightarrow 3$), 12.1 eV ($3 \longrightarrow 1$), 1.89 eV ($3 \longrightarrow 2$), 10.2 eV ($2 \longrightarrow 1$)
41. (a) 13.6 eV; (b) 3.40 eV
42. 4.1 m/s
43. (a) $n = 4$ to $n = 2$; (b) Balmer series
44. (a) 30.5 nm; (b) 291 nm; (c) 825 THz, 365 THz
45. (a) 13.6 eV; (b) −27.2 eV
46. (a) $n = 2$ to $n = 1$; (b) Lyman series
47. (a) 2.6 eV; (b) $n = 4$ to $n = 2$
49. 0.68
50. $n = 3$ to $n = 1$
52. 0.439
54. (a) n; (b) $2\ell + 1$; (c) n^2
55. (a) 0.0037; (b) 0.0054
56. $n \approx 4348$
58. (c) $(r^2/8a^3)(2 - r/a)^2 e^{-r/a}$
59. (a) $P_{210} = (r^4/8a^5)e^{-r/a}\cos^2\theta$; $P_{21+1} = P_{21-1} = (r^4/16a^5)e^{-r/a}\sin^2\theta$

Chapter 41

2. (a) 14; (b) 6; (c) 6; (d) 2
3. (a) 3; (b) 3
4. (a) 3.64×10^{-34} J · s; (b) 3.15×10^{-34} J · s
5. (a) 32; (b) 2; (c) 18; (d) 8
6. $n = 4$; $\ell = 3$; $m_\ell = +3, +2, +1, 0, -1, -2, -3$; $m_s = \pm\frac{1}{2}$
7. 24.1°
8. $\ell = 4$; $n \geq 5$; $m_s \pm \frac{1}{2}$
9. $n > 3$; $m_\ell = +3, +2, +1, 0, -1, -2, -3$; $m_s = +\frac{1}{2}, -\frac{1}{2}$
10. 50

11. (a) $\sqrt{12}\hbar$; (b)$\sqrt{12}\mu_B$;
(c)

m_ℓ	L_z	$\mu_{\text{orb},z}$	θ
-3	$-3\hbar$	$+3\mu_B$	150°
-2	$-2\hbar$	$+2\mu_B$	125°
-1	$-\hbar$	$+\mu_B$	107°
0	0	0	90°
$+1$	$+\hbar$	$-\mu_B$	73.2°
$+2$	$+2\hbar$	$-2\mu_B$	54.7°
$+3$	$+3\hbar$	$-3\mu_B$	30.0°

12. (a) 3; (b) 9; (c) 2; (d) 18; (e) 3
14. (a) 3×10^{74}; (b) 6×10^{74}; (c) 6×10^{-38} rad
15. 54.7° and 125°
16. (a) 58 μeV; (b) 14 GHz; (c) 2.1 cm, short radio wave region
17. 73 km/s^2
18. (a) 1.5×10^{-21} N; (b) 20 μm
19. 5.35 cm
20. 51 mT
21. (a) 2.13 meV; (b) 18 T
22. 19 mT
23. $44(h^2/8mL^2)$
24. $17.25(h^2/8mL^2)$
25. (a) $51(h^2/8mL^2)$; (b) $56(h^2/8mL^2)$; (c) $59(h^2/8mL^2)$
26. (a) $18.00(h^2/8mL^2)$; (b) $18.25(h^2/8mL^2)$; (c) $19.00(h^2/8mL^2)$
27. $42(h^2/8mL^2)$
28. (a) $45(h^2/8mL^2)$; (b) $47(h^2/8mL^2)$; (c) $48(h^2/8mL^2)$
30. selenium: 4p, 4 electrons; bromine: 4p, 5 electrons; krypton: 4p, 6 electrons
31. argon
32. $n = 1$, $\ell = 0$, $m_\ell = 0$, $m_s = \pm\frac{1}{2}$
33. (a) $(2, 0, 0, \pm\frac{1}{2})$; (b) $n = 2$, $\ell = 1$, $m_\ell = 1$, 0, or -1, $m_s = \pm\frac{1}{2}$
34. (a) 18 (36 if indistinguishability is not taken into account); (b) 6, the states where both electrons share the quantum numbers $(n, \ell, m_\ell, m_s) = (2, 1, 1, \frac{1}{2})$, $(2, 1, 1, -\frac{1}{2})$, $(2, 1, 0, \frac{1}{2})$,
$(2, 1, 0, -\frac{1}{2})$, $(2, 1, -1, \frac{1}{2})$, $(2, 1, -1, -\frac{1}{2})$
36. 12.4 kV
39. 49.6 pm, 99.2 pm
40. (a) 5.7 keV; (b) 87 pm, 14 keV; 220 pm, 5.7 keV
42. (a) 24.8 pm; (b) and (c) remain unchanged
43. (a) 35.4 pm, as for molybdenum; (b) 57 pm; (c) 50 pm
44. 6.44 keV
45. 9/16
46. 2.2 keV
48. (a) 19.7 keV, 17.5 keV; (b) Zr or Nb, Zr better
49. (a) 69.5 kV; (b) 17.9 pm; (c) K_α: 21.4 pm, K_β: 18.5 pm
50. 81 pm
51. (a) $(Z-1)^2/(Z'-1)^2$; (b) 57.5; (c) 2070
52. (b) 24 %, 15 %, 11 %, 7.9 %, 6.5 %, 4.7 %, 3.5 %, 2.5 %, 2.0 %, 1.5 %
53. (a) 6; (b) 3.2×10^6 years
54. 1.3×10^{15} moles
55. 9.1×10^{-7}
56. (a) 2.55 s; (b) 500 ps; (c) $(4.5 \times 10^{-4})°$ or 1.6" of arc
57. 10,000 K
58. -2.75×10^5 K
59. (a) 3.60 mm; (b) 5.25×10^{17}
60. $7.3 \times 10^{17}\ \text{s}^{-1}$
61. 4.7 km
62. 2×10^7
63. $2.0 \times 10^{16}\ \text{s}^{-1}$
64. 1.8 pm
65. (a) 3.03×10^5; (b) 1430 MHz; (d) 3.30×10^{-6}
66. (a) approximately none; (b) 68 J
67. (a) no; (b) 140 nm
68. (a) 7.33 μm; (b) $7.07 \times 10^5\ \text{W/m}^2$; (c) $2.49 \times 10^{10}\ \text{W/m}^2$
69. (a) 4.3 μm; (b) 10 μm; (c) infrared
70. (a) 6.9 μeV; (b) radio waves

Chapter 42

1. $8.43 \times 10^{28}\ \text{m}^{-3}$
3. 3490 atm
5. (a) $+8.0 \times 10^{-11}\ \Omega \cdot \text{m/K}$; (b) $-210\ \Omega \cdot \text{m/K}$
6. $5.90 \times 10^{28}\ \text{m}^{-3}$
7. (b) $1.52 \times 10^{28}\ \text{m}^{-3} \cdot \text{eV}^{-1}$
9. (a) 0; (b) 0.0955
10. $1.9 \times 10^{28}\ \text{m}^{-3} \cdot \text{eV}^{-1}$
12. 5.53 eV
13. 0.91
14. (a) 6.81 eV; (b) $1.77 \times 10^{28}\ \text{m}^{-3} \cdot \text{eV}^{-1}$; (c) $1.59 \times 10^{28}\ \text{m}^{-3} \cdot \text{eV}^{-1}$
15. (a) 2500 K; (b) 5300 K
17. (a) 90.0%; (b) 12.5%; (c) sodium
18. 1.36, 1.67, 0.90, 0.10, 0.00 $\times 10^{28}\ \text{m}^{-3} \cdot \text{eV}^{-1}$
19. (a) $2.7 \times 10^{25}\ \text{m}^{-3}$; (b) $8.43 \times 10^{28}\ \text{m}^{-3}$;

(c) 3100; (d) molecules: 3.3 nm, electrons: 0.228 nm
20. about 10^{-42}
21. (a) 1.0, 0.99, 0.50, 0.014, 2.5 $\times 10^{-17}$; (b) 700 K
23. 3
24. 57 meV above
25. (a) $5.86 \times 10^{28}\,\mathrm{m^{-3}}$; (b) 5.52 eV; (c) 1390 km/s; (d) 0.522 nm
26. (a) $1.31 \times 10^{29}\,\mathrm{m^{-3}}$; (b) 9.43 eV; (c) 1820 km/s; (d) 0.40 nm
27. (b) $1.80 \times 10^{28}\,\mathrm{m^{-3} \cdot eV^{-1}}$
30. 57.1 kJ
31. (a) 19.8 kJ; (b) 197 s
32. (a) 0.0055; (b) 0.018
33. 200° C
34. 0.029
35. (a) 225 nm; (b) ultraviolet
36. (a) gallium: $+3e$; arsenic: $+5e$; (b) 2
37. (a) 109.5°; (b) 235 pm
38. (a) 1.5×10^{-6}; (b) 1.5×10^{-6}
40. (a) n-type; (b) $5 \times 10^{21}\,\mathrm{m^{-3}}$; (c) 2.5×10^{5}
41. 0.22 μg
42. (a) 0.744 eV above; (b) 7.13×10^{-7}
43. (a) pure: 4.78×10^{-10}; doped: 0.0141; (b) 0.824
44. (b) 2.5×10^{8}
45. 6.02×10^{5}
46. opaque
47. 4.20 eV
48. 13 μm
49. (a) 5.0×10^{-17} F; (b) about $300e$

Chapter 43

1. 28.3 MeV
2. 15.8 fm
3. (a) 0.390 MeV; (b) 4.61 MeV
4. 27
6. 12 km
7. (a) six; (b) eight
8. (a) ^{142}Nd, ^{143}Nd, ^{146}Nd, ^{148}Nd, ^{150}Nd; (b) ^{97}Rb, ^{98}Sr, ^{99}Y, ^{100}Sr, ^{101}Nb, ^{102}Mo, ^{103}Tc, ^{105}Rh, ^{109}In, ^{110}Sn, ^{111}Sb, ^{112}Te; (c) ^{60}Zn, ^{60}Cu, ^{60}Ni, ^{60}Co, ^{60}Fe
10. (a) yttrium and iodine; (b) yttrium, 50; iodine, 74; (c) 19
11. (a) 1150 MeV; (b) 4.81 MeV/nucleon, 12.2 MeV/proton
12. (a) $2.3 \times 10^{17}\,\mathrm{kg/m^3}$ for each; (b) $1.0 \times 10^{25}\,\mathrm{C/m^3}$ for ^{55}Mn, $8.8 \times 10^{24}\,\mathrm{C/m^3}$ for ^{209}Bi
14. (b) 0.05% 0.50%, 0.81%, 0.83%, 0.81%, 0.78%, 0.74%, 0.72%, 0.71%
15. (a) 6.2 fm; (b) yes
16. 4×10^{-22} s
17. $K \approx 30$ MeV
18. (a) 1.000000 u, 11.90683 u, 236.2025 u
20. (a) 19.8 MeV, 6.26 MeV, 2.22 MeV; (b) 28.3 MeV; (c) 7.07 MeV
21. ^{25}Mg: 9.303%, ^{26}Mg: 11.71%
22. (a) +7.29 MeV; (b) +8.07 MeV; (c) −91.0 MeV
23. 1.6×10^{25} MeV
24. 1.0087 u
25. 7.92 MeV
26. (a) 1/4; (b) 1/8
27. 280 d
28. 3.0×10^{19}
29. (a) $7.6 \times 10^{16}\,\mathrm{s^{-1}}$; (b) $4.9 \times 10^{16}\,\mathrm{s^{-1}}$
30. (a) $4.8 \times 10^{-18}\,\mathrm{s^{-1}}$; (b) 4.6×10^{9} y
31. (a) 64.2 h; (b) 0.125; (c) 0.0749
32. (a) 5.04×10^{18}; (b) $4.60 \times 10^{6}\,\mathrm{s^{-1}}$
33. 5.3×10^{22}
34. 265 mg
35. (a) 2.0×10^{20}; (b) $2.8 \times 10^{9}\,\mathrm{s^{-1}}$
36. (a) 59.5 d; (b) 1.18
37. 209 d
38. 87.8 mg
39. 1.13×10^{11} y
42. 660 mg
43. (a) $8.88 \times 10^{10}\,\mathrm{s^{-1}}$; (b) $8.88 \times 10^{10}\,\mathrm{s^{-1}}$; (c) 1.19×10^{15}; (d) 0.111 μg
44. (a) $3.66 \times 10^{7}\,\mathrm{s^{-1}}$; (b) $t \gg 3.82$ d; (c) $3.66 \times 10^{7}\,\mathrm{s^{-1}}$; (d) 6.42 ng
45. 730 cm^2
47. Pu: 1.2×10^{-17}, Cm: $e^{-9173} \approx 0$
48. (a) 4.25 MeV; (b) −24.1 MeV; (c) 28.3 MeV
49. 4.269 MeV
50. $Q_3 = -9.50$ MeV, $Q_4 = 4.66$ MeV, $Q_5 = -1.30$ MeV
51. (a) 31.8 MeV, 5.98 MeV; (b) 86 MeV
52. (b) $4n+3$, $4n$, $4n+2$, $4n+3$, $4n$, $4n+1$, $4n+2$, $4n+1$, $4n+1$
53. ^{7}Li
54. (a) 900 fm; (b) 6.4 fm; (c) no; (d) yes
55. 1.21 MeV

57. 0.782 MeV
58. 600 keV
59. (b) 0.961 MeV
60. (b) 2.8×10^{13} W
61. 78.4 eV
62. 1600 y
63. (a) U: 1.06×10^{19}, Pb: 0.624×10^{19}; (b) 1.69×10^{19}; (c) 2.98×10^{9} y
64. 132 μg
65. 1.8 mg
66. 145 Bq = 3.92 nCi
67. 1.02 mg
68. 7.3 mSv
69. 13 mJ
70. (a) 18 mJ; (b) 2.9 mSv = 0.29 rem
71. (a) 6.3×10^{18}; (b) 2.5×10^{11}; (c) 0.20 J; (d) 2.3 mGy; (e) 30 mSv
72. 3.87×10^{10} K
73. (a) 6.6 MeV; (b) no
74. (a) ^{18}O, ^{60}Ni, ^{92}Mo, ^{144}Sm, ^{207}Pb; (b) ^{40}K, ^{91}Zr, ^{121}Sb, ^{143}Nd; (c) ^{13}C, ^{40}K, ^{49}Ti, ^{205}Tl, ^{207}Pb
75. (a) 25.4 MeV; (b) 12.8 MeV; (c) 25.0 MeV
76. (b) 1.00; (c) 70.8; (d) 0.0100; (e) 0.708; (f) no
77. 0.49
78. 9.0×10^{8} Bq
79. (a) β^- decay; (b) 8.2×10^{7}; (c) 1.2×10^{6}
80. (a) 1×10^{8}; (b) $(1 \times 10^{8})e^{-(\ln 2)(D-1996)/(30.2\,\text{y})}$, where D is the current year
81. 3.2×10^{12} Bq = 86 Ci
82. ^{225}Ao
83. 4.28×10^{9} y
84. 10^{13} atoms
85. 1.3×10^{-13} m
86. 7.31 $|rmMeV$
87. 3.2×10^{4} y
88. 4.9×10^{13} Bq

Chapter 44

1. (a) 2.6×10^{24}; (b) 8.2×10^{13} J; (c) 2.6×10^{4} y
2. by rows: ^{95}Sr, ^{95}Y, ^{134}Te; 3
3. 3.1×10^{10} s^{-1}
4. 4.54×10^{26} MeV
6. +5.00 MeV
7. −23.0 MeV
8. (a) 16 fissions/day; (b) 4.3×10^{8}
9. 181 MeV
11. (a) ^{153}Nd; (b) 110 MeV to ^{83}Ge, 60 MeV to ^{153}Nd; (c) 1.6×10^{7} m/s for ^{83}Ge, 8.7×10^{6} m/s for ^{153}Nd
12. (a) 10; (b) 226 MeV
13. (a) 252 MeV; (b) typical fission energy is 200 MeV
14. (a) +25%; (b) zero; (c) −36%
15. 461 kg
16. 617 kg
17. yes
18. (a) 44 kton
19. 557 W
20. (a) 1.2 MeV; (b) 3.2 kg
21. $^{238}\text{U} + n \rightarrow {}^{239}\text{U} \rightarrow {}^{239}\text{Np} + \text{e}$, $^{239}\text{Np} \rightarrow {}^{239}\text{Pu} + \text{e}$
23. (a) 84 kg; (b) 1.7×10^{25}; (c) 1.3×10^{25}
24. 1.6×10^{16}
25. 0.99938
26. 8030 MW
27. (b) 1.0, 0.89, 0.28, 0.019; (c) 8
28. 3.6×10^{9} y
29. (a) 75 kW; (b) 5800 kg
31. 1.7×10^{9} y
32. (a) 30 MeV; (b) 6 MeV
33. 170 keV
35. (a) 170 kV
36. 1.41 MeV
37. 0.151
38. (b) 500 km/s
41. (a) 3.1×10^{31} protons/m^3; (b) 1.2×10^{6} times
43. (a) 4.3×10^{9} kg/s; (b) 3.1×10^{-4}
44. (a) 4.0×10^{27} MeV; (b) 5.1×10^{26} MeV
45. (a) 1.83×10^{38} s^{-1}; (b) 8.25×10^{28} s^{-1}
47. (a) 4.1 eV/atom; (b) 9.0 MJ/kg; (c) 1500 y
48. 5×10^{9} y
49. 1.6×10^{8} y
50. (a) 6.3×10^{14} J/kg; (b) 6.2×10^{11} kg/s; (c) 4.3×10^{9} kg/s; (d) 15×10^{9} y
51. (a) 24.9 MeV; (b) 8.65 megaton TNT
53. 14.4 kW
54. $K_\alpha = 3.5$ MeV, $K_n = 14.1$ MeV

Chapter 45

1. 6.03×10^{-29} kg
2. 2.4×10^{-43}
3. 18.4 fm
4. $\pi^- \rightarrow \mu^- + \overline{\nu}$

5. 1.08×10^{42} J
6. one
7. 2.7 cm/s
8. 31 nm
9. 769 MeV
10. (a) 1.90×10^{-18} kg · m/s; (b) 9.90 m
13. (a) L_e, angular momentum; (b) L_μ, charge; (c) energy, L_μ
14. (a) $2e^+ + e^- + 5\nu + 4\overline{\nu}$; (b) boson, meson, $B = 0$
15. $Q = 0$, $B = -1$, $S = 0$
16. (b), (d)
17. (a) energy; (b) strangeness; (c) charge
18. (a) 605 MeV; (b) −181 MeV
19. 338 MeV
21. (a) K^+; (b) $\overline{n}$; (c) K^0
22. (a) 37.7 MeV; (b) 5.35 MeV; (c) 32.4 MeV
23. (a) $\overline{u}\overline{u}\overline{d}$; (b) $\overline{u}\overline{d}\overline{d}$
24. (a) n; (b) Σ^+; (c) Ξ^-
25. (a) not possible; (b) uuu
26. (a) sud; (b) uss
29. Σ^0, 7530 km/s
30. 1.6×10^{10} ly
31. 666 nm
32. 3.16×10^8 ly
33. (b) 4.5 H-atoms/m^3
34. (b) 0.934; (c) 1.50×10^{10} ly
35. (a) 256 μeV; (b) 4.84 mm
36. $102M$
37. (a) 122 m/s; (b) 246 y
38. (b) $2\pi r^{3/2}/\sqrt{GM}$
39. (b) 2.38×10^9 K
40. (a) 2.6 K; (b) 29 nm
41. (a) $0.785c$; (b) $0.993c$; (c) C2; (d) C1; (e) 51 ns; (f) 40 ns
42. $1A$, $2J$, $3I$, $4F$, $5G$, $6C$, $7H$, $8D$, $9E$
43. (c) $r\alpha/c + (r\alpha/c)^2 + (r\alpha/c)^3 + \ldots$; (d) $r\alpha/c$; (e) $\alpha = H$; (f) 7.4×10^8 ly; (g) 7.8×10^8 y; (h) 7.4×10^8 y; (i) 7.8×10^8 ly; (j) 1.2×10^9 ly; (k) 1.2×10^9 y; (l) 4.4×10^8 y
44. 13×10^9 y

SECTION SEVEN
COMPARISON OF PROBLEMS WITH THE FIFTH EDITION

In the table below, the left column (in bold type) of each group gives the numbers of problems and exercises in the current (sixth) edition of *Fundamentals of Physics* and for each entry the right column gives the number of the same problem or exercise in the fifth edition. Any problem that did not appear in that edition is labeled "new". Significant changes in a problem are noted here. If units have been switched from British to SI units, the problem is labeled "SI". Some of the exercises and problems in the sixth edition are from the supplemental problem book published to accompany the fifth edition.

Chapter 1

1E	1–5E	
2E	new	
3E	1–3E	
4E	new	
5E	1–6E	
6E	new	
7P	1–12P	
8P	new	
9P	1–14P	
10E	1–20E	
11E	1–19E	
12E	1–24E	
13P	1–30P	
14P	new	
15P	1–26P	
16P	1–27P	
17P	1–28P	
18P	new	
19E	1–34E	
20P	new	
21P	1–37E	
22P	1–35P	
23P	1–39P	
24P	1–38P	
25	new	
26	new	
27	1–40	
28	new	
29	new	

Chapter 2

1E	2–4E	
2E	new	
3E	2–8E	
4P	2–14P	
5P	2–11P	SI
6P	2–6P	SI
7P	2–12P	
8P	2–15P	
9P	1–29P	SI
10E	2–17E	
11E	2–16E	
12P	2–13P	
13P	2–19P	
14E	2–25E	
15E	2–26E	
16E	2–21E	
17E	2–20E	
18P	2–29P	
19P	2–27E	
20P	2–103	
21P	2–32P	
22E	2–35E	
23E	2–38E	
24E	2–33E	
25E	2–34E	
26E	2–102	
27E	2–36E	
28E	2–37E	
29E	2–39E	
30E	2–41E	
31E	2–44P	
32E	2–98	
33P	2–48P	
34P	2–58P	
35P	2–49P	
36P	2–56P	
37P	2–51P	
38P	2–57E	SI
39P	2–55P	
40E	2–62E	
41E	2–60E	
42E	2–64E	
43E	2–61E	
44E	2–65E	
45E	2–67E	
46P	2–74P	
47P	2–68P	
48P	2–111	

49P	2–69P
50P	2–97
51P	2–73P
52P	2–70P
53P	2–78P
54P	2–71P
55P	2–79P
56P	2–113
57P	2–85P
58P	2–76P
59P	2–87P
60P	2–83P
61P	2–88P
62P	2–77P
63P	2–75P
64P	2–84P
65P	2–90P

Chapter 3

1E	3–1E
2P	3–6P
3E	3–9E
4E	3–8E
5E	3–10E
6P	3–11E
7P	3–17P
8P	3–16P
9P	3–18P
10E	3–4E
11E	3–2E
12E	3–3E
13E	3–20E
14E	3–19E
15E	3–5P
16E	3–21E
17E	3–22E
18P	3–23E
19P	3–7P
20P	3–64
21P	3–27P
22P	3–66
23P	3–32P
24P	3–65
25P	3–33P
26P	3–63
27P	3–34P
28E	3–67
29E	3–42E
30E	3–46P
31P	3–48P
32P	3–49P
33P	3–55P
34P	3–61
35P	3–56P
36P	3–68
37P	3–58P
38P	3–51P

Chapter 4

1E	4–1E	
2E	4–2E	
3E	4–3E	
4P	3–29P	
5E	4–6E	
6E	4–8E	
7P	4–9E	
8P	4–104	
9E	4–11E	
10E	4–10E	
11E	4–12E	
12E	4–13E	
13P	4–15P	
14P	4–16P	
15P	4–14P	
16P	4–17P	
17E	4–19E	
18E	4–22E	
19E	4–24E	
20E	4–18E	
21E	4–21E	
22E	4–34P	
23E	4–26E	
24P	4–113	
25P	4–35P	
26P	4–109	
27P	4–36P	
28P	4–28E	
29P	4–37P	
30P	4–101	
31P	4–44P	
32P	4–108	
33P	4–48P	SI
34P	4–32P	
35P	4–50P	
36P	4–49P	
37P	4–51P	
38P	4–111	
39P	4–55P	SI
40P	4–53P	
41P	4–56P	
42E	4–58E	
43E	4–61E	
44E	4–63E	
45E	4–66E	
46P	4–114	
47P	4–67P	
48P	4–64E	
49P	4–68P	
50P	4–65E	
51P	4–71P	
52P	4–70P	
53E	4–75E	
54E	4–72E	
55P	4–73E	
56E	4–77E	
57E	4–79E	
58E	4–78E	
59P	4–82P	
60P	4–118	
61P	4–84P	
62P	4–86P	
63P	4–117	
64	new	

Chapter 5

1E	5–1E
2E	5–4E
3E	5–82
4E	5–5E
5E	5–6E
6P	5–10P
7P	5–8E
8P	5–11P
9E	5–18E
10E	5–78
11E	5–15E
12E	5–14E
13E	5–23E
14E	5–88
15E	5–27E
16E	5–89
17E	5–29E
18E	5–30E
19E	5–21E

20E	5–33E
21E	5–32E
22E	5–22E
23E	5–93
24P	5–94
25P	5–36P
26P	5–41P
27P	5–37P
28P	5–42P
29P	5–38P
30P	5–91
31P	5–40P
32P	5–45P
33P	5–43P
34P	5–49P
35P	5–51P
36P	5–48P
37P	5–54P
38P	5–39P
39P	5–77
40P	5–52P
41P	5–56P
42P	5–53P
43P	5–58P
44P	5–96
45P	5–60P
46P	5–66P
47P	5–64P
48P	5–69P
49P	5–68P
50P	5–74P
51P	5–73P
52P	5–63P
53P	5–70P
54P	5–65P
55P	5–50P
56P	5–62P

Chapter 6

1E	6–1E	
2E	6–3E	
3E	6–2E	
4E	6–5E	
5E	6–8E	
6E	6–13E	
7E	6–15E	
8E	6–12E	
9P	6–11E	
10P	6–76	
11P	6–18P	
12P	6–7E	
13P	6–23P	
14P	6–24P	
15P	6–25P	
16P	6–26P	
17P	6–29P	
18P	6–30P	
19P	6–31P	
20P	6–73	
21P	6–32P	
22P	6–33P	
23P	6–35P	
24P	6–75	
25P	6–38P	
26P	6–36P	
27P	6–39P	
28P	6–41P	
29P	6-40P	
30P	6–42P	
31P	6–43P	
32E	new	
33E	6–44E	
34P	6–45P	
35P	6–46P	
36E	6–50E	SI
37E	6–48E	
38E	6–91	
39E	6–49E	
40P	6–74	
41P	6–57P	
42P	6–61P	
43P	6–62P	
44P	6–69P	
45P	6–66P	
46P	6–92	
47P	6–70P	
48	new	

Chapter 7

1E	7–2E
2E	7–4E
3E	7–3E
4P	7–7P
5P	7–6P
6E	7–11E
7E	7–10E
8E	7–60
9E	7–9E
10P	7–17P
11P	7–16P
12P	7–58
13P	7–18P
14E	7–20E
15E	7–23E
16E	6–22E
17P	7–24P
18P	6–26P
19P	7–25P
20E	7–36E
21E	7–35E
22P	7–40P
23P	7–37P
24E	7–27E
25E	7–28E
26P	7–33P
27P	7–30P
28P	7–62
29P	7–31P
30E	7–41E
31E	7–45P
32E	7–44E
33P	7–48P
34P	7–65
35P	7–49P
36P	7–51P
37P	7–50P
38P	7–64
39P	7–46P
40P	7–47P

Chapter 8

1E	8–2E
2E	8–3E
3E	8–4E
4E	8–5E
5E	8–6E
6P	8–10P
7P	8–8P
8P	8–9P
9E	8–13E
10E	8–12E
11E	8–15E
12E	8–14E
13E	8–17E

14P	8–20P
15P	8–21P
16P	8–22P
17P	8–23P
18P	8–35P
19P	8–25P
20P	8–39P
21P	8–26P
22P	8–150
23P	8–33P
24P	8–110
25P	8–34P
26P	8–40P
27P	8–36P
28P	8–99
29P	8–41P
30P	8–100
31P	8–42P
32P	8–112
33P	8–43P
34P	8–113
35P	8–45P
36E	8–48E
37P	8–47P
38P	8–115
39E	8–50E
40E	8–51E
41P	8–54P
42P	8–52P
43E	8–66E
44E	8–68E
45E	8–102
46E	8–57E
47E	8–58E
48E	8–62E
49E	8–67E
50P	8–72P
51P	8–69P
52P	8–70P
53P	8–79P
54P	8–77P
55P	8–74P
56P	8–80P
57P	8–73P
58P	8–75P
59P	8–81P
60P	8–78P
61P	8–83P
62P	8–84P
63P	8–85P
64P	8–87P
65P	8–89P
66	8–68
67	8–136

Chapter 9

1E	9–1E
2E	9–2E
3E	9–3E
4E	9–4E
5E	9–5E
6P	9–7P
7P	9–8P
8P	9–9P
9P	9–11P
10E	9–12E
11E	9–13E
12E	9–14E
13P	9–18P
14P	9–19P
15P	9–21P
16P	9–82
17P	9–22P
18P	9–20P
19P	9–23P
20E	9–25E
21E	9–24E
22E	9–28E
23P	9–30P
24P	9–29P
25P	9–31P
26P	9–84
27E	9–34E
28E	9–35E
29E	9–37E
30P	9–78
31E	9–36E
32E	9–38E
33P	9–39P
34P	9–42P
35P	9–40P
36P	10–49P
37P	10–66P
38P	10–51P
39P	9–43P
40P	9–46P
41E	9–51E
42E	9–50E
43E	9–49E
44E	9–52E
45E	9–53E
46E	9–54E
47P	9–57P
48P	9–56P
49E	new
50E	9–59E
51E	9–62E
52E	9–63E
53E	9–65E
54E	9–67E
55E	9–69E
56P	9–71P
57P	9–73P
58P	9–74P
59	new

Chapter 10

1E	10–2E
2E	10–3E
3E	10–5E
4E	10–6E
5E	10–8E
6E	10–7E
7E	10–9E
8P	10–15P
9P	10–11P
10P	10–82
11P	10–12P
12P	10–16P
13P	10–13P
14P	10–19P
15P	10–24P
16P	10–20P
17P	10–21P
18P	10–22P
19P	10–26P
20E	10–42E
21E	10–41E
22E	10–44E
23E	10–40E
24E	10–45E
25P	10–47P
26P	10–50P
27P	10–52P
28P	10–46E

29P	10–53P
30P	10–89
31P	10–54P
32P	10–88
33P	10–55P
34P	10–56P
35E	10–28E
36E	10–30E
37E	10–31E
38E	10–27P
39E	10–32E
40P	10–34P
41P	10–33E
42P	new
43P	10–36P
44P	10–38P
45P	10–37P
46E	10–43E
47E	10–60P
48E	10–61E
49E	10–63E
50P	10–67P
51P	10–69P
52P	10–71P
53P	10–72P
54P	10–92
55P	10–68P
56	new
57	new

Chapter 11

1E	11–2E
2E	11–3E
3E	11–4E
4E	11–5E
5E	11–6E
6P	11–10P
7P	11–9P
8E	11–11E
9E	11–12E
10E	11–13E
11E	11–15E
12E	11–17E
13P	11–20P
14P	11–19P
15P	11–22P
16P	11–23P
17P	11–21P
18P	11–99
19E	11–25E
20E	11–26E
21E	11–27E
22E	11–28E
23E	11–32E
24E	11–33E
25P	11–38P
26P	11–35P
27P	11–40P
28P	11–36P
29P	11–41P
30P	11–96
31P	11–44P
32P	11–43P
33E	11–45E
34P	11–46P
35E	11–47E
36E	11–49E
37E	11–50E
38E	11–51E
39E	11–53E
40P	11–95
41P	11–56P
42P	11–57P
43P	11–58P
44P	11–59P
45E	11–61E
46E	11–60E
47P	11–63P
48P	11–64P
49E	11–66E
50E	11–65E
51E	11–69E
52E	11–67E
53P	11–70P
54P	11–105
55P	11–74P
56P	11–71P
57P	11–75P
58E	11–77E
59E	11–78E
60E	11–79E
61E	11–80E
62P	11–81P
63P	11–82P
64P	11–84P
65P	11–83P
66P	11–86P
67P	11–87P
68	11–31E
69	new
70	new

Chapter 12

1E	12–3E
2P	12–10P
3E	12–2E
4E	12–1E
5E	12–4E
6E	12–11E
7E	12–7E
8P	12–9P
9P	12–12P
10P	12–13P
11P	12–14P
12P	12–15P
13P	12–80
14P	12–16P
15E	12–17E
16P	12–18P
17E	12–20E
18E	12–21E
19E	12–22E
20P	12–23P
21P	12–26P
22P	12–24P
23E	12–27E
24E	12–29E
25E	12–31E
26P	12–34P
27P	12–33P
28P	12–83
29E	12–36E
30E	12–37E
31E	12–38E
32P	12–40P
33E	12–43E
34E	12–42E
35E	12–44E
36P	12–48P
37P	12–49P
38P	12–46P
39E	12–51E
40E	12–50E
41E	12–53E
42E	12–52E

43E	12–56E
44E	12–54E
45P	12–61P
46P	12–59P
47P	12–62P
48P	12–63P
49P	12–64P
50P	12–65P
51P	12–70P
52P	12–66P
53P	12–68P
54P	12–57E
55P	12–67P
56P	12–86
57P	12–87
58P	12–88
59P	12–69P

Chapter 13

1E	13–1E
2E	13–3E
3E	13–4E
4E	13–5E
5E	13–7E
6E	13–9E
7E	13–12E
8E	13–13E
9E	13–15E
10E	13–6E
11E	13–16E
12E	13–11E
13E	13–17E
14E	13–18E
15P	13–21P
16P	13–23P
17P	13–22P
18P	13–25P
19P	13–28P
20P	13–26P
21P	13–29P
22P	13–27P
23P	13–32P
24P	13–31P
25P	13–33P
26P	13–34P
27P	13–35P
28P	13–36P
29P	13–40P
30P	13–37P
31P	13–41P
32P	13–38P
33P	13–43P
34P	13–39P
35P	13–44P
36E	13–46E
37E	13–50E
38P	13–54P
39P	13–55P
40P	13–52P
41	new
42	new

Chapter 14

1E	14–1E
2E	14–2E
3E	14–4E
4E	14–3E
5P	14–5P
6E	14–7E
7E	14–6E
8P	14–92
9P	14–10P
10P	14–11P
11P	14–13P
12P	14–9P
13P	14–15P
14E	14–18E
15E	14–17P
16E	14–21E
17E	14–23E
18P	14–94
19P	14–25P
20P	14–27P
21P	14–26P
22E	14–28E
23P	14–31P
24P	14–32P
25P	14–33P
26P	14–34P
27E	14–35E
28E	14–36E
29E	14–37E
30E	14–39E
31P	14–43P
32P	14–46P
33P	14–44P
34P	14–96
35P	14–45P
36P	14–97
37P	14–48P
38P	14–51P
39P	14–49P
40E	14–54E
41E	14–55E
42E	14–56E
43E	14–57E
44E	14–58E
45E	14–59E
46E	14–62E
47E	14–61E
48E	14–64E
49E	14–65E
50E	14–66E
51P	14–68P
52P	14–99
53P	14–70P
54P	14–71P
55P	14–73P
56E	14–76E
57E	14–75E
58P	14–77P
59P	14–78P
60P	14–79P
61P	14–80P
62P	14–81P
63P	14–83P
64E	14–86E

Chapter 15

1E	15–3E	
2E	15–2E	
3E	15–5E	
4E	15–4E	
5E	15–6E	
6P	15–7E	SI
7P	15–8P	SI
8E	15–9E	
9E	15–11E	
10E	15–12E	
11E	15–14E	SI
12E	15–15E	
13E	15–16E	
14E	15–18E	
15P	15–19P	

16P	15–21P	
17P	15–22P	
18P	15–24P	
19P	15–24P	
20E	15–27E	
21E	15–28P	
22E	15–29E	SI
23E	15–30E	SI
24E	15–32E	SI
25E	15–34E	
26E	15–36E	SI
27E	15–37E	
28E	15–39E	
29P	15–41P	
30P	15–33E	
31P	15–42P	
32P	15–48P	
33P	15–43P	
34P	15–46P	
35P	15–44P	
36P	15–45P	SI
37P	15–51P	
38P	15–47P	
39E	15–55E	SI
40E	15–53E	
41P	15–56P	
42E	15–54E	
43E	15–58E	
44E	15–59E	
45E	15–61E	SI
46E	15–60E	SI
47E	15–64E	SI
48E	15–66E	
49E	15–67E	
50E	15–65E	
51P	15–69P	
52P	15–70P	
53P	15–74P	
54P	15–75P	
55P	15–77P	
56P	15–78P	
57P	15–79P	
58P	15–80P	
59	new	
60	new	
61	new	
62	new	

Chapter 16

1E	16–1E	
2E	16–2E	
3E	16–4E	
4E	16–6E	
5E	16–7E	
6E	16–8E	SI
7E	16–11E	
8E	16–12E	
9E	16–13E	
10E	16–14E	
11E	16–15E	
12E	16–16E	
13E	16–18E	
14P	16–19P	
15P	16–24P	
16P	16–25P	
17P	16–26P	
18P	16–27P	
19P	16–28P	
20P	16–29P	
21P	16–30P	
22P	16–32P	
23P	16–31P	
24P	16–33P	
25P	16–34P	
26P	16–22P	
27P	16–35P	
28P	16–36P	
29P	16–37P	
30P	16–38P	
31E	16–40E	
32E	16–41E	
33E	16–42E	
34E	16–44E	
35E	16–43E	
36E	16–46E	
37E	16–45E	
38P	16–48P	
39P	16–50P	
40E	16–52E	
41P	16–55P	
42E	16–57E	
43E	16–58E	
44E	16–61E	
45E	16–63E	
46E	16–65E	
47E	16–64E	
48E	16–66E	
49E	16–67E	
50E	16–68E	
51E	16–69E	
52P	16–74P	
53P	16–75P	
54P	16–77P	
55P	16–76P	
56P	16–78P	
57P	16–79P	
58P	16–80P	
59E	16–85E	
60E	16–84E	
61E	16–86E	
62P	16–89P	
63E	16–90E	
64P	16–91P	
65	new	

Chapter 17

1E	17–2E
2E	17–3E
3E	17–4E
4E	17–5E
5E	17–7E
6P	17–12P
7P	17–13P
8P	17–15P
9P	17–16P
10E	17–18E
11E	17–17E
12E	17–20E
13E	17–23E
14E	17–22E
15P	17–25P
16P	17–24E
17P	17–26P
18P	17–28P
19P	17–27P
20P	17–29P
21P	17–30P
22P	17–31P
23P	17–32P
24E	17–34E
25P	17–35P
26E	17–37E
27E	17–37E
28P	17–38P
29E	17–39E

30P	17–42P
31P	17–43P
32E	17–45E
33E	17–46E
34E	17–47E
35E	17–49E
36E	17–52E
37E	17–53E
38E	17–55E
39P	17–56P
40P	17–61P
41P	17–58P
42P	17–69
43P	17–62P
44P	17–64P
45P	17–63P
46P	17–68
47P	17–57P
48P	17–75
49P	17–65P
50P	17–76
51P	17–66P
52	new

Chapter 18

1E	18–1E	
2E	18–3E	
3E	18–4E	
4E	18–2E	
5P	18–10P	
6P	18–8E	
7P	18–11P	
8E	18–12E	
9E	18–14E	
10P	18–17P	
11P	18–18P	
12P	18–23E	
13P	18–21P	
14P	18–19P	
15P	18–22P	
16P	18–24P	
17E	18–28E	
18E	18–27E	
19E	18–29E	
20E	18–30E	
21E	18–31E	
22E	18–34E	
23E	18–35E	
24P	18–37P	
25P	18–36P	
26P	18–39P	
27P	18–38P	
28P	18–42P	
29P	18–45P	
30E	18–48E	
31E	18–50E	
32E	18–51E	
33E	18–52E	
34P	18–53P	
35P	18–55P	
36P	18–54P	
37P	18–56P	
38P	18–58P	
39P	18–59P	
40P	18–101	
41P	18–60P	
42E	18–62E	
43E	18–61E	
44P	18–64P	
45P	18–63P	
46E	18–66E	
47E	18–68E	
48E	18–69E	
49E	18–74P	
50E	18–78P	
51E	18–81P	
52P	18–82P	SI
53P	18–84P	
54P	18–85P	
55P	18–89P	
56P	18–87P	
57P	18–88P	
58P	18–72E	
59P	18–76P	
60P	18–75P	

Chapter 19

1E	19–3E
2E	19–2E
3E	19–4P
4E	19–5E
5E	19–9E
6E	19–7E
7P	19–11P
8P	19–12P
9P	19–10P
10E	19–14E
11E	19–15E
12E	19–17E
13E	19–16E
14E	19–24E
15E	19–23E
16E	19–30P
17E	19–26E
18P	19–28P
19P	19–29P
20P	19–32P
21P	19–33P
22P	19–34P
23P	19–35P
24P	19–103
25P	19–36P
26E	19–41E
27E	19–39E
28E	19–43E
29E	19–44E
30E	19–43E
31E	19–47E
32E	19–46E
33E	19–49E
34E	19–51E
35E	19–52E
36P	19–54E
37P	19–60P
38P	19–57P
39P	19–110
40P	19–59P
41P	19–62P
42P	19–63P
43P	19–65P
44P	19–66P
45P	19–67P
46P	19–64P
47P	19–68P
48E	19–73E
49E	19–71E
50E	19–74E
51E	19–76E
52E	19–75E
53P	19–78P
54E	19–79E
55E	19–80E
56E	19–82E
57E	19–83E
58E	19–86E

59E	19–84E	
60E	19–87E	
61P	19–88P	
62P	19–92P	
63P	19–91P	
64P	19–119	
65P	19–95P	
66P	19–96P	
67	new	
68	new	

Chapter 20

1E	20–2E	
2E	20–1E	
3P	20–3P	
4P	20–5P	
5E	20–7E	
6E	20–8E	
7E	20–10E	
8E	20–11E	SI
9E	20–9E	
10E	20–12E	
11P	20–14P	
12P	20–17P	
13P	20–15P	
14P	20–18P	
15P	20–21P	
16P	20–22P	
17P	20–25P	
18E	20–26E	
19E	20–27E	
20E	20–28E	
21E	20–29E	
22E	20–30E	
23P	20–33P	
24P	20–32P	
25E	20–34E	
26E	20–35E	
27P	20–39P	
28P	20–38P	
29P	20–40P	
30E	20–41E	
31E	20–42E	
32E	20–46P	
33E	20–43E	
34P	20–48P	
35P	20–45P	
36E	20–51E	
37E	20–50E	
38E	20–52E	
39P	20–55P	
40E	20–54E	
41P	20–56P	
42P	20–57P	
43P	20–59P	
44E	20–60E	
45E	20–61E	
46P	20–63P	
47P	20–65P	
48P	20–67P	
49P	20–66P	
50E	20–93	
51E	20–68E	
52E	20–69E	
53P	20–70P	
54E	20–72E	
55E	20–71E	
56E	20–74E	
57E	20–73E	
58E	20–75P	
59E	20–76E	
60P	20–80P	
61P	20–85P	

Chapter 21

1E	21–2E
2E	21–1E
3E	21–6E
4E	21–3E
5E	21–7E
6E	21–8E
7E	21–11E
8P	21–15P
9P	21–17P
10P	21–19P
11P	21–20P
12P	21–22P
13P	21–24P
14P	21–23P
15P	21–25P
16P	21–26P
17P	21–27P
18P	21–28P
19P	21–29P
20P	21–30P
21E	21–33E
22E	21–37E
23E	21–38E
24E	21–39E
25E	21–40E
26P	21–46P
27P	21–42P
28P	new
29P	21–45P
30P	21–47P
31P	21–49P
32P	21–48P
33P	21–50P
34E	21–78
35E	21–57E
36E	21–58E
37E	21–60P
38E	21–59E
39P	21–62P
40P	21–61P
41P	21–63P
42E	21–64W
43E	21–65E
44P	21–68P
45P	21–69P

Chapter 22

1E	22–4E
2E	22–3E
3E	22–5E
4E	22–7E
5P	22–10P
6P	22–11P
7P	22–13P
8P	22–8P
9P	22–16P
10P	22–14P
11P	22–17P
12P	22–15P
13P	22–18P
14P	22–19P
15P	22–20P
16P	22–21P
17P	22–22P
18E	22–23E
19E	22–25E
20E	22–26E
21E	22–27E
22E	22–29E

23E 22–28E
24E 22–31E
25P 22–32P
26P 22–34P
27P 22–35P
28P 22–36P
29E 22–39E
30 22–41

Chapter 23

1E 23–1E
2E 23–4E
3E 23–5E
4E 23–6E
5E 23–7E
6E 23–8E
7E 23–9E
8P 23–11P
9P 23–14P
10P 23–12P
11P 23–15P
12P 23–19P
13P 23–22P
14E 23–24E
15E 23–23E
16P 23–25P
17P 23–26P
18E 23–28E
19P 23–30P
20P 23–31P
21P 23–32P
22P 23–29P
23P 23–34P
24P 23–33P
25P 23–35P
26E 23–36E
27P 23–38P
28E 23–40E
29E 23–39E
30E 23–43E
31E 23–41E
32E 23–42E
33E 23–44E
34E 23–45E
35E 23–47E
36E 23–48E
37E 23–50E
38P 23–51P
39P 23–52P
40P 23–54P
41P 23–55P
42P 23–53P
43P 23–57P
44E 23–58E
45E 23–59E
46P 23–60P
47P 23–61P
48 new

Chapter 24

1E 24–1E
2E 24–2E
3E 24–3E
4E 24–5E
5E 24–7E
6E 24–10E
7P 24–12P
8P 24–61
9P 24–14P
10P 24–63
11P 24–15P
12E 24–17E
138 24–18E
14E 24–19E
15P 24–21P
16E 24–23E
17E 24–22E
18P 24–24P
19P 24–27P
20P 24–26P
21P 24–28P
22P 24–66
23P 24–29P
24P 24–67
25P 24–31P
26E 24–32E
27E 24–33E
28E 24–34E
29P 24–35P
30P 24–36P
31P 24–37P
32P 24–38P
33P 24–41P
34E 24–43E
35E 24–42E
36E 24–46E
37E 24–48E
38E 24–49E
39P 24–50P
40P 24–51P
41P 24–52P
42P 24–56P
43P 24–53P
44P 24–54P
45P 24–55P
46P 24–57P
47P 24–58P
48 new

Chapter 25

1E 25–2E
2E 25–1E
3P 25–3P
4E 25–6E
5E 25–9E
6E 25–8E
7P 25–11P
8P 25–12P
9P 25–13P
10P 25–10P
11P 25–14P
12E 25–18E
13E 25–15E
14E 25–20E
15E 25–21E
16E 25–23E
17P 25–26P
18P 25–25P
19P 25–27P
20P 25–30P
21P 25–28P
22E 25–33E
23P 25–34P
24E 25–35E
25E 25–37E
26E 25–36E
27E 25–38E
28P 25–41P
29P 25–42P
30E 25–43E
31E 25–46E
32E 25–47E
33P 25–49P
34P 25–51P

35P	25–50P
36E	25–52E
37E	25–56E
38E	25–58E
39P	25–60P
40P	25–61P
41P	25–62P
42P	25–63P
43P	25–64P
44P	25–65P
45P	25–66P
46P	25–70P
47P	25–71P
48P	25–67P
49P	25–73P
50E	25–74E
51E	25–76E
52E	24–77E
53P	25–80P
54P	25–81P
55P	25–79P
56P	25–82P
57	new

Chapter 26

1E	26–1E
2E	26–2E
3E	26–3E
4E	26–4E
5E	26–5E
6E	26–6E
7E	26–10E
8E	26–7E
9P	26–12P
10E	26–15E
11E	26–14E
12E	26–17E
13E	26–16E
14P	26–24P
15P	26–20P
16P	26–26P
17P	26–27P
18P	26–28P
19P	26–29P
20P	26–31P
21P	26–30P
22E	26–33E
23E	26–35E
24E	26–38E
25E	26–39E
26P	26–42P
27P	26–43P
28P	26–45P
29P	26–46P
30P	26–47P
31P	26–48P
32P	26–41P
33P	26–49P
34E	26–52E
35E	26–53E
36E	26–55E
37E	26–56E
38P	26–58P
39P	26–57P
40P	26–63P
41P	26–64P
42P	26–65P
43E	26–66E
44E	26–67E
45P	26–68P
46P	26–69P
47P	26–72P
48	new

Chapter 27

1E	27–2E
2P	27–3P
3P	27–4P
4E	27–7E
5E	27–6E
6E	27–5E
7E	27–8E
8P	27–11P
9P	27–12P
10P	27–15P
11P	27–13P
12E	27–18E
13E	27–17E
14E	27–16E
15E	27–19E
16E	27–21E
17E	27–20E
18E	27–23E
19E	27–27E
20E	27–28E
21P	27–29P
22P	27–33P
23P	27–34P
24P	27–36P
25P	27–35P
26P	27–38P
27P	27–40P
28P	27–39P
29P	27–41P
30P	27–42P
31E	27–44E
32E	27–43E
33E	27–47E
34E	27–45E
35E	27–49E
36E	27–48E
37P	27–56P
38P	27–54P
39P	27–57P
40P	27–55P
41P	27–59P
42	new
43	new
44	new
45	new

Chapter 28

1E	28–1E
2E	28–3E
3E	28–4E
4E	28–5E
5E	28–6E
6E	28–7E
7E	28–9E
8E	28–8E
9E	28–11E
10E	28–10E
11E	28–13E
12P	28–16P
13P	28–17P
14P	28–21P
15P	28–22P
16P	28–23P
17P	28–24P
18E	28–29E
19E	28–27E
20E	28–31E
21E	28–32E
22E	28–33E

23E	28–34E
24E	28–35E
25E	28–36E
26P	28–39P
27P	28–41P
28P	28–47P
29P	28–42P
30P	28–44P
31P	28–48P
32P	28–49P
33P	28–43P
34P	28–50P
35P	28–51P
36E	28–53E
37P	28–55P
38P	28–57P
39P	28–58P
40P	28–60P
41P	28–61P
42P	28–62P
43P	28–63P
44E	28–65E
45E	28–67E
46E	28–66E
47E	28–68E
48P	28–73P
49P	28–69P
50P	28–74P
51P	28–70P
52P	28–71P
53P	28–72E
54P	28–75P
55P	28–77P
56	new
57	new
58	new

Chapter 29

1E	29–2E
2E	29–3E
3E	29–4E
4P	29–6P
5P	29–5P
6E	29–8E
7E	29–9E
8E	29–10E
9P	29–12P
10P	29–11P
11P	29–13P
12E	29–14E
13P	29–17P
14P	29–18P
15E	29–20E
16E	29–19E
17E	29–23E
18E	29–24E
19E	29–26E
20E	29–25E
21E	29–27E
22P	29–28P
23P	29–29P
24P	29–31P
25P	29–34P
26P	29–32P
27P	29–36P
28P	29–82
29P	29–37P
30E	29–40E
31P	29–44P
32P	29–42P
33E	29–45E
34E	29–47E
35E	29–46E
36P	29–48P
37P	29–53P part added
38P	29–52P
39E	29–55E
40E	29–54E
41E	29–56E
42P	29–57P
43P	29–60P
44P	29–58P
45P	29–61P
46P	29–59P
47P	29–62P
48E	29–64E
49E	29–63E
50E	29–65E
51E	29–66E
52E	29–67E
53E	29–68E
54P	29–70P
55P	29–69P

Chapter 30

1E	30–3E
2E	30–4E
3E	30–7E
4E	30–6E
5E	30–8E
6E	30–9E
7P	30–16E
8P	30–12P
9P	30–13P
10P	30–11P
11P	30–17P
12P	30–19P
13P	30–18P
14P	30–20P
15P	30–21P
16P	30–23P
17P	30–22P
18P	30–24P
19P	30–26P
20P	30–25P
21E	30–27E
22E	30–28E
23E	30–29E
24E	30–30E
25P	30–34P
26P	30–35P
27P	30–36P
28P	30–39P
29P	30–38P
30E	30–41E
31E	30–40E
32E	30–43E
33P	30–45P
34P	30–44P
35P	30–48P
36P	30–81
37P	30–50P
38P	30–49P
39P	30–51P
40E	30–54E
41E	30–53E
42E	30–55E
43E	30–56E
44P	30–58P
45P	30–59P
46P	30–61P
47P	30–60P
48E	30–63E

49E	30–62E
50E	30–64E
51E	30–65E
52E	30–66E
53P	30–68P
54P	30–71P
55P	30–72P
56P	30–67P
57P	30–69P
58P	30–70P

Chapter 31

1E	31–3E
2E	31–2E
3E	31–5E
4E	31–6E
5E	31–8E
6P	31–9P
7P	31–10P
8P	31–14P
9P	31–18P
10P	31–16P
11P	31–19P
12P	31–17P
13P	31–20P
14P	31–21P
15P	31–22P
16P	31–23P
17P	31–25P
18P	31–24P
19P	31–26P
20P	31–112
21P	31–28P
22P	31–27P
23P	31–30P
24P	31–29P
25E	31–32E
26E	31–31E
27E	31–33E
28P	31–35P
29P	31–34E
30P	31–36P
31P	31–39P
32E	31–41P
33E	31–40E
34P	31–42P
35P	31–44P
36E	31–46E
37E	31–45E
38P	31–49P
39P	31–50P
40E	31–51E
41E	31–52E
42P	31–55P
43P	31–56P
44P	31–57P
45E	31–59E
46E	31–58E
47E	31–60E
48E	31–62E
49E	31–63E
50P	31–65P
51P	31–67P
52P	31–68P
53P	31–69P
54P	31–70P
55P	31–73P
56E	31–75E
57E	31–78P
58E	31–77E
59E	31–79P
60P	31–80P
61P	31–82P
62E	31–84E
63E	31–83E
64E	31–86E
65E	31–85E
66E	31–89E
67P	31–91P
68E	31–94E
69E	31–93E
70E	31–95E
71P	31–96P
72P	31–97P
73P	31–99P
74P	31–98P
75P	31–100P

Chapter 32

1E	32–1E
2E	32–2E
3P	32–3P
4E	32–6E
5E	32–5E
6P	32–8P
7P	32–9P
8P	32–10P
9E	32–12E
10E	32–11E
11E	32–14E
12E	32–18E
13P	32–19P
14E	32–20E
15E	32–21E
16E	32–22E
17E	32–23E
18E	32–24E
19P	32–26P
20E	32–28E
21E	32–29E
22E	32-31E
23E	32–30E
24P	32–33P
25P	32–32P
26E	32–36E
27E	32–37E
28P	32–39P
29E	32–40E
30E	32–41E
31E	32–42E
32E	32–43E
33P	32–47P
34P	32–45P
35P	32–48P
36P	32–56
37P	32–49P
38P	32–58

Chapter 33

1E	33–1E
2E	33–3E
3E	33–4E
4E	33–5E
5P	33–6P
6E	33–7E
7P	33–8P
8E	33–9E
9E	33–11E
10E	33–14E
11P	33–15P
12P	33–17P
13P	33–18P
14P	33–19P
15P	33–24P

16P	33–22P
17P	33–25P
18P	33–23P
19P	33–26P
20P	33–28P
21P	33–29P
22P	33–27P
23P	33–30P
24E	33–32E
25E	33–31E
26P	33–34P
27P	33–33P
28P	33–35P
29P	33–37P
30E	33–38E
31E	33–39E
32E	33–40E
33E	33–43E
34P	33–44P
35P	33–46P
36P	33–45P
37E	33–49E
38E	33–50E
39E	33–51E
40P	33–53P
41P	33–55P
42P	33–57P
43P	33–56P
44P	33–58P
45P	33–59P
46P	33–64P
47P	33–67P
48P	33–65P
49E	33–72E
50E	33–70E
51E	33–69E
52E	33–71E
53E	33–73E
54E	33–74E
55E	33–76E
56P	33–80P
57P	33–82P
58P	33–81P
59P	33–83P
60P	33–84P
61P	33–85P
62E	33–86E
63E	33–87E
64E	33–88E
65P	33–90P

Chapter 34

1E	34–2E
2E	34–1E
3E	34–4E
4E	34–6E
5P	34–7P
6E	34–8E
7E	34–9E
8E	34–11E
9E	34–12E
10E	34–15E
11E	34–16E
12E	34–17E
13E	34–20E
14E	34–23E
15E	34–24E
16P	34–27P
17P	34–28P
18P	34–29P
19P	34–31P
20E	34–33E
21E	34–34E
22E	34–36E
23E	34–37E
24P	34–39P
25P	34–38P
26P	34–40P
27P	34–42P
28P	34–44P
29P	34–45P
30P	34–46P
31P	34–47P
32E	34–48E
33E	34–49E
34E	34–52E
35E	34–53P
36P	34–55P
37P	34–56P
38P	34–57P
39P	34–58P
40P	34–60P
41P	34–59P
42E	34–61E
43E	34–62E
44E	34–63E
45E	34–64E
46P	34–109
47P	34–65P
48P	34–68P
49P	34–67P
50P	34–70P
51P	34–71P
52E	34–74E
53E	34–75E
54E	34–77E
55E	34–76E
56P	34–79P
57P	34–78P
58P	34–81P
59P	34–80P
60E	34–85E
61E	34–86E
62E	34–87E
63E	new
64E	new
65P	new
66P	new

Chapter 35

1E	35–3E
2E	35–4E
3E	35–6E
4P	35–7P
5P	35–9P
6P	35–8P
7P	35–11P
8P	35–12P
9E	35–15E
10P	35–16P
11P	35–17P
12P	35–18P
13P	35–19P
14P	35–20P
15P	35–21P
16P	35–22P
17E	35–24E
18E	35–28E
19E	35–27E
20E	35–29E
21E	35–31E
22E	35–32E
23P	35–42P
24P	35–34P
25P	35–43P

26P 35–70
27P 35–35P
28P 35–39P
29P 35–40P
30P 35–38P
31P 35–44P
32E 35–47E
33E 35–46E
34P 35–48P
35P 35–50P
36P 35–53P
37P 35–52P

Chapter 36

1E 36–1E
2E 36–2E
3E 36–4E
4E 36–5E
5P 36–9P
6P 36–11P
7P 36–12P
8P 36–13P
9P 36–15P
10P 36–14P
11E 36–16E
12E 36–17E
13E 36–19E
14E 36–22E
15E 36–21E
16E 36–24E
17E 36–25E
18E 36–26E
19P 36–27P
20P 36–29P
21P 36–33P
22P 36–30P
23E 36–36E
24E 36–35E
25E 36–38P
26E 36–37E
27P 36–39P
28P 36–40P
29P 36–41P
30E 36–42E
31E 36–44E
32E 36–43E
33E 36–45E
34E 36–47E
35E 36–49E
36P 36–51P
37E 36–50E
38P 36–52P
39P 36–53P
40P 36–54P
41P 36–57P
42P 36–55P
43P 36–62P
44P 36–63P
45P 36–65P
46P 36–58P
47P 36–66P
48P 36–67P
49P 36–68P
50P 36–69P
51P 36–70P
52P 36–71P
53P 36–72P
54E 36–74E
55E 36–73E
56P 36–76P
57P 36–75P
58P 36–77P
59 new
60 new

Chapter 37

1E 37–3E
2E 37–2E
3E 37–4E
4E 37–5E
5E 37–7E
6P 37–9P
7P 37–8P
8E 37–11E
9E 37–13P
10E 37–12E
11P 37–15P
12P 37–14P
13P 37–16P
14E 37–18E
15E 37–19E
16E 37–20E
17E 37–21E
18E 37–22E
19E 37–24E
20E 37–26E
21P 37–29P
22P 37–31P
23P 37–33P
24P 37–35P
25P 37–34P
26P 37–36P
27E 37–37E
28E 37–38E
29P 37–40P
30P 37–41P
31P 37–42P
32P 37–43P
33E 37–45E
34E 37–47E
35E 37–48E
36E 37–50E
37P 37–52P
38P 37–53P
39P 37–54P
40P 37–55P
41P 37–56P
42P 37–57P
43P 37–62P
44P 37–61P
45P 37–17P
46E 37–63E
47E 37–65E
48E 37–64E
49E 37–68E
50E 37–67E
51P 37–72P
52P 37–71P
53E 37–73E
54E 37–76E
55E 37–78E
56E 37–77E
57P 37–79P
58P 37–81P
59P 37–82P
60P 37–80P
61P 37–83P
62 new
63 new
64 new

Chapter 38

1E 38–2E
2E 38–1E

3E	38–5E
4E	38–4E
5P	38–6P
6P	38–8P
7E	38–9E
8E	38–12E
9E	38–11E
10E	38–10E
11E	38–14E
12P	38–17P
13P	38–15P
14E	38–19E
15E	38–18E
16E	38–20E
17E	38–21E
18P	38–24P
19P	38–23P
20P	38–25P
21E	38–26E
22E	38–27E
23E	38–29E
24E	38–30E
25P	38–31P
26P	38–34P
27E	38–35E
28E	18-90E
29E	18-91E
30E	18-93E
31P	38–38P
32E	38–40E
33E	38–42E
34E	38–43E
35E	38–44E
36E	10–74E
37P	38–47E
38P	38–48P
39P	38–52P
40P	38–49P
41P	38–53P
42P	38–56P
43P	38–55P
44P	38–58P
45P	38–57P
46P	38–60P
47P	38–61P
48P	38–62P
49P	38–63P
50P	38–64P
51P	10-76P
52P	38–66P
53P	38–67P

Chapter 39

1E	39–3E
2E	39–4E
3E	39–1E
4E	39–2E
5E	39–7E
6E	39–8E
7E	39–5E
8E	39–6E
9E	39–9E
10P	39–10P
11P	39–13P
12P	39–14P
13P	39–11P
14P	39–12P
15P	39–15P
16E	39–18E
17E	39–17E
18E	39–16E
19E	39–20E
20E	39–19E
21E	39–21E
22P	39–23P
23P	39–22P
24P	39–24P
25P	39–25P
26P	39–27P
27P	39–26P
28P	39–28P
29P	39–29P
30P	39–30P
31E	39–33E
32E	39–32E
33E	39–31E
34P	39–36P
35P	39–35P
36P	39–34P
37P	39–37P
38P	39–40P
39P	39–39P
40P	39–38P
41P	39–41P
42P	39–44P
43P	39–43P
44P	39–42P
45P	39–47P
46P	39–48P
47P	39–45P
48P	39–46P
49E	39–50P
50E	39–49E
51E	39–51E
52E	39–52E
53P	39–53P
54P	39–56P
55P	39–55P
56P	39–54P
57P	39–58P
58P	39–57P
59P	39–60P
60P	39–59P
61P	39–61P
62P	39–62P
63P	39–63P
64P	39–64P
65P	39–65P
66E	39–66E
67P	39–67P
68P	39–68P
69P	39–69P
70P	39–71P
71P	39–70P
72P	39–72P
73P	39–73P
74E	39–74E
75E	39–76E
76E	39–75E
77P	39–77P
78P	39–78P
79P	39–79E
80P	39–80P
81P	39–82P
82P	39–81P
83P	39–83P

Chapter 40

1E	40–2E
2E	40–1E
3E	40–4E
4E	40–3E
5E	40–8E
6E	40–5E
7E	40–6E

8P 40–12P
9P 40–11P
10P 40–10P
11P new
12P new
13P 40–14P revised
14E new
15E 40–9E revised
16P new
17P 40–13P
18E 40–18E
19E 40–17E
20E 40–16E
21P 40–19P
22P 40–21P
23P 40–20P
24E new
25E new
26P new
27P new
28P new
29P new
30E 40–22E
31E 40–23E
32E 40–24E
33E 40–28E
34E 40–26E
35E 40–27E
36E 40–25E
37E 40–29E
38E 40–32E
39E 40–31E
40E 40–30E
41P 40–33P
42P 40–36P
43P 40–37P
44P 40–34P
45P 40–35P
46P 40–38P
47P 40–39P
48P 40–40P
49P 40–42P
50P 40–41P
51P 40–44P
52P 40–43P
53P 40–45P
54P 40–46P
55P 40–48P
56P 40–47P
57P 40–49P
58P 40–50P
59P 40–51P

Chapter 41

1E 41–1E
2E 41–2E
3E 41–5E
4E 41–4E
5E 41–3E
6E 41–7E
7E 41–6E
8E 41–9E
9E 41–8E
10E 41–10E
11P 41–12P
12E 41–11E
13P 41–14P
14P 41–13P
15E 41–15E
16E 41–17E
17E 41–16E
18P 41–18P
19E 41–19E
20E 41–20E
21E 41–22E
22E 41–21E
23E new
24E new
25P new
26P new
27P new
28P new
29E 41–23E
30E 41–24P
31E 41–25P
32E 41–26P
33P 41–27P
34P 41–28P
35P 41–29P
36E 41–32E
37E 41–31E
38E 41–30E
39P 41–34P
40P 41–33P
41P 41–35P
42P 41–36E
43P 41–37E
44P 41–38E
45P 41–40E
46P 41–39P
47E 41–41P
48P 41–43P
49E 41–42P
50P 41–44P
51P 41–45P
52P 41–46P
53E 41–47E
54E 41–48E
55E 41–50E
56E 41–49E
57E 41–51E
58E 41–52E
59E 41–54E
60E 41–53E
61E 41–57E
62E 41–56E
63E 41–55E
64P 41–59P
65P 41–58P
66P 41–60P
67P 41–62P
68P 41–61P
69 41–63
70 41–64

Chapter 42

1E 42–1E
2E 42–3E
3E 42–2E
4E 42–6E
5E 42–5E
6E 42–4E
7E 42–8E
8E 42–7E
9E 42–9E
10E 42–10E
11E 42–11E
12E 42–12E editorial change
13E 42–13E
14P 42–14P
15P 42–17P
16P 42–16P
17P 42–15P editorial change
18P 42–20P
19P 42–19P

20P 42–18P
21P 42–21P
22P 42–23P
23P 42–22P
24P 42–26P
25P 42–25P
26P 42–24P
27P 42–27P
28P 42–27P
29P 42–29P
30P 42–30P
31P 42–31P
32P 42–32P
33P 42–33P
34P 42–34P
35E 42–37P
36P 42–36P
37P 42–35P
38P 42–38P
39P 42–39P
40P 42–40P
41P 42–41P
42P 42–43P
43P 42–42P
44E 42–45P
4EP 42–44P editorial change
46P 42–47P
47P 42–46P
48P 42–48P
49P 42–49P

Chapter 43

1E 43–2E
2E 43–1E
3P 43–3P
4E 43–6E
5E 43–7E
6E 43–4E
7E 43–5E
8E 43–8E
9E 43–11E
10E 43–10E revised
11E 43–9E
12E 43–12E
13E 43–13E
14E 43–14E
15E 43–16E
16E 43–15E
17E 43–17E
18E 43–18E
19P 43–19P
20P 43–21P
21P 43–20P editorial change
22P 43–24P
23P 43–23P
24P 43–22P
25P 43–25P editorial change
26E 43–28E
27E 43–27E
28E 43–26E
29E 43–29E
30E 43–31E
31E 43–30E
32E 43–32E
33E 43–33E
34P 43–35P
35P 43–34P
36P 43–36P
37P 43–37P
38P 43–39P
39P 43–38P
40P 43–42P
41P 43–43P
42P 43–40P
43P 43–44P
44P 43–45P
45P 43–41P
46E 43–46E
47E 43–47E
48P 43–49P
49P 43–48P
50P 43–51P
51P 43–50P
52E 43–54E
53E 43–52E
54E 43–56E
55E 43–53E
56P 43–57P
57P 43–55E
58P 43–58P
59P 43–59P
60P 43–60P
61P 43–61P
62E 43–63E
63E 43–62E
64P 43–64P
65P 43–65P
66E 43–66E
67E 43–67E
68E 43–68E
69P 43–70P
70E 43–69E
71P 43–71P
72E 43–73E
73E 43–72E
74E 43–74E
75P 43–75P
76 43–80
77 new
78 new
79 new
80 new
81 new
82 new
83 new
84 new
85 new
86 new
87 new
88 new

Chapter 44

1E 44–1E
2E 44–4E
3E 44–3E
4E 44–2E
5E 44–5E
6E 44–7E
7E 44–6E
8E 44–9E
9E 44–8E
10P 44–10P
11P 44–12P
12P 44–11P
13P 44–13P
14P 44–14P
15E 44–15E
16E 44–16E
17E 44–17E
18P 44–21P
19P 44–18P
20P 44–19P
21P 44–20P
22P 44–23P
23P 44–22P

24P	44–25P
25P	44–24P
26P	44–26P
27P	44–27P
28E	44–28E
29E	44–29E
30P	44–31P
31P	44–30P
32E	44–33E
33E	44–32E
34E	44–34E
35E	44–35E
36P	44–37P
37P	44–36P
38P	44–38P
39E	44–40E
40E	44–39E
41E	44–41E
42P	44–42P
43P	44–44P
44P	44–43P
45P	44–45P
46P	44–47P
47P	44–46P
48P	44–48P
49P	44–50P
50P	44–49P
51P	44–51P
52E	44–52E
53P	44–54P
54P	44–53P

Chapter 45

1E	45–1E
2E	45–3E
3E	45–2E
4E	45–4E
5E	45–5E
6P	45–8P
7P	45–7P
8P	45–6P
9P	45–10P
10P	45–9P
11P	45–11P
12E	45–12E
13E	45–13E
14P	45–14P
15E	45–15E
16E	45–16E
17E	45–17E
18E	45–18E
19E	45–19E
20P	45–21P
21P	45–20P
22P	45–22P
23E	45–23E
24E	45–24E
25E	45–26E
26E	45–25E
27E	45–27E
28P	45–28P
29P	45–29P
30E	45–30E
31E	45–31E
32E	45–32E
33P	45–34P
34P	45–33P
35P	45–35P
36E	45–36E
37P	45–37P
38P	45–38P
39E	45–40E
40E	45–39E
41	45–41
42	new
43	45–42
44	new

SECTION EIGHT
COMPUTER PROJECTS

Computational physics has become a part of many introductory courses. Its inclusion serves two purposes: students use their programming skills and they learn to apply the fundamental principles they are studying to situations that are more complicated than those presented in their physics text. The act of writing and debugging a computer program often helps students understand more precisely the equations they are working with. Perhaps more importantly from a pedagogical viewpoint, a computer can be programmed to solve the same problem over and over with different input data. With proper instruction, a student can use this technique to develop insight into the physics being studied.

To help in these endeavors, some computer projects are suggested on the following pages. In addition, outlines of programs are supplied to help with the work on the projects. The projects and programs are keyed to the chapter of *Fundamentals of Physics* in which the physical principles used are discussed. You or the students must flesh out the programs to make them compatible with the programming language or spreadsheet that is used and to make the input and output routines work on the specific computer that is used. You may want to revise some projects to bring them more in line with your goals for the course.

Many of the projects require graphs to be drawn. Beginners may want to use a spreadsheet with graphing capability or else a simple BASIC or Pascal program to generate the required data, then draw the graphs manually. More advanced programmers will want to include programming steps to draw the plots on a monitor screen. Either technique is satisfactory.

Projects may be assigned as homework or carried out in groups during the laboratory portion of the course. The latter is especially beneficial since the projects can then be used to generate class discussions of the physics.

Chapter 2

A computer can be used to calculate the position and velocity of an object when its acceleration is a known function of time. Divide the time axis into a large number of small intervals, each of duration Δt. If Δt is sufficiently small the acceleration can be approximated by a constant in each interval, perhaps with a different value in different intervals. If v_b is the velocity at the beginning of an interval then $v_e = v_b + a\Delta t$ can be used to calculate the velocity at the end of the interval. If a is the average acceleration for the interval then this expression is exact. In most cases, however, the average acceleration is not known and so must be approximated. For the projects in this section it is approximated by $[a(t_b) + a(t_e)]/2$, where $a(t_b)$ is the acceleration at the beginning of the interval and $a(t_e)$ is the acceleration at the end. For each of the projects below the function $a(t)$ must be supplied.

If x_b is the coordinate of the object at the beginning of the interval, then the coordinate at the end of the interval is given by $x_e = x_b + v\Delta t$, where v is the average velocity in the interval. Approximate v by $(v_b + v_e)/2$. Smaller values for Δt make the approximations for v_e and x_e better. Δt, however, cannot be taken to be so small that significance is lost when the computer sums the terms in these equations. Note that the coordinate and velocity at the end of any interval are the coordinate and velocity at the beginning of the next interval. A skeleton program might look like this:

```
input initial values: t_0, x_0, v_0
input final time and interval width: t_f, Δt
set t_b = t_0, x_b = x_0, and v_b = v_0
calculate acceleration at beginning of first interval: a_b = a(t_b)
begin loop over intervals
      calculate time at end of interval: t_e = t_b + Δt
      calculate acceleration at end of interval: a_e = a(t_e)
      calculate "average" acceleration in interval: a = (a_b + a_e)/2
      calculate velocity at end of interval: v_e = v_b + aΔt
      calculate "average" velocity in interval: v = (v_b + v_e)/2
      calculate coordinate at end of interval: x_e = x_b + vΔt
    * if t_e ≥ t_f then
            print or display t_b, x_b, v_b
            print or display t_e, x_e, v_e
            exit loop
      end of if statement
      set t_b = t_e, x_b = x_e, v_b = v_e, a_b = a_e in preparation for next interval
end loop over intervals
stop
```

The line marked with an asterisk will be different for different applications. You will change this line, for example, if you want to display results and exit the loop when the velocity or coordinate have certain values.

Because Δt is arbitrary, the end of the last interval may not correspond exactly to t_f. The first line of output corresponds to a time just before t_f and the second line corresponds to a time just after. You can force $t = t_f$ at the end of the last interval by asking the computer to recalculate Δt near the beginning of the program. First use $(t_f - t_0)/\Delta t$ to estimate the number of intervals, then round the result to the nearest integer N. Finally take $\Delta t = (t_f - t_0)/N$.

Most programming languages allow you to write a separate section of the program to define the function $a(t)$. Then, the lines implementing $a_b = a(t_b)$ and $a_e = a(t_e)$ simply refer to the function definition. This is more efficient than writing the instructions for $a(t)$ twice, once for $t = t_b$ and once for $t = t_e$.

PROJECT 1. Test the program on a problem you can analyze analytically. Suppose the acceleration in m/s^2 is given by $12t$, for t in seconds. Take the initial coordinate (at $t = 0$) to be 5.0 m and the initial velocity to be -120 m/s. Find the coordinate and velocity at $t = 4.5$ s.

Start with $\Delta t = 0.5$ s and carry out the computation. Repeat several times, each time halving Δt, until you get the same answers to three significant figures on successive trials. Work the problem analytically to obtain an exact solution and carefully compare the answers with those obtained by the program. Check your program code if they differ.

PROJECT 2. Now use the program to determine when and where the particle is instantaneously stopped. You want the program to stop when the velocity changes sign in some interval. Use $v_b v_e \le 0$ as the condition for displaying results and exiting the loop. The correct answer is between the two results displayed. If they are the same to within the number of significant figures you desire you are finished. If they are not, run the program again with a smaller value of Δt. Try to obtain three significant figure accuracy.

Now find when the particle is at $x = -100$ m and its velocity when it is there. The condition for displaying results and exiting the loop is now $(x_b+100)(x_e+100) \le 0$. Obtain three significant figure accuracy.

PROJECT 3. Try a more complicated example. Suppose the particle starts at $x = 0$ with a velocity of -120 m/s and has an acceleration that is given in m/s^2 by $a(t) = 30e^{-t/8}$, where t is the time in seconds. Find the time at which it is instantaneously at rest and its position at that time. [ans: 5.55 s; −295 m]

You can use the program to generate a list of values ready to be plotted by hand or to plot values on the monitor screen. Suppose the time interval desired between displayed points is Δt_d. Input the value of Δt_d just after the values for t_f and Δt are read, then set $t_d = t_0 + \Delta t_d$. Replace the last statements of the program, from the if statement, with

```
        if t_e ≥ t_d then
            print, display, or plot x_e or v_e
            increment t_d by Δt_d
        end of if statement
        set t_b = t_e, x_b = x_e, v_b = v_e, a_b = a_e in preparation for next interval
        if t_e ≥ t_f then exit loop
    end loop over intervals
    stop
```

Because the computer may produce values of t_e that are in error in the last place carried, the set of points you get may be somewhat different from what you anticipate. If this is intolerable, round the values of t_e so the position of the last non-zero digit is a few less than the number of figures carried by the machine.

PROJECT 4. A particle starts from rest at the origin and has an acceleration that is given in m/s^2 by $a(t) = 30te^{-t}$, where t is the time in seconds. Draw graphs of its coordinate and velocity as functions of time from $t = 0$ to $t = 5$ s. Plot points every 0.5 s. You should obtain two significant figure accuracy with $N = 100$.

The acceleration is quite small near $t = 0$ and increases as t increases. This is the effect of the factor t. It reaches a maximum at $t = 1$ s, then decreases. This is the effect of the exponential factor. Your graph of $v(t)$ should have its greatest slope in the vicinity of $t = 1$ s, then approach a line with a slope of zero. The velocity has become nearly constant. What do you think the limiting value of the velocity is?

The graph of $x(t)$ should have a slope of zero at $t = 0$ (the initial velocity is zero) but it soon curves upward and eventually approaches a straight line, the slope of which is the limiting value of the velocity.

Chapter 4

The program outlined for Chapter 2 projects can be revised to investigate two-dimensional motion. You must now supply two components of the acceleration as functions of t and have the computer carry out calculations for each of the two components of the velocity and position vectors. The outline of a sample program might be:

```
    input initial values: t_0, x_0, y_0, v_0x, v_0y
    input final time and interval width: t_f, Δt
    set t_b = t_0, x_b = x_0, y_b = y_0, v_xb = v_0x, v_yb = v_0y
    calculate acceleration at beginning of first interval: a_xb = a_x(t_b), a_yb = a_y(t_b)
    begin loop over intervals
        calculate time at end of interval: t_e = t_b + Δt
        calculate acceleration at end of interval: a_xe = a_x(t_e), a_ye = a_y(t_e)
```

```
    calculate "average" acceleration: a_x = (a_xb + a_xe)/2, a_y = (a_yb + a_ye)/2
    calculate velocity at end of interval: v_xe = v_xb + a_x Δt, v_ye = v_yb + a_y Δt
    calculate "average" velocity : v_x = (v_xb + v_xe)/2, v_y = (v_yb + v_ye)/2
    calculate coordinates at end of interval: x_e = x_b + v_x Δt, y_e = y_b + v_y Δt
  * if t_e ≥ t_f then
        print or display t_b, x_b, y_b, v_xb, v_yb
        print or display t_e, x_e, y_e, v_xe, v_ye
       exit loop
    end of if statement
    set t_b = t_e, x_b = x_e, y_b = y_e, v_xb = v_xe, v_yb = v_ye, a_xb = a_xe, a_yb = a_ye
end loop over intervals
stop
```

As before, the line with the asterisk may be changed for other applications.

Instead of v_{0x} and v_{0y} you may wish to input the initial speed v_0 and angle ϕ_0 between the velocity and the x axis, then have the computer calculate v_{0x} and v_{0y} using $v_{0x} = v_0 \cos\phi_0$ and $v_{0y} = v_0 \sin\phi_0$. You may also wish to define $a_x(t)$ and $a_y(t)$ in a separate section of the program.

PROJECT 1. Start with a projectile motion problem that can be solved analytically. Suppose a projectile is fired over level ground at 350 m/s, at an angle of 25° above the horizontal. When does it reach the highest point? How high and how far down range is the highest point? When does it hit the ground and what is the range? Compare your answers with the analytic solutions.

If the x axis is horizontal and the y axis is positive in the upward direction then the components of the acceleration are given by $a_x = 0$ and $a_y = -9.8\,\text{m/s}^2$. To find the highest point exit the loop when $v_{ye} \le 0$. To find the range, exit the loop when $y_e \le 0$. You must experiment a little to find an appropriate value for Δt. Start with 0.1 s and reduce it by a factor of 5 in successive calculations. Stop when you get the same results to three significant figures.

PROJECT 2. Now suppose the acceleration of the projectile has a horizontal component that varies with time, as a rocket might have. Take $a_x = 3.0t\,\text{m/s}^2$ and $a_y = -9.8\,\text{m/s}^2$, where t is in seconds. Use the same initial conditions (an initial velocity of 350 m/s, 25° above the horizontal) and find the time the projectile reaches its highest point and the coordinates of the highest point, assuming it is fired from the origin. Find the range over level ground and the velocity of the object just before it lands. [ans: highest point: $t = 15.1\,\text{s}$, $x = 6.51 \times 10^3\,\text{m}$, $y = 1.12 \times 10^3\,\text{m}$; range: $t = 30.2\,\text{s}$, $x = 2.33 \times 10^4\,\text{m}$, $v_x = 1.68 \times 10^3\,\text{m/s}$, $v_y = -148\,\text{m/s}$]

PROJECT 3. Now suppose the horizontal component of the acceleration is given in m/s^2 by $a_x(t) = 30te^{-t}$ while the y component is still $a_y = -9.8\,\text{m/s}^2$. The x axis is horizontal and the y axis is positive in the upward direction. The initial velocity is still 350 m/s, 25° above the horizontal. At what time does the object reach the highest point on its trajectory and what are the coordinates of that point? At what time does it return to the level of the firing point and what is the range? What are the components of its velocity just before landing? [ans: highest point: $t = 15.1\,\text{s}$, $x = 5.18 \times 10^3\,\text{m}$, $y = 1.12 \times 10^3\,\text{m}$; range: $t = 30.2\,\text{s}$, $x = 1.04 \times 10^4\,\text{m}$; velocity: $v_x = 347\,\text{m/s}$, $v_y = -148\,\text{m/s}$]

Chapter 6

The computer program outlined for the projects of Chapter 2 must be revised if the acceleration is a function of position or velocity. The average acceleration in an interval cannot be approximated by $a = (a_b + a_e)/2$ because a_e cannot be computed until v_e or x_e are known. You can, however, use the acceleration at the beginning of the interval and write $v_e = v_b + a_b\Delta t$. This is usually a poor approximation to the average acceleration, so compared to the program of Chapter 2, much smaller intervals must be used to obtain the same accuracy.

Errors arise when a large number of intervals are used because the computer normally carries only a small number of significant figures (eight or ten) and the last is often in error. These so-called truncation errors accumulate. You can decrease the effect significantly by carrying out the calculation in double precision.

The outline of a possible program for one-dimensional motion is:

```
input initial values: t0, x0, v0
input final time and interval width: tf, Δt
set tb = t0, xb = x0, vb = v0
begin loop over intervals
    calculate acceleration at beginning of interval: ab = a(tb)
    calculate velocity at end of interval: ve = vb + abΔt
    calculate "average" velocity: v = (vb + ve)/2
    calculate coordinate at end of interval: xe = xb + vΔt
    calculate time at end of interval: te = tb + Δt
  * if te ≥ tf then
        print or display tb, xb, vb
        print or display te, xe, ve
        exit loop
    end of if statement
    set tb = te, xb = xe, vb = ve
end loop over intervals
stop
```

You may want to revise the program so it displays values for a sequence of times. See the Computer Projects section for Chapter 2.

Sometimes air resistance must be taken into account when an object moves in the air. Consider an object that is moving vertically and take its acceleration to be given by $a = -g - (b/m)v$, where the positive direction is upward, m is the mass of the object, and b is a constant that depends on the interaction of the object with the air.

PROJECT 1. Suppose an object with $(b/m) = 0.100\,\mathrm{s}^{-1}$ is fired upward from ground level with an initial speed of 100 m/s. On separate graphs plot the coordinate and velocity every 0.5 s from $t = 0$ to $t = 17$ s. First test the program to see what interval width it needs to obtain two significant figure accuracy over that range of time.

How much time does the projectile take to get to the highest point on its trajectory and how high is that point? How long does it take to get back to the ground and what is its velocity just before it reaches the ground? You might use your graph to obtain approximate answers, then use the program to refine them. [ans: highest point: 7.03 s, 311 m; ground: 16.2 s, 58.9 m/s]

Notice that the projectile takes longer to fall from the highest point than it does to reach that point and that it returns to ground level with a speed that is less than the firing speed. Compare your answers with those you would obtain if the object were in free fall. The time to reach the highest point is ______ and the highest point is ______ with air resistance than without. The total flight time is ______ and the speed on impact is ______ with air resistance than without.

PROJECT 2. Here's a problem for which the acceleration depends on position. Starting from rest at $x = 0$, a 3.5 kg box is dragged along the ground in the positive x direction by a constant force of 12 N acting horizontally. The ground is rougher toward larger x and the coefficient of kinetic friction increases according to $\mu_k = 0.070\sqrt{x}$. Use the program to find its position and velocity every second from $t = 0$ to the time it comes to rest again. At what time does the box come to rest? How far has it been dragged? [ans: 11.4 s; 56.2 m]

What is the maximum speed of the box? When does it have this speed? Where is it when it has this speed? [ans: 7.56 m/s; 4.84 s; 25.0 m from the starting point]

For the special case of an acceleration that is linear in the velocity, $a = (a_b + a_e)/2$ *can* be used to calculate the velocity at the end of the interval. Here's how. The acceleration at the beginning of the interval is given by $a_b = -g - (b/m)v_b$ and the acceleration at the end of the interval is given by $a_e = -g - (b/m)v_e$, so $a = (a_b + a_e)/2 = -g - (b/2m)v_b - (b/2m)v_e$. The velocity at the end of the interval is given by $v_e = v_b + a\Delta t = v_b - g\Delta t - (b\Delta t/2m)v_b - (b\Delta t/2m)v_e$. Solve this expression for v_e. The result is $v_e = [v_b(1 - b\Delta t/2m) - g\Delta t]/(1 + b\Delta t/2m)$. If $h = b\Delta t/2m$ this becomes $v_e = [v_b(1 - h) - g\Delta t]/(1 + h)$. An outline of a program is:

```
input initial values: t0, x0, v0
input final time and interval width: tf, Δt
calculate parameter: h = bΔt/2m
set tb = t0, xb = x0, vb = v0
begin loop over intervals
     calculate velocity at end of interval: ve = [vb(1 − h) − gΔt]/(1 + h)
     calculate "average" velocity: v = (vb + ve)/2
     calculate coordinate at end of interval: xe = xb + vΔt
     calculate time at end of interval: te = tb + Δt
   * if te ≥ tf then
          print or display tb, xb, vb
          print or display te, xe, ve
          exit loop
     end of if statement
     set tb = te, xb = xe, vb = ve
end loop over intervals
stop
```

PROJECT 3. Use this program to find the position and velocity of the projectile of the first project at the end of 17 s. Notice that far fewer intervals are required to obtain two significant figure accuracy.

Now use the program to investigate the influence of the air on a projectile fired straight upward. Make a table of the time to reach the highest point, the coordinate of the highest point, the time to reach the ground, and the velocity just before reaching ground, all as functions of the resistance parameter b/m. Try $b/m = 1.00\,\text{s}^{-1}$, $0.100\,\text{s}^{-1}$, and $0.0100\,\text{s}^{-1}$.

This exercise is designed to give you some idea of the effect of air resistance. A javelin has a small value of b/m, a basketball has a larger value, and a ping-pong ball has a still larger value. The larger b/m the greater the effects.

You might try to find a value for the resistance coefficient b/m of a ping-pong ball or other object. Shoot it into the air with a spring gun, measure the initial speed with a photogate timer, and time its return to the firing level. Then try various values of b in your computer program until you find the value that reproduces the experimentally determined time of flight.

PROJECT 4. You can use the program to investigate the approach to terminal velocity. Suppose an object with $b/m = 0.100\,\mathrm{s}^{-1}$ is dropped from a high cliff. Use the program to plot its velocity every 2 s over the first minute of its fall. The speed should approach the value $mg/b = 98\,\mathrm{m/s}$. Repeat for objects with $b/m = 0.0500\,\mathrm{s}^{-1}$ and $0.500\,\mathrm{s}^{-1}$.

The program can be modified to deal with a projectile moving in two dimensions, and subjected to a drag force that is proportional to its velocity. The acceleration components are now given by $a_x = -(b/m)v_x$ and $a_y = -g - (b/m)v_y$. The outline of a program is:

```
input initial values: t0, x0, y0, v0x, v0y
input final time and interval width: tf, Δt
calculate parameter: h = bΔt/2m
set tb = t0, xb = x0, yb = y0, vxb = v0x, vyb = v0y
begin loop over intervals
      calculate velocity at end of interval:
            vxe = vxb(1 − h)/(1 + h)
            vye = [vyb(1 − h) − gΔt]/(1 + h)
      calculate "average" velocity: vx = (vxb + vxe)/2, vy = (vyb + vye)/2
      calculate coordinates at end of interval: xe = xb + vxΔt, ye = yb + vyΔt
      calculate time at end of interval: te = tb + Δt
    * if te ≥ tf then
            print or display tb, xb, yb, vxb, vyb
            print or display te, xe, ye, vxe, vye
            exit loop
      end of if statement
      set tb = te, xb = xe, yb = ye, vxb = vxe, vyb = vye
end loop over intervals
stop
```

PROJECT 5. A projectile with $b/m = 0.100\,\mathrm{s}^{-1}$ is fired over level ground with an initial velocity of 100 m/s, 45° above the horizontal. Find its coordinates every 0.5 s from $t = 0$ (the time of firing) to $t = 12.5\,\mathrm{s}$. Plot its trajectory. First find the value of Δt required to obtain three significant figure accuracy over the entire time interval.

Find the time the projectile reaches the highest point and the coordinates of the highest point. Take the x axis to be horizontal and the y axis to be positive in the upward direction. [ans: $t = 5.43\,\mathrm{s}$, $x = 296\,\mathrm{m}$, $y = 175\,\mathrm{m}$]

Find the time of flight and range. What are its velocity components just before it lands? [ans: $t = 12.1\,\mathrm{s}$, $x = 495\,\mathrm{m}$, $v_x = 21.2\,\mathrm{m/s}$, $v_y = -47.5\,\mathrm{m/s}$]

PROJECT 6. If air resistance is significant, maximum range is obtained for a firing angle that is different from 45°. Consider a projectile with $b/m = 0.100\,\mathrm{s}^{-1}$, fired with an initial speed of 100 m/s over level ground. Find the range to three significant figures for firing angles of 30°, 35°, and 40°. You already know the range for 45°. The firing angle for maximum range is between ______ and ______. You may wish to refine the interval to obtain the firing angle to two significant figures. [ans: 510 m; 519 m; 514 m]

PROJECT 7. Suppose the projectile of the last project ($b/m = 0.100$) is fired from the edge of a high cliff instead of over level ground. Assume the same initial conditions (fired from the origin with a speed of 100 m/s, 45° above the horizontal) and plot its trajectory for the first minute of its flight.

Notice that the trajectory is not symmetric about the highest point but is blunted in the forward direction. Near the end of the time interval the projectile is falling nearly straight down. Both the horizontal and vertical components of the velocity approach limiting values. The limiting value of the horizontal component is ______ and the limiting value of the vertical component is ______.

Chapter 7

A force with x component $F(x)$ does work given by $W = \int_{x_i}^{x_f} F(x)\,dx$ as the object on which it acts moves from x_i to x_f along the x axis. A computer can be used to evaluate integrals of this form. One of the simplest techniques to use is Simpson's rule. An interval of the x axis, from x_b to x_e, is divided into two equal parts of width Δx. Let F_b be the value of the force at the beginning of the interval ($x = x_b$), F_m be the value at the middle ($x = x_b + \Delta x$), and F_e be the value of the end ($x = x_e$). Then according to Simpson's rule the integral over the interval can be approximated by

$$\int_{x_b}^{x_e} F(x)\,dx = \frac{\Delta x}{3}(F_b + 4F_m + F_e)\,.$$

This expression can be derived easily by fitting the function $F(x)$ to a quadratic of the form $F(x) = A_0 + A_1(x - x_b) + A_2(x - x_b)^2$. The coefficients are chosen so the quadratic yields the correct values of the function at $x = x_b$, $x = x_m$, and $x = x_e$. They are $A_0 = F_b$, $A_1 = -(F_e - 4F_m + 3F_b)/2\Delta x$, and $A_2 = (F_e - 2F_m + F_b)/2(\Delta x)^2$. The quadratic can be integrated easily. The approximation becomes better as the interval becomes narrower.

To evaluate an integral for an extended portion of the x axis divide the region from x_i to x_f into N intervals, where N is an even number. The interval width is given by $\Delta x = (x_f - x_i)/N$. Label the points x_0 ($= x_i$), x_1, x_2, ..., x_N ($= x_f$) and the corresponding values of the force F_0, F_1, F_2, ..., F_N. Apply the formula given above to each pair of intervals; for example, x_0 is the first point, x_1 is the second point, and x_2 is the third point in the first application; x_2 is the first point, x_3 is the second point, and x_4 is the third point in the next application. Except for x_0 and x_N, each point with an even label enters the final formula twice: once as the first point in an integration and once as a third point. Each point with an odd label enters as a midpoint of an integration. Thus

$$\int_{x_i}^{x_f} F(x)\,dx = \frac{\Delta x}{3}\Big[(F_N - F_0) + 2(F_0 + F_2 + F_4 + \dots + F_{N-2}) + 4(F_1 + F_3 + \dots + F_{N-1})\Big]$$

F_0 is subtracted at the beginning of the equation because it is also included in the sum over values with even labels. There it is multiplied by 2 but it should be included only once. F_N is not included in any of the sums but it must be included once in the final equation, so it is added at the beginning.

Write a computer program to evaluate a work integral for one-dimensional motion. Input the lower and upper limits of the integral and the number of intervals. You can force N to be an even integer by dividing the value read by 2, rounding the result to the nearest integer, then multiplying by 2. Calculate Δx. Now write a loop to sum all the force values with even labels and sum all the force values with odd labels. Instructions in the loop will be executed $N/2$ times. An outline might be:

```
input limits of integral: x_i, x_f
input number of intervals: N
replace N with nearest even integer
calculate interval width: Δx = (x_f − x_i)/N
initialize quantity to hold sum of values with even labels: S_e = 0
initialize quantity to hold sum of values with odd labels: S_o = 0
set x = x_i
begin loop over intervals: counter runs from 1 to N/2
      calculate F(x) and add it to the sum of values with even labels:
         replace S_e with S_e + F(x)
      increment x by Δx
      calculate F(x) and add it to the sum of values with odd labels:
         replace S_o with S_o + F(x)
      increment x by Δx
end loop
calculate force for upper and lower limits: F_0 = F(x_i), F_N = F(x_f)
evaluate integral: W = (Δx/3)(F_N − F_0 + 2S_e + 4S_o)
display result
stop
```

You may want to define the force function $F(x)$ in a separate section of the program.

PROJECT 1. This problem can be solved analytically. Use it to check your program. A 2.00 kg block moves along the x axis, subjected to the force $\vec{F} = (6 - 4x)\,\hat{\imath}$, where $\vec{F}$ is in newtons and x is in meters. What work is done by this force as the block moves from $x = 0$ to $x = 1.00$ m? from $x = 0$ to $x = 5.00$ m? Try $N = 2$ and double its value on successive runs until you get the same first three significant figures. Check the answers by working the problem analytically. [ans: 4.00 J; −20.0 J]

If $\vec{F}$ is the total force acting on the block, then the work that it does is equal to the change in the kinetic energy of the block. Suppose the block has a speed of 6.00 m/s when it is at $x = 0$. What is its speed when it is at $x = 1.00$ m? at 5.00 m? [ans: 6.32 m/s; 4.00 m/s]

PROJECT 2. A certain non-ideal spring exerts a force that is given by $F(x) = -250x - 125xe^{-x}$, where x is its extension (if positive) or its compression (if negative). A 2.3 kg block is attached to one end and the other end is held fixed. The block is pulled out so the spring is extended by 25 cm, then it is released from rest. How much work does the spring do on the block as the block moves from its initial position to the position for which the spring is neither extended or compressed? If the force of the spring is the only force acting on the block, what is its speed as it passes this point? [ans: 11.1 J; 3.11 m/s]

The block continues to move, compressing the spring. By how much is it compressed when the block comes to rest instantaneously? Since the block is at rest both initially and finally the total work done by the spring over the entire trip is 0. Use trial and error to find the final value of x. Alternatively you might imbed the program in a loop in which x_f is incremented each time around and search for the values of x_f that straddle $W = 0$. [ans: −0.237 m]

If the force is given as a function of time and you are asked for the work it does over a given time interval, you must change your strategy somewhat. If the force is the total force, you can use one of the programs of Chapters 2 or 4 to find the velocity. The total work done, of course, is just the change in kinetic energy. If the force is not the total force, you can use the power equation $dW/dt = Fv$ to write $W = \int Fv\,dt$. The program is the same as before except the integrand is now Fv. You must know the velocity as a function of time to use this method.

PROJECT 3. The position as a function of time for a 25-kg crate sliding across the floor (along the x axis) is given by $x(t) = 3te^{-0.60t}$, where x is in meters and t is in seconds. Numerically evaluate the integral for the work done by the total force during the first 2.0 s. You must find $v(t)$ and $F(t)$ by differentiating $x(t)$ and using $F = ma$. Select an integration interval for three significant figure accuracy. [ans: -112 J]

Use the expression you found for $v(t)$ to calculate the change in the kinetic energy during the first 2.0 s. Your result should agree with the result of the numerical integration.

PROJECT 4. A 2.0-kg block starts from rest at the origin and moves along the x axis. The total force acting on it is given by $F(t) = (8.0 + 12t^2)$, where F is in newtons and t is in seconds. Use analytical means to show that $v(t) = 4.0t + 2.0t^3$ and $x(t) = 2.0t^2 + 0.50t^4$.

One of the forces acting on the block is given by $F_1(t) = 3.0t^2$. Find the work done by F_1 during the first 3.0 s of the motion. Do this by analytical evaluation of $\int Fv\,dt$ and by numerical integration. [ans: 972 J]

PROJECT 5. The velocity of a body with mass m dropped from rest near the surface of the earth is given by $v(t) = (mg/b)(1 - e^{-bt/m})$, where b is the drag coefficient and down is taken to be the positive direction. Suppose an object with mass $m = 5.0$ kg and drag coefficient $b = 75$ kg/s is dropped from the edge of a high cliff. Use numerical integration to find the work done by gravity during the first 5.0 s. [ans: 5280 J]

The work done by gravity, of course, is given by mgs, where s is the distance fallen. The expression for $v(t)$ can be integrated with respect to time, with the result $s(t) = (mg/b)[t - (m/b) + (m/b)e^{-bt/m}]$. Use this expression to calculate the distance fallen in 5.0 s, then use $W = mgs$ to calculate the work done by gravity. You should obtain agreement with your numerical calculation.

If the object moves in two dimensions, the integral for the work becomes $W = \int \vec{F} \cdot d\vec{s} = \int (F_x\,dx + F_y\,dy)$, where $d\vec{s} = \hat{\imath}\,dx + \hat{\jmath}\,dy$ is an infinitesimal segment of the path. It is a vector tangent to the path. Because the object is on a specified path in the xy plane, the infinitesimals dx and dy are related to each other. You must know the path to determine the relationship.

There are several ways of specifying a path. One way is to give the functional relationship between the coordinates of points on the path: $y = g(x)$, where $g(x)$ is a specified function. Then, $dy = (dg/dx)\,dx$ and the work integral becomes $W = \int [F_x + (dg/dx)F_y]\,dx$. For example, a straight line with slope S and intercept A is given by $y = A + Sx$ and the work is given by $W = \int (F_x + SF_y)\,dx$. The limits of integration are the x coordinates of the points at the beginning and end of the path.

You must use the functional relationship between x and y in another way. The components of the force will be given as functions of both x and y but there can be only a single variable of integration. You must substitute the function $g(x)$ wherever y occurs in the expressions for the force components. The following is an example.

PROJECT 6. A particle moves along a straight line from $x = 1.5$ m, $y = 2.0$ m to $x = 3.0$ m, $y = 5.0$ m. One of the forces acting on it is given by $\vec{F} = 3x^2y^2\,\hat{\imath} + 2x^3y\,\hat{\jmath}$. What work does this force do?

The path is given by $y = -1.0 + 2.0x$, so $dy = 2.0\,dx$. The force components are given by $F_x = 3x^2y^2 = 3x^2(-1 + 2x)^2$ and $F_y = 2x^3y = 2x^3(-1 + 2x)$. The work integral is $\int_{1.5}^{3.0}[3x^2(-1+2x)^2 + 4x^3(-1+2x)]\,\mathrm{d}x$. The integration program can now be used to calculate the work done. [ans: 662 J]

The method does not work if the path is parallel to the y axis and it requires a large number of intervals if the path is nearly parallel to the y axis. Then, you should use y as the variable of integration. The expression for work becomes $\int[(\mathrm{d}x/\mathrm{d}y)F_x + F_y]\,\mathrm{d}y$.

A path can also be specified by giving both x and y as functions of some parameter. The parameter need not have any physical significance. For the straight line path of the previous project, you might write $x = 1.5 + 1.5s$ and $y = 2.0 + 3.0s$. When the parameter $s = 0$, then $x = 1.5$ and $y = 2.0$. When $s = 1$, then $x = 3.0$ and $y = 5.0$. Now, $\mathrm{d}x = 1.5\,\mathrm{d}s$ and $\mathrm{d}y = 3.0\,\mathrm{d}s$. The integral for the work becomes $W = \int_0^1[(\mathrm{d}x/\mathrm{d}s)F_x + (\mathrm{d}y/\mathrm{d}s)F_y]\,\mathrm{d}s$. You must now substitute for both x and y in terms of s in the expressions for F_x and F_y.

If the path is an arc of a circle, you might use the angle between the x axis and the radial line to the point as a parameter. Thus $x = R\cos\theta$ and $y = R\sin\theta$, where R is radius of the path. The integral for the work becomes $W = \int[-F_x\sin\theta + F_y\cos\theta]R\,\mathrm{d}\theta$. Notice that $-F_x\sin\theta + F_y\cos\theta$ is the component of the force along a line that is tangent to the path and $R\,\mathrm{d}\theta$ is the length of a path segment that subtends the angle $\mathrm{d}\theta$ at the center of the circle.

PROJECT 7. A particle travels counterclockwise around a circular path with a radius of 2.0 m, centered at the origin. What work is done by the force $\vec{F} = 3x^2y^2\,\hat{\imath} + 2x^3y\,\hat{\jmath}$ as the particle goes from $x = 2.0$ m, $y = 0$ to $x = -\sqrt{2.0}$ m, $y = \sqrt{2.0}$ m? This is an arc of $3\pi/4$ radians or 135°. [ans: −5.66 J]

Chapter 8

The force $\vec{F} = 3x^2y^2\,\hat{\imath} + 2x^3y\,\hat{\jmath}$ is a conservative force. You can see what this means by computing the work it does as the particle on which it acts goes via different paths from the same initial point to the same final point.

PROJECT 1. Use the program of the previous chapter to calculate the work done by the force $\vec{F} = 3x^2y^2\,\hat{\imath} + 2x^3y\,\hat{\jmath}$ as the particle goes from $x = 2.0$ m, $y = 0$ to $x = 4.0$ m, $y = 2.0$ m along each of the following paths.

a. Along the x axis from $x = 2.0$ m to $x = 4.0$ m, then along the line $x = 4.0$ m from $y = 0$ to $y = 2.0$ m. Carry out the integration in two parts, one for each segment.
b. Along the line $x = 2.0$ m from $y = 0$ to $y = 2.0$ m, then along the line $y = 2.0$ m from $x = 2.0$ m to $x = 4.0$ m.
c. Along the straight line that joins the to points. You might take $x = 2.0 + 2.0s$ and $y = 2.0s$, where s is a parameter that varies between 0 and 1.
d. Along the perimeter of a 2.0-m radius circle centered at $x = 4.0$ m, $y = 0$. For this path $x = 4.0 - 2.0\cos\theta$ and $y = 2.0\sin\theta$, where θ varies from 0 to $\pi/2$ radians (90°).

Notice that in every case the work done is 256 J.

PROJECT 2. Since the work done by the force of the previous project is independent of the path, a potential energy function is associated with it. The value of this function at any point is the negative of work done by the force as the particle moves from some reference point to the point in question. The table below is actually a grid of points in the xy plane. The x coordinate is given at the top and the y coordinate is given at the left side. Take the origin to be the reference point ($U = 0$ there) and calculate the potential energy associated with each of the other points. You may want to automate the process by reading in the coordinates x_f and y_f of the point. Use the path given by $x = x_f s$ and $y = y_f s$, where s is a parameter than varies from 0 to 1. Since $dx/ds = x_f$ and $dy/ds = y_f$ the potential energy is given by $U(x_f, y_f) = -\int_0^1 (F_x x_f + F_y y_f)\,ds$. You must write the force components in terms of s. Fill in the table below with values of U.

$y(m)$ \ $x(m)$	0	1	2	3	4
4					
3					
2					
1					
0	0				

The analytic form for the potential energy function is $U(x, y) = -x^3y^2$. You can easily check this since the derivative of U with respect to x must give the negative of the x component of the force and the derivative of U with respect to y must give the negative of the y component. Use the exact analytic expression to check the results of your numerical calculations. You might construct a program consisting of two loops, one over values of the x coordinate and the other over values of the y coordinate, to compute and display values of U.

Use the table to compute the following quantities:

a. The change in the potential energy when the particle moves from $x = 2\,\text{m}$, $y = 1\,\text{m}$ to $x = 3\,\text{m}$, $y = 3\,\text{m}$. [ans: $-235\,\text{J}$]

b. The work done by the force when the particle moves from $x = 3\,\text{m}$, $y = 4\,\text{m}$ to $x = 1\,\text{m}$, $y = 1\,\text{m}$. [ans: $-431\,\text{J}$]

c. The change in the kinetic energy of the particle when it moves from $x = 3\,\text{m}$, $y = 1\,\text{m}$ to $x = 2\,\text{m}$, $y = 4\,\text{m}$. Assume only one force acts on it. [ans: $+101\,\text{J}$]

PROJECT 3. By way of contrast, the force $\vec{F} = (3x^2 - 6)y^2\,\hat{\imath} + 2x^3y\,\hat{\jmath}$ is not conservative. Use the program to calculate the work done by the force as the particle goes from $x = 2.0\,\text{m}$, $y = 0$ to $x = 4.0\,\text{m}$, $y = 2.0\,\text{m}$ along each of the following paths.

a. Along the x axis from $x = 2.0\,\text{m}$ to $x = 4.0\,\text{m}$, then along the line $x = 4.0\,\text{m}$ from $y = 0$ to $y = 2.0\,\text{m}$.

b. Along the line $x = 2.0\,\text{m}$ from $y = 0$ to $y = 2.0\,\text{m}$, then along the line $y = 2.0\,\text{m}$ from $x = 2.0\,\text{m}$ to $x = 4.0\,\text{m}$.

c. Along the straight line that joins the to points.

d. Along the perimeter of a 2.0-m radius circle centered at $x = 4.0\,\text{m}$, $y = 0$.

Notice that the work done is different for different paths. You can easily see that it is impossible to assign a potential energy to each point in space so that the work done by the force equals the difference in the values assigned to the end points of the path. [ans: 256 J; 208 J; 240 J; 224 J]

Chapter 9

A computer can be used to follow individual objects in a system as they interact with each other and to follow the center of mass of the system. As an example, consider two carts connected by an ideal spring on a horizontal frictionless air track. If the spring has a natural length ℓ_0 (when it is neither compressed or extended), then when its length is ℓ it exerts a force of magnitude $k|\ell - \ell_0|$ on each of the carts. Here k is the force constant for the spring. If the spring is extended ($\ell > \ell_0$), then both forces are toward the center of the spring. If the spring is compressed ($\ell < \ell_0$), then both forces are away from the center of the spring.

Let x_1 be the coordinate of one cart and x_2 be the coordinate of the other and take $x_2 > x_1$. Then, $\ell = x_2 - x_1$. The force on cart 1 is given by $F_1 = k(x_2 - x_1 - \ell_0)$ and the force on cart 2 is given by $F_2 = -k(x_2 - x_1 - \ell_0)$.

A program that will calculate the coordinate and velocity of each cart as functions of time can be modeled after the first program of Chapter 6. In addition, the program calculates the coordinate and velocity of the center of mass. Here's an outline.

input initial values: t_0, x_{10}, v_{10}, x_{20}, v_{20}
input final time and interval width: t_f, Δt
input display interval: Δt_d
set $t_b = t_0$, $t_d = t_0 + \Delta t_d$, $x_{1b} = x_{10}$, $v_{1b} = v_{10}$, $x_{2b} = x_{20}$, $v_{2b} = v_{20}$
calculate coordinate and velocity of center of mass:
  $x_{\text{cm}} = (m_1 x_{1b} + m_2 x_{2b})/(m_1 + m_2)$
  $v_{\text{cm}} = (m_1 v_{1b} + m_2 v_{2b})/(m_1 + m_2)$
display x_{1b}, x_{2b}, x_{cm}
begin loop over intervals
  calculate force on 1 at beginning of interval: $F_1 = k(x_{2b} - x_{1b} - \ell_0)$
  calculate accelerations at beginning of interval: $a_1 = F_1/m_1$, $a_2 = -F_1/m_2$
  calculate velocities at end of interval: $v_{1e} = v_{1b} + a_1\Delta t$, $v_{2e} = v_{2b} + a_2\Delta t$
  calculate "average" velocities: $v_1 = (v_{1b} + v_{1e})/2$, $v_2 = (v_{2b} + v_{2e})/2$
  calculate coordinates at end of interval: $x_{1e} = x_{1b} + v_1\Delta t$, $x_{2e} = x_{2b} + v_2\Delta t$
  calculate time at end of interval: $t_e = t_b + \Delta t$
  if $t_e \geq t_d$ then
    calculate coordinate and velocity of center of mass:
      $x_{\text{cm}} = (m_1 x_{1e} + m_2 x_{2e})/(m_1 + m_2)$
      $v_{\text{cm}} = (m_1 v_{1e} + m_2 v_{2e})/(m_1 + m_2)$
   * display x_{1e}, x_{2e}, x_{cm}
    increment t_d by Δt_d
  end of if statement
  if $t_e \geq t_f$ then exit loop
  set $t_b = t_f$, $x_{1b} = x_{1e}$, $v_{1b} = v_{1e}$, $x_{2b} = x_{2e}$, $v_{2b} = v_{2e}$
end loop over intervals
stop

For some applications the line marked with an asterisk is replaced by: display v_{1e}, v_{2e}, v_{cm}. You will need to use a large number of extremely small intervals, so truncation errors may be significant. If possible, use double precision variables.

PROJECT 1. Take $m_1 = 250\,\text{g}$, $m_2 = 600\,\text{g}$, $\ell_0 = 0.50\,\text{m}$, and $k = 5.00\,\text{N/m}$. Initially ($t = 0$), cart 1 is at rest at the origin and cart 2 is at rest at $x_2 = 0.70\,\text{m}$. Plot the coordinates of the carts and the coordinate of the center of mass every 0.10 s from $t = 0$ to $t = 2.0\,\text{s}$. You will need to take $\Delta t = 0.02\,\text{s}$ to obtain two significant figure accuracy.

Notice that the center of mass is initially at rest and remains at rest even though the carts oscillate back and forth. Now repeat the calculation with $v_{20} = 0.30\,\text{m/s}$. The carts again oscillate but as they do the center of mass moves with constant velocity in the positive x direction. Check that the calculation gives the same result as $v_{\text{cm}} = (m_1 v_{10} + m_2 v_{20})/(m_1 + m_2)$ for the velocity of the center of mass.

For both of the situations considered, the total momentum of the system, given by $P = (m_1 + m_2)v_{\text{cm}}$, is conserved. Now, suppose an external force of 0.50 N is applied to the first cart in the positive x direction. The acceleration of that cart is computed using $a_1 = (F_1 + 0.50)/m_1$, where F_1 is still the force of the spring on cart 1. The acceleration of cart 2 is $a_2 = -F_1/m_2$, as before. Plot the coordinates of the carts and the coordinate of the center of mass every 0.10 s from $t = 0$ to $t = 2.0\,\text{s}$. The curve representing the coordinate of the center of mass should be parabolic. In fact, $x_{\text{cm}} = \frac{1}{2}a_{\text{cm}}t^2$, where $a_{\text{cm}} = F_e/(m_1 + m_2)$. Here F_e is the external force, 0.50 N. Plot v_{cm} as a function of time and verify that it is a straight line with the correct slope.

A computer program can also be used to investigate the motion of a rocket. If the rocket and fuel have mass $M(t)$ and the fuel is ejected with velocity $\mathbf{u}$, measured relative to the rocket, then the acceleration of the rocket is given by

$$M\frac{d\mathbf{v}}{dt} = \mathbf{F} + \frac{dM}{dt}\mathbf{u},$$

where $\mathbf{F}$ is the external force acting on the rocket (the force of gravity, for example).

First, consider a rocket that is fired straight upward and assume the acceleration due to gravity is uniform. If the positive y axis is upward, then $\mathbf{u} = -u\,\mathbf{j}$, $\mathbf{v} = v\,\mathbf{j}$, and $\mathbf{F} = -Mg\,\mathbf{j}$. If the fuel is expended uniformly, you may write $M(t) = M_0 - Kt$, where K is a constant. The differential equation becomes

$$(M_0 - Kt)\frac{dv}{dt} = uK - (M_0 - Kt)g\,.$$

This expression gives the acceleration as a function of time and can be used in conjunction with the computer program of Chapter 2 to calculate the position and velocity of the rocket at any time until the rocket runs out of fuel.

The differential equation above can also be solved analytically to obtain

$$v(t) = v_0 - gt + u\ln\frac{M_0}{M_0 - Kt},$$

where v_0 is the velocity at $t = 0$. You will use this expression in the next computer project to check your program.

PROJECT 2. Consider a rocket that carries 80% of its original mass as fuel and ejects it uniformly for 5.0 s, at which time the fuel is gone and the engine shuts off. The rocket starts from rest and the speed of the fuel relative to the rocket is $u = 5000\,\text{m/s}$. Use the program of Chapter 2 to find the speed and altitude of the rocket at burnout. First, you need to find a value for K. At the end of 5.0 s, M is the mass of the rocket alone and this is $0.20M_0$. Thus, $M_0 - 5.0K = 0.20M_0$, so $K = 0.16M_0$. While fuel is being expended, the acceleration of the rocket is given by $a(t) = -g + 0.16u/(1 - 0.16t)$. Use this expression in the program. Check your answer by using the analytic expression above to calculate the speed. [ans: $8.00 \times 10^3\,\text{m/s}$; $1.48 \times 10^4\,\text{m}$]

The next project deals with a rocket that moves in two dimensions. If the acceleration due to gravity is uniform, then

$$\frac{d\mathbf{v}}{dt} = -\frac{K\mathbf{u}}{M_0 - Kt} - g\mathbf{j},$$

where the y axis is positive in the upward direction. Since $\mathbf{u}$ is always opposite to $\mathbf{v}$, we may write $\mathbf{u} = -u\mathbf{v}/v$, where the unit vector $\mathbf{v}/v$ is in the direction of $\mathbf{v}$. Thus,

$$\frac{d\mathbf{v}}{dt} = \left[\frac{Ku}{M_0 - Kt}\right]\frac{\mathbf{v}}{v} - g\mathbf{j}.$$

This equation holds while fuel is being ejected. After burnout, the rocket becomes an ordinary projectile and $d\mathbf{v}/dt = -g\mathbf{j}$.

Write a program to investigate the motion of a rocket fired from rest over level ground. Since each component of the acceleration depends on both components of the velocity, you cannot use $\mathbf{s} = (\mathbf{a}_b + \mathbf{a}_e)/2$ to approximate the average acceleration in an interval. Use instead the acceleration at the beginning of the interval. For the first interval the program will have trouble with the unit vector $\mathbf{v}/v$ since it involves division by 0. For that interval replace v_x/v by $\cos\phi_0$ and v_y/v by $\sin\phi_0$, where ϕ_0 is the firing angle. At the end of the interval, after the velocity components have been computed, calculate $v = \sqrt{v_{xe}^2 + v_{ye}^2}$ and the ratios v_{xe}/v and v_{ye}/v, in preparation for the next interval. This can be handled using the unit vector components α_x and α_y as you will see in the outline below.

The program should provide for the possibility of different interval widths before and after burnout. Before burnout the acceleration may be great and both the speed and direction of travel may change rapidly. You will want to use a small interval. After burnout you will want to use a larger interval to save time. Here's an outline of a program.

input burnout time: t_{bo}
input firing angle: ϕ_0
input interval widths: $(\Delta t)_b$ for $t < t_{bo}$, $(\Delta t)_a$ for $t > t_{bo}$
set interval width: $\Delta t = (\Delta t)_b$
calculate unit vector components: $\alpha_x = \cos\phi_0$, $\alpha_y = \sin\phi_0$
set initial values: $t_b = 0$, $x_b = 0$, $v_{xb} = 0$, $y_b = 0$, $v_{yb} = 0$
begin loop over intervals
    calculate x component of acceleration at beginning of interval:
        if $t_b < t_{bo}$ then $a_x = \alpha_x Ku/(M_0 - Kt_b)$
        if $t_b \geq t_{bo}$ then $a_x = 0$
    calculate y component of acceleration at beginning of interval:
        if $t_b < t_{bo}$ then $a_y = -g + \alpha_y Ku/(M_0 - Kt_b)$
        if $t_b \geq t_{bo}$ then $a_y = -9.8$
    calculate velocity at end of interval: $v_{xe} = v_{xb} + a_x\Delta t$, $v_{ye} = v_{yb} + a_y\Delta t$
    calculate "average" velocity: $v_x = (v_{xb} + v_{xe})/2$, $v_y = (v_{yb} + v_{ye})/2$
    calculate x coordinates at end of interval: $x_e = x_b + v_x\Delta t$, $y_e = y_b + v_y\Delta t$
    calculate time at end of interval: $t_e = t_b + \Delta t$
  * if $t_e \geq t_f$ then
            display or print t_b, x_b, y_b, v_{xb}, v_{yb}
            display or print t_e, x_e, y_e, v_{xe}, v_{ye}
            exit loop
    end of if statement
    set $t_b = t_e$, $x_b = x_e$, $v_{xb} = v_{xe}$, $y_b = y_e$, $v_{yb} = v_{ye}$
    if $t_b < t_{bo}$ set $\Delta t = (\Delta t)_b$, if $t_b \geq t_{bo}$ set $\Delta t = (\Delta t)_a$

calculate unit vector components:

$v = \sqrt{v_{xe}^2 + v_{ye}^2}$, $\alpha_x = v_{xe}/v$, $\alpha_y = v_{ye}/v$

end loop over intervals when $y_e < 0$

stop

The line marked with an asterisk will be different for different applications. To calculate the coordinates and velocity for the highest point on the trajectory, end the loop when $v_{ye} \leq 0$. To calculate the range over level ground, end the loop when $y_e \leq 0$.

PROJECT 3. A toy rocket has a mass (without fuel) of 250 g. It carries 10 g of fuel and is fired from rest at 45° above the horizontal. The fuel leaves the rocket at 4000 m/s and burns out after 0.200 s of flight. Use the program to find the coordinates and velocity components at burnout ($t = 0.20$ s). For two significant figure accuracy use $\Delta t = 0.001$ s before burnout and $\Delta t = 0.01$ s after burnout. [ans: $x = 12.0$ m, $y = 10.7$ m, $v_x = 121$ m/s, $v_y = 108$ m/s]

Use the program to find the time the rocket reaches the highest point on its trajectory. Find its coordinates and speed when it is there. [ans: $t = 11.2$ s, $x = 1.35 \times 10^3$ m, $y = 605$ m, $v = 121$ m/s]

Use the program to find the time the rocket lands if it is fired over level ground. Find its coordinates and speed just before it lands. [ans: $t = 22.3$ s, $x = 2.70 \times 10^3$ m, $v_x = 121$ m/s, $v_y = -109$ m/s]

Chapter 10

Many details of a two-body collision are controlled by the impulse exerted by each object on the other. Consider the situation in which an object of mass m_1, moving with velocity v_{1b}, impinges on an object of mass m_2, initially at rest. Suppose that after the collision both objects move along the line of incidence. Let J be the impulse object 1 exerts on object 2. Then, the velocity of object 2 after the collision is given by $v_{2a} = J/m_2$. Object 2 exerts an impulse $-J$ on object 1 and its velocity after the collision is given by $v_{1a} = v_{1b} - J/m_1$. Both final velocities can be computed if the initial velocity of object 1 and the impulse J are known (along with the masses, of course).

Once the velocities are known, the change in the kinetic energy of the system can be found. Before the collision the kinetic energy is $K_b = \frac{1}{2}m_1v_{1b}^2$ and after the collision it is $K_a = \frac{1}{2}m_1v_{1a}^2 + \frac{1}{2}m_2v_{2a}^2$. The kinetic energy lost during the collision is given by $Q = K_b - K_a$. A computer can be used to investigate the outcome of a one-dimensional collision as a function of the impulse.

Write a program that calculates the final velocities and the change in kinetic energy for a given range of impulses. Input the masses and the initial velocity of object 1. Also input the limits of the range of impulses to be considered. The outline of a program is:

input masses, initial velocity: m_1, m_2, v_{1b}
calculate initial kinetic energy: $K_b = \frac{1}{2}m_1v_{1b}^2$
input limits of range of impulse: J_i, J_f
number of values of J: $N = 21$
calculate impulse increment: $\Delta J = (J_f - J_i)/(N - 1)$
set $J = J_i$
begin loop over impulse values; counter runs from 1 to N
    calculate final velocities: $v_{1a} = v_{1b} - J/m_1$, $v_{2a} = J/m_2$
    calculate final kinetic energy: $K_a = \frac{1}{2}m_1v_{1a}^2 + \frac{1}{2}m_2v_{2a}^2$
    calculate kinetic energy loss: $Q = K_b - K_a$
    print or display J, v_{1a}, v_{2a}, Q

increment J: replace J with $J + \Delta J$
end loop
stop

The number N was chosen to be 21 because 21 lines nicely fill the usual monitor screen. You may want to use a different value. Both J_i and J_f are included in the list of values considered.

PROJECT 1. An air-track cart of mass $m_1 = 250\,\text{g}$ impinges at $3.0\,\text{m/s}$ on a cart of mass $m_2 = 600\,\text{g}$, initially at rest. Use the computer program to list the final velocities and kinetic energy loss for values of the impulse J from 0 to $2\,\text{kg·m/s}$.

From the table you generated, pick out the two values of the impulse that straddle the condition $v_{1a} = v_{2a}$. This is the completely inelastic collision. Note that Q is a maximum in this region of the table. Rerun the program with a narrower range of impulse values. Repeat if necessary and find J, v_{1a}, and Q for a completely inelastic collision to three significant figure accuracy. [ans: $0.529\,\text{kg·m/s}$; $0.882\,\text{m/s}$; $0.794\,\text{J}$]

From the original table ($J = 0$ to $2\,\text{kg·m/s}$) pick out the two values that straddle the condition $v_{1a} = 0$. Find to three significant figures the impulse, the final velocity of cart 2, and the kinetic energy loss. [ans: $0.750\,\text{kg·m/s}$; $1.250\,\text{m/s}$; $0.656\,\text{J}$]

From the original table pick out the two values that straddle the condition $Q = 0$. This is the elastic collision. Find to three significant figures the impulse and final velocities. [ans: $1.06\,\text{kg·m/s}$; $-1.24\,\text{m/s}$; $1.77\,\text{m/s}$]

PROJECT 2. Now repeat the calculations for two carts with the same mass. Take $m_1 = m_2 = 250\,\text{g}$ and $v_{1b} = 3.0\,\text{m/s}$. Notice that for this special case the impulse that stops the incident cart is the same as the impulse that produces an elastic collision. [ans: completely inelastic: $J = 0.375\,\text{kg·m/s}$, $v_{1a} = v_{2a} = 1.50\,\text{m/s}$, $Q = 0.563\,\text{J}$; elastic: $J = 0.750\,\text{kg·m/s}$, $v_{1a} = 0$, $v_{2a} = 3.00\,\text{m/s}$, $Q = 0$]

Two-dimensional collisions are somewhat more difficult to analyze. Consider a collision between two pucks on an air table. One puck, with mass m_1 and speed v_{1b}, is incident along the x axis on a puck with mass m_2, initially at rest. Suppose the impulse exerted by the moving puck on the puck at rest has magnitude J and makes the angle α with the x axis. Puck 2 leaves the collision along the line of the impulse; its velocity makes the angle $\theta_2 = \alpha$ with the x axis. Its speed after the collision is given by $v_{2a} = J/m_2$. The impulse-momentum equations for the incident puck are $-J\cos\alpha = m_1 v_{1a}\cos\theta_1 - m_1 v_{1b}$ and $-J\sin\alpha = m_1 v_{1a}\sin\theta_1$. These can be solved for v_{1a} and θ_1, with the result $v_{1a} = \sqrt{(J^2 + m_1^2 v_{1b}^2 - 2m_1 v_{1b} J\cos\alpha)}\big/ m_1$ and $\theta_1 = -\arctan[J\sin\alpha/(m_1 v_{1b} - J\cos\alpha)]$. To conserve momentum the sign of θ_1 must be opposite that of θ_2; that is, the two pucks must leave the collision on opposite sides of the line of incidence. If the sign of θ_1 produced by the computer is not correct, add 180° to it or subtract 180° from it.

Write a program that accepts values for the masses, initial velocity of the incident puck, and angle of the impulse with the x axis, then calculates the magnitudes and directions of the velocities after the collision and the loss in kinetic energy, all as functions of the magnitude of the impulse. An outline might be:

input masses, initial velocity: m_1, m_2, v_{1b}
calculate initial kinetic energy: $K_a = \frac{1}{2}m_1 v_{1a}^2$
input angle of impulse: α
input limits of range of impulse: J_i, J_f
number of values of J: $N = 21$
calculate impulse increment: $\Delta J = (J_f - J_i)/(N-1)$
set $J = J_i$

begin loop over impulse values; counter runs from 1 to $N+1$

calculate final speeds:

$v_{1a} = \sqrt{(J^2 + m_1^2 v_{1b}^2 - 2m_1 v_{1b} J \cos\alpha)} \Big/ m_1$

$v_{2a} = J/m_2$

calculate angles:

$\theta_1 = -\arctan[J \sin\alpha/(m_1 v_{1b} - J\cos\alpha)]$

$\theta_2 = \alpha$

check sign of θ_1 and correct if necessary

calculate final kinetic energy: $K_a = \frac{1}{2}m_1 v_{1a}^2 + \frac{1}{2}m_2 v_{2a}^2$

calculate kinetic energy loss: $Q = K_b - K_a$

print or display J, v_{1a}, v_{2a}, θ_1, θ_2, Q

increment J: replace J with $J + \Delta J$

end loop

stop

PROJECT 3. Puck 1 has a mass of 250 g and is incident at 3.0 m/s on puck 2, which has a mass of 600 g and is initially at rest. The impulse of puck 1 on puck 2 makes an angle of 20° with the axis of incidence (the x axis). Use the program to construct a table of the final speeds, angles of motion, and kinetic energy losses for impulses with magnitudes in the range from 0 to 2 kg·m/s.

Use the table to find two values of the impulse magnitude that straddle the value for an elastic collision. Rerun the program to find the impulse for an elastic collision to three significant figures. What are the final speeds and angle of motion then? [ans: $J = 0.995$ kg·m/s; $\mathbf{v}_{1a} = 1.55$ m/s at −119°; $\mathbf{v}_{2a} = 1.66$ m/s at 20°]

Now find the magnitude of the impulse for which puck 2 is scattered through 90°. What are the velocities of the pucks after the collision and what kinetic energy is lost? [ans: $J = 0.798$ kg·m/s; $\mathbf{v}_{1a} = 1.09$ m/s at −90°; $\mathbf{v}_{2a} = 1.33$ m/s at 20°; $Q = 1.33$ J]

Now suppose the impulse makes an angle of 60° with the line of incidence. What magnitude of impulse produces an elastic collision? What are the final velocities? [ans: $J = 0.529$ kg·m/s; $\mathbf{v}_{1a} = 2.67$ m/s at −43.4°; $\mathbf{v}_{2a} = 0.882$ m/s at 60°]

For what magnitude of impulse is puck 1 scattered through 90°? What are the final velocities and the kinetic energy loss? [ans: $J = 1.50$ kg·m/s; $\mathbf{v}_{1a} = 5.20$ m/s at −90°; $\mathbf{v}_{2a} = 2.50$ m/s at 60°; $Q = -4.13$ J]

The negative value of Q in the last case means kinetic energy must be *added* during the collision (in an explosion, for example).

PROJECT 4. What happens if the incident puck has greater mass than the target puck? Take $m_1 = 600$ g, $v_{1b} = 3.0$ m/s, and $m_2 = 250$ g. For $\alpha = 20°$ and 60° calculate the magnitude of the impulse that will produce an elastic collision and find the velocities of the pucks after such a collision. [ans: 20°: $J = 0.995$ kg·m/s, $\mathbf{v}_{1a} = 1.55$ m/s at −21.5°, $\mathbf{v}_{2a} = 3.78$ m/s at 20°; 60°: $J = 0.529$ kg·m/s, $\mathbf{v}_{1a} = 2.67$ m/s at −16.6°, $\mathbf{v}_{2a} = 2.12$ m/s at 60°]

PROJECT 5. If the masses are equal, the pucks always leave an elastic collision with their lines of motion making an angle of 90° with each other. Use the program to verify this for $m_1 = m_2 = 250$ g, $v_{1b} = 3.0$ m/s, and $\alpha = 20°$. Also verify it for $\alpha = 60°$. [ans: 20°: $J = 0.705$ kg·m/s, $\mathbf{v}_{1a} = 1.03$ m/s at −70.0°, $\mathbf{v}_{2a} = 2.82$ m/s at 20°; 60°: $J = 0.375$ kg·m/s, $\mathbf{v}_{1a} = 2.60$ m/s at −30.0°, $\mathbf{v}_{2a} = 1.50$ m/s at 60°]

Chapter 12

Conservation of angular momentum problems usually involve a system of two or more parts that exert torques on each other. The net external torque, however, vanishes. The angular momentum of each part changes, but the sum of the changes vanishes. You can use a computer to examine details of the motion.

Consider two wheels that are free to rotate on the same axle and that exert torques on each other. If the torque on wheel 1 is τ, then the angular velocity ω_1 of that wheel obeys $\tau = I_1\, d\omega_1/dt$, where I_1 is the rotational inertia of the wheel. The torque on wheel 2 is $-\tau$ and the angular velocity of that wheel obeys $-\tau = I_2\, d\omega_2/dt$. If τ is given as a function of time, a computer can be used to solve for θ_1, ω_1, θ_2, and ω_2, all as functions of time. Here's an outline, modeled after the program given in Chapter 2. The program also contains instructions to compute the angular momenta of the wheels and the total angular momentum of the system so you can check on angular momentum conservation.

> input initial angular velocities: ω_{10}, ω_{20}
> input final time, interval width: t_f, Δt
> input display interval: Δt_d
> set $t_b = 0$, $t_d = \Delta t_d$, $\theta_{1b} = 0$, $\omega_{1b} = \omega_{10}$, $\omega_{2b} = \omega_{20}$
> calculate torque on 1: $\tau_b = \tau(t_b)$
> **begin loop** over intervals
>   calculate time at end of interval: $t_e = t_b + \Delta t$
>   calculate torque on 1 at end of interval: $\tau_e = \tau(t_e)$
>   calculate "average" torque: $\tau = (\tau_b + \tau_e)/2$
>   calculate angular velocities at end of interval:
> $\omega_{1e} = \omega_{1b} + \tau\Delta t/I_1$
> $\omega_{2e} = \omega_{2b} - \tau\Delta t/I_2$
>   calculate "average" angular velocities:
> $\omega_1 = (\omega_{1b} + \omega_{1e})/2$
> $\omega_2 = (\omega_{2b} + \omega_{2e})/2$
>   calculate angular positions at end of interval:
> $\theta_{1e} = \theta_{1b} + \omega_1\Delta t$
> $\theta_{2e} = \theta_{2b} + \omega_2\Delta t$
>   if $t_e \geq t_d$ then
>   calculate angular momenta: $L_1 = I_1\omega_{1e}$, $L_2 = I_2\omega_{2e}$, $L = L_1 + L_2$
>   print or display t_e, θ_{1e}, ω_{1e}, θ_{2e}, ω_{2e}, L_1, L_2, L
>   increment t_d by Δt_d
>   end of if statement
>   if $t_e \geq t_f$ then exit loop
>   set $t_b = t_e$, $\theta_{1b} = \theta_{1e}$, $\omega_{1b} = \omega_{1e}$, $\theta_{2b} = \theta_{2e}$, $\omega_{2b} = \omega_{2e}$, $\tau_b = \tau_e$
> **end loop** over intervals
> stop

PROJECT 1. Suppose wheel 1 has a rotational inertia of 2.5 kg·m^2 and is initially rotating at 100 rad/s. Wheel 2 has a rotational inertia of 1.5 kg·m^2 and is initially at rest. The wheels come in contact and exert torques on each other, the torque of wheel 2 on wheel 1 being given by $\tau(t) = -20te^{-t/2}$, in N·m for t in seconds. Use the program to plot the angular velocities of the wheels from $t = 0$ to $t = 15$ s. Use an integration interval of 0.005 s and a display interval of 0.5 s.

What are the initial values of the angular momenta of the wheels and the total angular momentum of the system? What are the values after 15 s? [ans: $L_1 = 250\,\mathrm{kg{\cdot}m^2/s}$, $L_2 = 0$, $L = 250\,\mathrm{kg{\cdot}m^2/s}$; $L_1 = 170\,\mathrm{kg{\cdot}m^2/s}$, $L_2 = 80\,\mathrm{kg{\cdot}m^2/s}$, $L = 250\,\mathrm{kg{\cdot}m^2/s}$]

Suppose that in addition to the torque that wheel 2 exerts on wheel 1 an external agent exerts a torque of 8.0 N·m on wheel 1, in the direction of its angular velocity. What now are the angular momenta of the wheels and the total angular momentum of the system after 15 s? The torque on wheel 1 is $8.0 - 20te^{-t/2}$ and the torque on wheel 2 is $20te^{-t/2}$. [ans: $L_1 = 290\,\mathrm{kg{\cdot}m^2/s}$, $L_2 = 80\,\mathrm{kg{\cdot}m^2/s}$, $L = 370\,\mathrm{kg{\cdot}m^2/s}$]

Notice that the angular momenta of the individual wheels change with time in both cases. The total angular momentum, however, is conserved in the first situation but not in the second. Since $\mathrm{d}L/\mathrm{d}t = \tau_{\text{ext}}$ and in the second situation $\tau_{\text{ext}} = 8.0$ N·m the total angular momentum at the end of 15 s should be $L_f = 8.0 \times 15 = 120\,\mathrm{kg{\cdot}m^2/s}$. Is this result produced by the program?

PROJECT 2. Conservation of angular momentum is sometimes demonstrated by considering the inelastic "collision" between two wheels on the same axle. Suppose wheel 1 of the previous project starts with an angular velocity of 100 rad/s and exerts a constant torque of 8.0 N·m on wheel 2, which starts from rest. Wheel 2, of course, exerts a torque of −8.0 N·m on wheel 1. Use the program to find the time for which $\omega_1 = \omega_2$ and the value of the angular velocity then. Through what angle has each wheel rotated by the time they reach the same angular velocity? Compare the value obtained for the final angular velocity with that predicted by $I_1\omega_{10} = I_1\omega_1 + I_2\omega_2$. [ans: 11.7 s, 62.5 rad/s, 952 rad, 366 rad]

The time for the wheels to reach the same angular velocity and the angle through which they turn depend on the torque they exert on each other but the final value of the angular velocity does not. To verify this statement repeat the calculation for the same initial conditions but for a torque on wheel 1 that is given by $\tau = -20te^{-t/3}$, in N·m for t in seconds. [ans: 5.24 s, 62.5 rad/s, 438 rad, 144 rad]

PROJECT 3. The kinetic energy of the wheels is not conserved during the inelastic "collision" of the previous project. You can see this by calculating the initial and final kinetic energies, but you can also use a computer to calculate the work done by each torque. If a wheel turns through the small angle $\Delta\theta$ while the torque τ acts on it, the work done by the torque is given by $\Delta W = \tau\Delta\theta$. Modify the program so it computes the work done on each wheel. Just before the loop over intervals set $W_1 = 0$ and $W_2 = 0$. These variables will be used to sum the work done in the intervals. Just before displaying results have the computer increment W_1 by $\tau(\theta_{1e} - \theta_{1b})$ and W_2 by $-\tau(\theta_{2e} - \theta_{2b})$. Here τ is the average torque exerted by wheel 2 on wheel 1. Also have the computer calculate the total work $W = W_1 + W_2$ and display the result.

Take the initial conditions to be as before ($\omega_{10} = 100$ rad/s, $\omega_{20} = 0$). Take the torque of wheel 1 on wheel 2 to be a constant 8.0 N·m. Find the work done on each wheel and the total work done from $t = 0$ to the time when the wheels have the same angular velocity. Compare the total work done to the change in the total kinetic energy: $\Delta K = \frac{1}{2}(I_1 + I_2)\omega_f^2 - \frac{1}{2}I_1\omega_{10}^2$, where ω_f is the final angular velocity. [ans: -7.62×10^3 J, 2.93×10^3 J, -4.69×10^3 J]

Now repeat the calculation for a torque on wheel 1 that is given by $\tau = -20te^{-t/3}$, in N·m for t in seconds. You should get the same answers.

Can the "collision" be elastic? Take the torque on wheel 1 to be −8.0 N·m and calculate the total work done as a function of time. Search for a time such that the total work is zero. You might use the condition $W \geq 0$ to exit the loop over intervals. If the wheels no longer interact after this time, the collision is elastic. What are the final angular velocities of the wheels? [ans: 23.4 s, 25.0 rad/s, 125 rad/s]

The text showed you how to analyze the change in angular velocity of a spinning ice skater as she brings her arms toward the center of her torso from an outstretched position. Before she brings them in, her rotational inertia is I_i and her angular velocity is ω_i. After she brings them in, her rotational inertia is I_f and her angular velocity is ω_f. Conservation of angular momentum leads to $I_i\omega_i = I_f\omega_f$, so $\omega_f = (I_i/I_f)\omega_i$.

Here you will use a computer program to investigate some of the details. In particular, you will be able to follow her angular velocity as she brings her arms in. Let I_T be the rotational inertia of her torso alone and let I_A be the rotational inertia of her arms, including any weights she might hold to heighten the effect. I_A is a function of time, determined by the rate with which she moves her arms. Angular momentum is conserved at each stage of the process so

$$\omega(t) = \frac{I_{Ai} + I_T}{I_A(t) + I_T}\omega_i \,,$$

where I_{Ai} is the initial value of the rotational inertia of her arms.

If $\omega(t)$ is known, a computer program can be used to calculate the angle $\theta(t)$ through which the skater has rotated since $t = 0$. The angle θ_e at the end of any interval of width Δt is related to the angle θ_b at the beginning by $\theta_e = \theta_b + \omega\Delta t$, where ω is the average angular velocity in the interval. As usual, we approximate the average angular velocity by $(\omega_b + \omega_e)/2$. Here is an outline of a program.

input final time and interval width: t_f, Δt
input display interval: Δt_d
set $t_b = 0$, $t_d = \Delta t_d$, $\theta_b = 0$
calculate angular velocity at beginning of first interval: $\omega_b = \omega(t_b)$
begin loop over intervals
    calculate time at end of interval: $t_e = t_b + \Delta t$
    calculate angular velocity at end of interval: $\omega_e = \omega(t_e)$
    calculate "average" angular velocity: $\omega = (\omega_b + \omega_e)/2$
    calculate angle at end of interval: $\theta_e = \theta_b + \omega\Delta t$
    if $t_e \geq t_d$ then
        calculate torque at end of interval: τ_e
        print or display t_e, θ_e, ω_e
        increment t_d by Δt_d
    end of if statement
    if $t_e \geq t_f$ then exit loop
    set $t_b = t_e$, $\theta_b = \theta_e$, $\omega_b = \omega_e$ in preparation for next interval
end loop over intervals
stop

Both the angular momentum of the skater's torso and the angular momentum of her arms change. Her torso exerts an torque on her arms and her arms exert a torque of equal magnitude but opposite direction on her torso. If τ represents the torque on her torso, then $\tau = I_T\,\mathrm{d}\omega/\mathrm{d}t$. Differentiation of the expression above for $\omega(t)$ produces

$$\tau = -\frac{I_T\omega_i(I_{Ai} + I_T)}{(I_A + I_T)^2}\frac{\mathrm{d}I_A}{\mathrm{d}t}$$

Use this expression in the program to calculate the torque just before t_e, θ_e, and ω_e are displayed, then display it along with the other quantities. A line for the instruction has already been indicated in the outline above.

PROJECT 4. A skater is initially turning at 12 rad/s. Her torso has a rotational inertia of 0.65 kg·m^2. She carries weights in her hands so that as she pulls her arms in their rotational inertia changes from 0.50 kg·m^2 to 0.04 kg·m^2. Suppose she moves her arms in such a way that their rotational inertia changes uniformly over 2.0 s. That is, $I_A(t) = I_{Ai} - Kt$, where $I_{Ai} = 0.50$ kg·m^2 and K is a constant chosen so $I_A = 0.04$ kg·m2 when $t = 2.0$ s.

Use the program to plot her angular velocity and the torque acting on her torso as functions of time from $t = 0$ to $t = 2.0$ s. What is her final angular velocity? Through what angle did she rotate during the process? At what time did she exert the maximum torque on her arms and what was the magnitude of that torque? [ans: 20.0 rad/s; 30.7 rad (4.88 rev); 1.73 N·m at 2.0 s]

Suppose now that all of the rotational inertia of her arms is actually due to the weights she is carrying and that she pulls them in at a constant rate. If m is the mass of one weight and r is it distance from the axis of rotation, then the rotational inertia of both weights together is $I_A = 2mr^2$. Since the weights are brought in at a constant rate, we may write $r = r_i + Kt$, where r_i is the initial distance and K is a constant. Take $m = 1.0$ kg and pick values of r_i and K so $I_{Ai} = 0.50$ kg·m^2 and $I_A = 0.05$ kg·m^2 at the end of 2.0 s. Then, use the program to plot her angular velocity and the torque acting on her torso as functions of time from $t = 0$ to $t = 2.0$ s. What is her final angular velocity? Through what angle did she rotate during the process? At what time did she exert the maximum torque on her arms and what was the magnitude of that torque? [ans: 20.0 rad/s; 32.2 rad (5.12 rev); 2.82 N·m at 0.95 s]

Chapter 14

In this section a computer is used to investigate satellite motion. A massive central body (the Sun or Earth, for example) is at the origin and is assumed to remain motionless. If the coordinates of the satellite are x and y, then the force exerted by the central body on the satellite has components that are given by $F_x = -GMmx/r^3$ and $F_y = -GMmy/r^3$, where G is the universal gravitational constant (6.67×10^{-11} m^3/s^3·kg), M is the mass of the central body, m is the mass of the satellite, and $r = \sqrt{x^2 + y^2}$ is the center-to-center distance of the satellite from the central body. These expressions assume both the central body and satellite have spherically symmetric mass distributions. Notice that the force is directed radially from the center of the satellite toward the center of the central body.

The acceleration of the satellite has components that are given by $a_x = -GMx/r^3$ and $a_y = -GMy/r^3$. Since the force depends on the coordinates of the satellite, you cannot use $\mathbf{a} = (\mathbf{a}_b + \mathbf{a}_e)/2$ to approximate the average acceleration in an interval. The acceleration at the beginning of an interval is an extremely poor approximation to the average acceleration and if it is used the intervals must be quite narrow and the running time must be quite long. Here you will learn a trick to circumvent the problem.

The program outlined below calculates the coordinates and acceleration at the beginning of each interval and the velocity at the midpoint of each interval. The acceleration at the beginning of an interval and the velocity at the midpoint of the previous interval are used to compute the velocity at the midpoint of the interval, then that velocity and the coordinates at the beginning of the interval are used to compute the coordinates at the end of the interval.

To start, the velocity half an interval before t_0 must be computed using the acceleration and velocity at t_0. When the computer is asked to display results, in the loop over intervals, x_b and y_b are the coordinates at the beginning of the interval while v_{xm} and v_{ym} are the velocity components for the midpoint of the interval. We wish to display the velocity components for the beginning and end of the interval, so the acceleration for the beginning is used to calculate these quantities before printing. Here's the outline.

input initial conditions: x_0, y_0, v_{x0}, v_{y0}
input final time and interval width: t_f, Δt
input display interval: Δt_d
calculate acceleration at t_0: a_x, a_y
calculate velocity at $t_0 - \Delta t/2$:
  $v_{xm} = v_{x0} - a_x \Delta t/2$
  $v_{ym} = v_{y0} - a_y \Delta t/2$
set $t_b = t_0$, $t_d = t_0 + \Delta t_d$, $x_b = x_0$, $y_b = y_0$
begin loop over intervals
  calculate acceleration at beginning of interval: a_x, a_y
  calculate velocity at midpoint of interval:
    replace v_{xm} with $v_{xm} + a_x \Delta t$
    replace v_{ym} with $v_{ym} + a_y \Delta t$
  calculate coordinates at end of interval:
    $x_e = x_b + v_{xm} \Delta t$
    $y_e = y_b + v_{ym} \Delta t$
  calculate time at end of interval: $t_e = t_b + \Delta t$
  if $t \geq t_d$ then
    calculate velocity at beginning of interval:
      $v_{xb} = v_{xm} - a_x \Delta t/2$
      $v_{yb} = v_{ym} - a_y \Delta t/2$
    print or display t_b, x_b, y_b, v_{xb}, v_{yb}
    calculate velocity at end of interval:
      $v_{xe} = v_{xm} + a_x \Delta t/2$
      $v_{ye} = v_{ym} + a_y \Delta t/2$
    print or display t_e, x_e, y_e, v_{xe}, v_{ye}
    set $t_d = t_d + \Delta t_d$
  end of if statement
  if $t_e \geq t_f$ then exit loop
  set $t_b = t_e$, $x_b = x_e$, $y_b = y_e$
end of loop over intervals
stop

To write an efficient program define the constant C by $C = -GM\Delta t$, then use $a_x = Cx_b/r^3$ and $a_y = Cy_b/r^3$ to compute the acceleration components.

In the following project you will use the program to verify Kepler's laws of planetary motion. First you will verify the conservation of energy and angular momentum.

PROJECT 1. Consider a satellite in orbit around Earth ($M = 5.98 \times 10^{24}$ kg). At $t = 0$ it is at $x = 7.2 \times 10^6$ m, $y = 0$ and has velocity components $v_{x0} = 0$, $v_{y0} = 9.0 \times 10^3$ m/s. Use the program with $\Delta t = 10$ s to plot the position for every 500 s from $t = 0$ to $t = 1.55 \times 10^4$ s.

Modify the program so it computes the total mechanical energy and angular momentum, both per unit satellite mass, for each point displayed. The total mechanical energy per unit mass is given by $E = \frac{1}{2}(v_{xe}^2 + v_{ye}^2) - GM/r$ and the angular momentum per unit mass is given by $\ell = x_e v_{ye} - y_e v_{xe}$. The results should be constant to two or three significant figures.

Now use the program to prove the orbit is an ellipse. An ellipse is a geometric figure such that the sum of the distances from any point on the figure to two fixed points, called foci, is the same as for any other point. For the orbit you generated above, one focus is at the origin and the other is on the x axis the same distance from the geometric center as the first. First locate the geometric center of the orbit. Use the coordinates generated by the program to find the x coordinates of the points where the orbit crosses the x axis, then calculate half their sum. Let this coordinate be $-x_c$. The second focus is at $x_F = -2x_c$.

Modify the program so it computes the sum of the distances from the point displayed to the foci. It is given by $\sqrt{x_e^2 + y_e^2} + \sqrt{(x_e - x_F)^2 + y_e^2}$. Check that this sum is the same for all displayed points to three or more significant figures.

Kepler's second law says that the vector from the central body to the satellite sweeps out equal areas in equal times. Let $\mathbf{r}_b$ be the position vector at the beginning of an interval and let $\mathbf{r}_e$ be the position vector at the end of the interval. The triangle defined by these two vectors has an area that is given by $A = \frac{1}{2}|\mathbf{r}_b \times \mathbf{r}_e| = \frac{1}{2}|x_b y_e - x_e y_b|$. This is the area swept out by the position vector in time Δt. Have the computer calculate A just before it displays results and ask it to display A along with the other quantities. Check to see if it is constant to a reasonable number of significant figures.

You should recognize that this calculation essentially repeats the calculation above to check on the constancy of the angular momentum. Since $\mathbf{r}_e = \mathbf{r}_b + \mathbf{v}\Delta t$, $A = \frac{1}{2}|\mathbf{r}_b \times \mathbf{v}|\Delta t$. Since the angular momentum is given by $\boldsymbol{\ell} = m\mathbf{r} \times \mathbf{v}$, this becomes $A = \frac{1}{2}\ell\Delta t$. A is constant because ℓ is constant.

Kepler's third law tells us that the period T is related to the length a of the semimajor axis by $T^2 = (4\pi^2/GM)a^3$. For the orbit generated above, the length of the semimajor axis is half the distance between the two points where the orbit crosses the x axis. Calculate this distance and compute the right side of the third law equation. Use the program to find the period. It is, for example, twice the time between two successive crossings of the x axis. Compute the left side of the third law equation and compare the value with the value you obtained for the right side.

PROJECT 2. The length of the semimajor axis of an elliptical orbit depends only on the total mechanical energy. Suppose the satellite of the previous project is started from $x = 5.5 \times 10^6$ m, $y = 0$ with a velocity in the positive y direction. Select the initial speed so its total mechanical energy is the same as before. Plot the trajectory and find the length of the semimajor axis. Compare it with the length of the semimajor axis for the previous project.

The length a of the semimajor axis is related to the total mechanical energy E by

$$E = -\frac{GMm}{2a}.$$

Suppose the satellite is started at $x = 7.2 \times 10^6$ m, $y = 0$ with a velocity in the positive y direction. What initial speed should it have if its orbit is to be circular? For a circular orbit, the semimajor axis is the same as the radius and for an initial position on the x axis it must be the same as x_0. Equate $-GMm/2x_0$ to $\frac{1}{2}mv_{y0}^2 - GMm/x_0$ and solve for v_{y0}. Use the program to plot the orbit. You might have the computer calculate $r^2 = x_e^2 + y_e^2$ for every displayed point and see if this quantity remains constant. Compared to the orbit of the first project is the total mechanical energy greater or less for the circular orbit?

If the total mechanical energy is negative, the satellite is bound and its orbit is an ellipse. This is so for the satellites studied above. If the total mechanical energy is zero, the orbit is a parabola and if the total mechanical energy is positive, the orbit is a hyperbola. In either of these cases the satellite is not bound and it eventually escapes from the gravitational pull of the central body. Here's an example.

PROJECT 3. Consider a spacecraft in the gravitational field of Earth, with initial coordinates $x_0 = 7.2 \times 10^6\,\mathrm{m}$, $y_0 = 0$. Take its initial velocity to be $v_{x0} = 0$, $v_{y0} = 1.2 \times 10^4\,\mathrm{m/s}$. Use the program to plot its position every 300 s from $t = 0$ to $t = 4500\,\mathrm{s}$. Also calculate the total mechanical energy for these times and note it is constant (within the limits of the calculation, of course) and positive. Notice that the trajectory tends to become a straight line as the spacecraft recedes from the central body. Find the angle this line makes with the x axis. [ans: 51°]

What happens to the line if the initial speed is increased? Use the program to verify your conjecture.

Chapter 16

Here numerical integration is used to investigate the oscillation of a mass m on an ideal spring with spring constant k. The acceleration of the mass is given by $a = -(k/m)x$, where x is the coordinate of the mass, measured from the equilibrium point. The motion is periodic with an angular frequency that is given by $\omega = \sqrt{k/m}$, a period that is given by $2\pi\sqrt{m/k}$, and an amplitude that is given by $x_m = \sqrt{x_0^2 + (v_0/\omega)^2}$, where x_0 is the initial coordinate and v_0 is the initial speed.

The first program discussed in the Computer Projects section of Chapter 6 can be used, but the interval width Δt must be quite small and the number of intervals correspondingly large to avoid accumulating errors. The accuracy can be improved considerably by using more terms in the equations for the velocity and coordinate at the end of an interval. Including terms that are proportional to $(\Delta t)^2$, the velocity is given by $v_e = v_b + a_b\Delta t + \frac{1}{2}(da/dt)(\Delta t)^2$, where the derivative da/dt is evaluated for the beginning of the interval. Since $a = -(k/m)x$, $da/dt = -(k/m)v_b$. The derivative of the acceleration with respect to time is sometimes called the "jerk". Define $j_e = -(k/m)v_e$ and use $v_e = (\frac{1}{2}j_e\Delta t + a_e)\Delta t + v_b$ to calculate the velocity at the end of an interval. Some factoring has been carried out to save computational time and to reduce truncation errors. Similarly $x_e = x_b + v_b\Delta t + \frac{1}{2}a_b(\Delta t)^2$, which is best written $x_e = (\frac{1}{2}a_b\Delta t + v_b)\Delta t + x_b$. You must change appropriate lines of the program.

PROJECT 1. Consider a 2.0-kg mass attached to an ideal spring with spring constant $k = 350\,\mathrm{N/m}$. It is released from rest at $x = 0.070\,\mathrm{m}$. Take $\Delta t = 0.001\,\mathrm{s}$ and plot the coordinate for every 0.05 s from $t = 0$ (the time of release) to $t = 1\,\mathrm{s}$. For use later have the computer calculate the potential energy ($U = \frac{1}{2}kx^2$), the kinetic energy ($K = \frac{1}{2}mv^2$), and the total mechanical energy ($E = K + U$) for each point plotted.

Notice that the graph predicts an oscillatory motion. Use the graph to estimate the amplitude and compare the value with 0.070 m. Use the graph to estimate the period and compare the value with $2\pi\sqrt{m/k}$.

Also notice that at times when the spring has its greatest extension (0.070 m) the speed is zero, the kinetic energy is zero, the potential energy is maximum, and the magnitude of the acceleration is maximum. Also notice that when the spring is neither extended or compressed ($x = 0$) the speed is maximum, the kinetic energy is maximum, the potential energy is zero, and the acceleration is zero. Finally, notice that the total mechanical energy is constant. Your results may show some fluctuation after the third or fourth significant figure but this is due to computational errors.

The coordinate as a function of time can be written $x(t) = x_m \cos(\omega t + \phi)$, where ϕ is the phase constant. For any value of ϕ the initial coordinate is given by $x_0 = x_m \cos\phi$ and the initial velocity is given by $v_0 = -\omega x_m \sin\phi$. For the initial conditions you used above (x_0 positive and $v_0 = 0$), the phase constant is zero. Other initial conditions result in different values for the phase constant. If the amplitude is the same, a different phase constant produces a function $x(t)$ that is shifted along the time axis relative to the function for $\phi = 0$.

Suppose the phase constant is $\pi/6$ rad (30°). Take the amplitude to be $x_m = 0.070\,\text{m}$ and calculate the initial coordinate and velocity. Now use the program to plot $x(t)$ for the first second of the motion and find the time at which the first maximum ($x = 0.070\,\text{m}$) occurs. Since the first maximum occurs at $t = 0$ for $\phi = 0$, the result you obtain is the amount by which the plot is shifted.

A phase constant of 2π radians (360°) corresponds to a shift equal to one period of the motion. A phase constant of $\pi/6$ radians corresponds to one twelfth of a period. Does this agree with your result?

PROJECT 2. Suppose the mass is also subjected to a resistive force, proportional to its velocity. Then, the acceleration is given by $a = -(k/m)x - (b/m)v$, where b is the drag coefficient. Take $m = 2.0\,\text{kg}$, $k = 350\,\text{N/m}$, $b = 2.8\,\text{kg/s}$, $x_0 = 0.070\,\text{m}$, and $v_0 = 0$. Use the computer program, with $\Delta t = 0.001\,\text{s}$, to plot the coordinate at intervals of 0.050 s from $t = 0$ (when the mass is released) to $t = 1.0\,\text{s}$. Also calculate the potential energy, the kinetic energy, and the total mechanical energy for these times. The rate of change of the acceleration is given by $j = -(k/m)v - (b/m)a$.

Notice that the amplitude decreases as time goes on. This decrease is related, of course, to the decrease in total mechanical energy. The resistive force does negative work on the mass. Plot the total mechanical energy as a function of time.

Does the drag force also change the period of the motion? Measure the period as the time between successive maxima and compare the result with $2\pi\sqrt{m/k}$.

PROJECT 3. The oscillator of the previous project is said to execute damped harmonic motion. The motion is said to be *underdamped* because the mass continues to oscillate. If the value of b is increased sufficiently, no oscillations occur and the motion is said to be *overdamped.*

Take $m = 2.0\,\text{kg}$, $k = 350\,\text{N/m}$, $b = 90\,\text{kg/s}$, $x_0 = 0.070\,\text{m}$, and $v_0 = 0$. Plot $x(t)$ every 0.050 s from $t = 0$ (when the mass is released) to $t = 1.0\,\text{s}$.

The motion of the mass is changed by an external force. Consider a force given by $F_m \cos(\omega'' t)$, where F_m and ω'' are constants. The angular frequency of the impressed force should not be confused with the natural angular frequency $\omega = \sqrt{k/m}$ of the mass and spring alone. The acceleration of the mass is now given by $a = -(k/m)x + (F_m/m)\cos(\omega'' t)$ and its derivative is given by $j = -(k/m)v - (F_m\omega''/m)\sin(\omega'' t)$. Change the program accordingly. The following project is designed to help you investigate forced oscillations without damping.

PROJECT 4. Take $m = 2.0\,\text{kg}$, $k = 350\,\text{N/m}$, $b = 0$, $x_0 = 0.070\,\text{m}$, and $v_0 = 0$. The natural angular frequency is $\omega = \sqrt{350/2} = 13.2\,\text{rad/s}$, corresponding to a period of 0.475 s. For each of the following impressed forces of the form $F_m \cos(\omega'' t)$, plot $x(t)$ from $t = 0$ to $t = 1.0\,\text{s}$. Use an integration interval of $\Delta t = 0.001\,\text{s}$ and a display interval of $\Delta t_d = 0.050\,\text{s}$. (a.) $F_m = 18\,\text{N}$, $\omega'' = 35\,\text{rad/s}$ (b.) $F_m = 18\,\text{N}$, $\omega'' = 15\,\text{rad/s}$.

For the conditions of part a, the motion is nearly sinusoidal with a period of about 0.47 s, the natural period. The influence of the impressed force is seen in deviations from a sinusoidal shape. When the impressed frequency is closer to the natural frequency, as in part b, the influence of the impressed force is more pronounced. The amplitude grows with time, then levels off. If the impressed frequency is made exactly equal to the natural frequency, the amplitude grows without bound.

The increase in amplitude can easily be accounted for in terms of the work done by the impressed force. If ω'' is different from ω, the impressed force is in the same direction as the velocity of the mass over some portions of the motion and in the opposite direction over other portions. If the frequencies are very different, as in part a, the net work done over a time period that is long compared with the period is almost zero. On the other hand, when the two frequencies are nearly the same, as in part b, the impressed force and velocity are in the same direction over a large portion of the motion and the amplitude grows with time. As time goes on, the impressed force and velocity are in the same direction over smaller portions of the motion and in opposite directions over larger portions. Eventually, the net work vanishes over a long time interval and the amplitude becomes constant.

You can easily verify these assertions by plotting the total mechanical energy as a function of time. Modify the program so it calculates and lists or plots the total mechanical energy every 0.05 s from $t = 0$ to $t = 2.0$ s. Note intervals during which the impressed force does positive work (the energy increases) and intervals during which it does negative work (the energy decreases).

PROJECT 5. Now investigate the motion when a damping force is present. Consider the same oscillator ($m = 2.0$ kg, $k = 350$ N/m), subjected to an impressed force with $F_m = 18$ N and $\omega'' = 35$ rad/s, but with the drag coefficient $b = 15$ kg/s. Use the same initial conditions ($x_0 = 0.070$ m, $v_0 = 0$) and integration interval ($\Delta t = 0.001$ s). Plot $x(t)$ every 0.050 s from $t = 0$ to $t = 2.0$ s.

Notice that motion starts out very much the same as when there is no damping. Its period is nearly the natural period. Now, however, this motion is quickly damped and what remains is a sinusoidal motion at the frequency of the impressed force. Verify that the period is about 0.18 s, corresponding to an angular frequency of 35 rad/s. The amplitude eventually becomes constant. The energy supplied by the impressed force is dissipated by damping.

Repeat the calculation for an impressed angular frequency of 15 rad/s, near the natural frequency. Note that the final amplitude is much larger than when the impressed frequency was far from the natural frequency.

A simple pendulum consists of a mass m at the end of a light rod of length ℓ, free to swing in a uniform gravitational field. The angular position is given by the angle $\theta(t)$, measured from the vertical. If g is the acceleration due to gravity, then θ obeys the differential equation

$$\frac{d^2\theta}{dt^2} = -\frac{g}{\ell}\sin\theta .$$

You have learned that if θ is always small and is measured in radians, then $\sin\theta$ may be replaced by θ itself and the solution to the resulting differential equation is $\theta(t) = \theta_m \sin(\omega t + \phi)$, where the angular frequency is given by $\omega = \sqrt{g/\ell}$.

Use a computer to investigate the motion when the amplitude is not small. You may want to change the program so it is written in terms of θ and its derivative Ω ($= d\theta/dt$): simply replace x with θ and v with Ω. The angular acceleration, which replaces a, is $\alpha = -(g/\ell)\sin\theta$ and its derivative is $j = -(g\Omega/\ell)\cos\theta$.

PROJECT 6. First use the program to check the small angle approximation. A simple pendulum with a length of 1.2 m is pulled aside 10° (0.175 rad) and released from rest. What is its period? Use an integration interval of 0.001 s and search for the second time after starting that Ω is zero. Compare the result with $2\pi\sqrt{\ell/g}$. [ans: 2.20 s]

Repeat the calculation for an initial angular displacement of 45° (0.785 rad). [ans: 2.29 s]

Repeat the calculation for an initial angular displacement of 75° (1.31 rad). [ans: 2.46 s]

Now check to see how close the motion is to simple harmonic. Take the initial angular displacement to be 75° and use the program to generate a table of $\theta(t)$ for the first 2.5 s of the motion. You might obtain values at intervals of 0.1 s. At the same time, have the program generate values of $\theta_m \cos(\omega t)$, where the value of ω is calculated from the value you found for the period.

The conversion of kinetic energy to potential energy and back again to kinetic energy is quite similar to the conversion that takes place for a spring-mass system, although the mechanism is the work done by the force of gravity, not the force of a spring. Suppose the pendulum is released from rest with an initial angular displacement of 75°. For each of the displayed points have the computer calculate the kinetic energy per unit mass ($\frac{1}{2}\ell^2\Omega^2$), the potential energy per unit mass $[g\ell(1 - \cos\theta)]$, and their sum.

Chapter 17

Start with a rather simple program to plot the displacement $y(x,t) = y_m \sin(kx - \omega t)$ for a sinusoidal traveling wave at a given time. Input the amplitude y_m, wavelength λ, frequency f, and time t. Have the computer calculate values of $y(x,t)$ for x from 0 to some final value x_f, at intervals of width Δx. Here's an outline.

```
input amplitude, wavelength, frequency: y_m, λ, f
input final value of x, interval width: x_f, Δx
input time: t
calculate angular wave number, angular frequency: k = 2π/λ, ω = 2πf
set x = 0
begin loop over intervals
     calculate y(x,t): y = y_m sin(kx − ωt)
     display or plot x, y
     increment x by Δx
if x > x_f then end loop over intervals
stop
```

A reasonably good graph can be obtained if Δx is chosen to be about $\lambda/20$ and x_f is chosen to be about 2λ. If you program the computer to plot the wave on your monitor, you might have the program calculate values of Δx and x_f rather than read them. If you are plotting by hand, you will want values that are 1, 2, or 5 times a power of ten.

PROJECT 1. Take $y_m = 2.0$ mm, $\lambda = 5.0$ cm, and $f = 10$ Hz. Make a graph of the wave at $t = 0$ and note the position of the maximum nearest the origin.

Now make a graph of the wave at $t = 4.0 \times 10^{-2}$ s and again note the position of the maximum nearest the origin. Measure the distance this maximum has moved in 4.0×10^{-2} s and use the value you obtain to calculate the wave speed. Compare your answer with $v = \lambda f$. The time was chosen so that no other maximum appears between the origin and the maximum you are following.

Two sinusoidal waves with the same frequency and wavelength, traveling in the same direction, sum to form another sinusoidal wave. You can use a computer program to plot the sum of the waves, then read the resultant amplitude and phase from the graph. Suppose $y_1(x,t) = y_{1m}\sin(kx - \omega t)$ and $y_2(x,t) = y_{2m}\sin(kx - \omega t - \phi)$. Then, the resultant wave is given by $y(x,t) = y_{1m}\sin(kx - \omega t) + y_{2m}\sin(kx - \omega t - \phi)$. Here's the outline of a program.

> input amplitudes, wavelength, frequency, phase constant: y_{1m}, y_{2m}, λ, f, ϕ
> input final value of x, interval width: x_f, Δx
> input time: t
> calculate angular wave number, angular frequency: $k = 2\pi/\lambda$, $\omega = 2\pi f$
> set $x = 0$
> **begin loop** over intervals
>   calculate $y(x,t)$: $y = y_{1m}\sin(kx - \omega t) + y_{2m}\sin(kx - \omega t - \phi)$
>   display or plot x, y
>   increment x by Δx
> if $x > x_f$ then **end loop** over intervals
> stop

PROJECT 2. No matter what the amplitudes and the phase difference, the resultant wave is sinusoidal and moves with same speed as the constituent waves. Take $y_{1m} = 2.0\,\text{mm}$, $y_{2m} = 4.0\,\text{mm}$, $\lambda = 5.0\,\text{cm}$, $f = 10\,\text{Hz}$, and $\phi = 65°$. Plot the resultant wave from $x = 0$ to $x = 10\,\text{cm}$ for $t = 0$. Then, plot the resultant wave for $t = 4.0 \times 10^{-2}\,\text{s}$ and calculate the wave speed. You should get the same answer as you obtained for the first project.

PROJECT 3. The amplitude of the resultant wave depends on the phase difference ϕ. First consider two waves with the same amplitude. Then, the resultant amplitude is given by $y_m = 2y_{1m}\cos(\phi/2)$. Take $y_{1m} = y_{2m} = 2.0\,\text{mm}$, $\lambda = 5.0\,\text{cm}$, $f = 10\,\text{Hz}$, and $\phi = 65°$. Plot the resultant wave from $x = 0$ to $x = 10\,\text{cm}$ for $t = 0$. Measure the amplitude and compare the result with the calculated value.

The phase constant α of the resultant wave also depends on the phases of the constituent waves. For equal amplitudes, $\alpha = \phi/2$. Use your graph to find α. At $t = 0$ the first constituent wave y_1 is zero and has positive slope at the origin. Use your graph to find the coordinate where the resultant wave is zero and has positive slope. The ratio of this coordinate to a wavelength is the same as the ratio of the phase constant to 360° or 2π rad. The coordinate should be $(32.5/360)\lambda = 0.45\,\text{cm}$. Is it?

What happens when the amplitudes are not equal? Take $y_{1m} = 2.0\,\text{mm}$, $y_{2m} = 4.0\,\text{mm}$, $\lambda = 5.0\,\text{cm}$, and $f = 10\,\text{Hz}$. For each of the following values of ϕ, find the amplitude and phase constant of the resultant wave: (a.) 0; (b.) 30°; (c.) 45°; (d.) 90°; (e.) 180°. [ans: 6.0 mm, 0; 5.8 mm, 20°; 5.6 mm, 30°; 4.5 mm, 63°; 2.0 mm, 180°]

Your program can also be used to investigate standing waves. A standing wave is composed of two traveling waves with the same amplitude, frequency, and wavelength, but traveling in opposite directions. Revise the program so only one amplitude is read and so $y = y_m\sin(kx - \omega t) + y_m\sin(kx + \omega t - \phi)$ is used to calculate the resultant wave.

PROJECT 4. Take $y_m = 2.0\,\text{mm}$, $\lambda = 5.0\,\text{cm}$, and $f = 10\,\text{Hz}$. Make a graph of the wave at $t = 0$ and note the position of the maximum nearest the origin. Now make a graph of the wave at $t = 0.020\,\text{s}$ and again note the position of the maximum nearest the origin. During this time each of the traveling waves moved 1.0 cm, one fifth of a wavelength. But the standing wave maximum did not move.

Notice that the maximum displacement is less than at $t = 0$. In fact, all parts of the string move together toward $y = 0$. At $t = 0.025$ s the displacement is everywhere 0. Use the program to verify this. Also plot the wave for $t = 0.05$ s, one half period after the start.

PROJECT 5. Identify the nodes on the graph you made as part of the last project for $t = 0$ and verify that they are half a wavelength apart. How do the phases of the traveling waves affect the positions of the nodes? For the same wavelength and frequency as the last project and for each of the following values of ϕ, find the coordinate of the node nearest the origin and verify that the node separation is the same: (a.) 0; (b.) 45°; (c.) 90°. [ans: 0; 0.31 cm; 0.63 cm]

Chapter 18

Modify the program of the last chapter to investigate beats. Two waves with slightly different frequencies are summed: $y(x,t) = y_{1m}\sin(k_1 x - \omega_1 t) + y_{2m}\sin(k_2 x - \omega_2 t - \phi)$. Once a value is chosen for x this can be written $y = y_{1m}\sin(\omega_1 t - \phi_1) + y_{2m}\sin(\omega_2 t - \phi_2)$, where $\phi_1 = k_1 x - \pi$ and $\phi_2 = k_2 x + \phi - \pi$. For the first few projects, choose x and ϕ so both ϕ_1 and ϕ_2 vanish. Then, $y = y_{1m}\sin(\omega_1 t) + y_{2m}\sin(\omega_2 t)$. Input the amplitudes and frequencies, then have the computer generate a plot of y for times from $t = 0$ to $t = t_f$, with an interval of Δt. Here's an outline.

> input amplitudes, frequencies: y_{1m}, y_{2m}, f_1, f_2
> input final value of t, interval width: t_f, Δt
> calculate angular frequencies: $\omega_1 = 2\pi f_1$, $\omega_2 = 2\pi f_2$
> set $t = 0$
> **begin loop** over intervals
>   calculate y: $y = y_{1m}\sin(\omega_1 t) + y_{2m}\sin(\omega_2 t)$
>   plot y
>   increment t by Δt
> if $t > t_f$ then **end loop** over intervals
> stop

Reasonable graphs are produced if Δt is taken to be about $1/50(f_1 + f_2)$ and t_f is taken to be about $2/|f_1 - f_2|$. The displacement y oscillates with a frequency of $(f_1 + f_2)/2$ so this value of Δt produces 100 points per period of oscillation. The beat frequency is $|f_1 - f_2|$ so this value of t_f lets you see two periods of the beat. Since a large number of points are generated, you should program the computer to produce the graph on a monitor screen rather than plot by hand.

PROJECT 1. First take $y_1 = y_2 = 2.0$ mm, $f_1 = 100$ Hz, and $f_2 = 110$ Hz. The graph shows a rapidly varying oscillation inside a more slowly varying envelope. Measure the period of the rapid oscillation and verify that it is $2/(f_1 + f_2)$. Measure the period of the envelope and verify that it is $1/|f_1 - f_2|$.

Now try $y_{1m} = y_{2m} = 2.0$ mm, $f_1 = 100$ Hz, and $f_2 = 105$ Hz. Verify that the period of the rapid oscillation is nearly the same and that the period of the envelope has doubled.

Finally try $y_{1m} = y_{2m} = 2.0$ mm, $f_1 = 500$ Hz, and $f_2 = 510$ Hz. Verify that the period of the rapid oscillation has decreased by a factor of about 5 while the beat period is the same as for the first case above ($f_1 = 100$ Hz, $f_2 = 110$ Hz).

PROJECT 2. What happens if the amplitudes are different? Try $y_{1m} = 2.0\,\text{mm}$, $y_{2m} = 4.0\,\text{mm}$, $f_1 = 100\,\text{Hz}$, and $f_2 = 110\,\text{Hz}$. Pay special attention to the regions between beats and tell how these compare to the same regions when the amplitudes are the same: ________________________________

Now increase the amplitude of the second wave to $y_{2m} = 6.0\,\text{mm}$. Does the result substantiate your statement?

PROJECT 3. Now investigate the influence of a phase difference. Rewrite the program instruction for the calculation of y so it reads $y = y_{1m}\sin(\omega_1 t) + y_{2m}\sin(\omega_2 t - \phi)$ and add an input statement for ϕ near the beginning of the program. Take $y_{1m} = y_{2m} = 2.0\,\text{mm}$, $f_1 = 100\,\text{Hz}$, and $f_2 = 110\,\text{Hz}$. Plot the displacement as a function of time for $\phi = 30°$, $60°$, $90°$, and $180°$. You already have a graph for $\phi = 0$. As you look at the graphs, pay particular attention to the position of the beat maxima along the time axis. What is the influence of the phase? ________________________________

Chapter 23

A computer can be programmed to calculate the electric field due to a line of charge if the charge distribution along the line is known. Suppose the charge is on the x axis from x_0 to x_f and the charge density is given by the known function $\lambda(x)$. Place the y axis so the field point is in the xy plane and has coordinates x and y. Then, the electric field at that point is given by

$$E_x = \frac{1}{4\pi\epsilon_0}\int_{x_0}^{x_f} \frac{\lambda(x')(x-x')\,dx'}{[(x-x')^2+y^2]^{3/2}},$$

$$E_y = \frac{y}{4\pi\epsilon_0}\int_{x_0}^{x_f} \frac{\lambda(x')\,dx'}{[(x-x')^2+y^2]^{3/2}}.$$

The z component is zero.

The Simpson's rule program discussed in Chapter 7 can be used to carry out the integrations. For most problems, however, this technique requires a large number of intervals because the functions $(x-x')/\left[(x-x')^2+y^2\right]^{3/2}$ and $1/\left[(x-x')^2+y^2\right]^{3/2}$ in the integrands vary rapidly with x'. On the other hand, λ usually varies much more slowly. In order to avoid this problem divide the x axis from x_0 to x_f into N segments, each of width $\Delta x'$, and approximate λ by a constant in each segment. The integral over each segment can be evaluated analytically. The indefinite integrals are $\int (x-x')\left[(x-x')^2+y^2\right]^{-3/2}\,dx' = \left[(x-x')^2+y^2\right]^{-1/2}$ and $\int \left[(x-x')^2+y^2\right]^{-3/2}\,dx' = -(1/y^2)(x-x')\left[(x-x')^2+y^2\right]^{-1/2}$. Let an interval start at x_b and evaluate λ at the midpoint of the interval: $\lambda_m = \lambda(x_b + \Delta x/2)$. Then, the components of the electric field are given approximately by

$$E_x = \frac{1}{4\pi\epsilon_0}\sum_{i=1}^{N}\lambda_m\left\{\frac{1}{[(x-x_b-\Delta x)^2+y^2]^{1/2}} - \frac{1}{[(x-x_b)^2+y^2]^{1/2}}\right\},$$

$$E_y = -\frac{1}{4\pi\epsilon_0}\frac{1}{y}\sum_{i=1}^{N}\lambda_m\left\{\frac{(x-x_b-\Delta x)}{[(x-x_b-\Delta x)^2+y^2]^{1/2}} - \frac{(x-x_b)}{[(x-x_b)^2+y^2]^{-1/2}}\right\}.$$

Write a program to calculate the components of the electric field produced by a line of charge. Input the coordinates x and y and the limits of integration x_0 and x_f, then have the program evaluate the sums and multiply the results by $1/4\pi\epsilon_0$ or $(1/4\pi\epsilon_0)(1/y)$, as appropriate. You will need to supply a programming line to define the function $\lambda(x)$. Here's an outline.

input number of intervals, limits of integration: N, x_0, x_f
calculate segment length: $\Delta x = (x_f - x_0)/N$
input coordinates: x, y
set $S_x = 0$, $S_y = 0$
set $x_b = x_0$
calculate $f_{1x} = 1/\left[(x - x_b)^2 + y^2\right]^{1/2}$
calculate $f_{1y} = (x - x_b)/\left[(x - x_b)^2 + y^2\right]^{1/2}$
begin loop over intervals: counter runs from 1 to N
    set $x_e = x_b + \Delta x$
    calculate $f_{2x} = 1/\left[(x - x_e)^2 + y^2\right]^{1/2}$
    calculate $f_{2y} = (x - x_e)/\left[(x - x_e)^2 + y^2\right]^{1/2}$
    calculate linear charge density at center of segment: λ_m
    $S_x = S_x + \lambda_m(f_{2x} - f_{1x})$
    $S_y = S_y + \lambda_m(f_{2y} - f_{1y})$
    replace x_b with x_e, f_{1x} with f_{2x}, f_{1y} with f_{2y}
end loop
calculate field components: $E_x = (1/4\pi\epsilon_0)S_x$, $E_y = -(1/4\pi\epsilon_0)(1/y)S_y$
display E_x, E_y
go back to enter another set of coordinates or quit

Test the program by considering a line of uniform charge density.

PROJECT 1. A line of charge runs along the x axis from the origin to $x = 0.10\,\text{m}$. Suppose the line contains 5.5×10^{-9} C of charge, distributed uniformly (λ = constant). Use the program to calculate the electric field components at $x = 0$, $y = 0.05\,\text{m}$. Evaluate the analytic expressions and compare answers.

Now suppose the same total charge is distributed on the same line with a linear density that is given by $\lambda = Ax^2$, where A is a constant. First show that $\lambda = 1.65 \times 10^{-5}x^2$ C/m for x in meters. Then, use the program to find the electric field components at points along the line $y = 0.050\,\text{m}$. Take points every 0.020 m from $x = -0.060\,\text{m}$ to $x = 0.100\,\text{m}$.

Estimate the value of x for which the electric field is in the y direction. Explain why it is not on the center line of the wire, $x = 0.050\,\text{m}$.

At points far away from the line of charge the field tends to become like that of a point charge. Use the program to calculate the field components along the line $x = 0$. Take $y = 0.10$, 1.0, 10, 100, and 1000 m. Modify the program so it also calculates the electric field at the same points for a point charge $q = 5.5 \times 10^{-9}$ C at the origin. Compare the fields of the line and point charge. Does the field of the line become more like that of the point charge at far away points?

Given a charge distribution that creates an electric field, a computer can be used to plot the electric field lines. We consider a distribution of point charges. It is usual to start at a point close to one of the charges, where the field line is along the line that joins the point and the charge. The electric field at the point is calculated and the result is used to locate a neighboring point on the same field line. It is a short distance away in the direction of the electric field vector. The process is then continued to locate other points on the same field line.

For simplicity we will deal with charges, fields, and field lines in the x, y plane. Suppose the electric field at a point with coordinates x and y has components E_x and E_y and we wish to find another point on the field line through x, y. $(E_x/E)\mathbf{i}+(E_y/E)\mathbf{j}$ is a unit vector tangent to the field line and $x + (E_x/E)\Delta s$, $y + (E_y/E)\Delta s$ are the coordinates of a point on the line Δs distant from

x, y. The following is the outline of a program that calculates a sequence of points on a single field line. You must supply the coordinates x_0, y_0 of the first point, the distance Δs between points, and the program instructions to calculate the electric field.

> input distance between points, number of points: Δs, N
> input the coordinates of the first point: x_0, y_0
> set $x_b = x_0$, $y_b = y_0$
> plot x_b, y_b
> **begin loop** over points; counter runs from 1 to N
>   calculate field components at x_b, y_b: E_x, E_y
>   calculate magnitude of field at x_b, y_b: $E = \sqrt{E_x^2 + E_y^2}$
>   calculate coordinates of new point:
> $x_e = x_b + (E_x/E)\Delta s$
> $y_e = y_b + (E_y/E)\Delta s$
>   plot x_e, y_e
>   set $x_b = x_e$, $y_b = y_e$
> **end loop** over points
> go back to get another starting point or quit

You may want to provide instructions so the field lines are plotted on the monitor screen. Alternatively, you may have the computer display the coordinates of the points so you can plot the line by hand. You should realize that the lines generated are approximate but the approximation becomes more accurate as Δs is made smaller. Do not make Δs so small that significance is lost in the calculation. If you plot by hand, you will not want to display and plot every point, particularly if Δs is small. Add program instructions so that only every 10 or 20 calculated points are displayed.

You might also want to stop plotting when the line gets close to a charge. One way to do this is to save the coordinates of the charges, then check x_e and y_e to see if they are near a corresponding charge coordinate. If they are, have the computer go to the last line of the program.

PROJECT 2. Check the program by considering a dipole. Charge $q_1 = 7.1 \times 10^{-9}$ C is located at the origin and charge $q_2 = -7.1 \times 10^{-9}$ C is located on the y axis at $y = -0.40$ m. Use the program to plot 2 field lines. Start one at $x = 1.0 \times 10^{-3}$ m, $y = 1.0 \times 10^{-3}$ m and the second at $x = 1.0 \times 10^{-3}$ m, $y = -1.0 \times 10^{-3}$ m. Take $\Delta s = 0.005$ m and plot about 100 points on each line. Do these lines look like the lines of a dipole?

PROJECT 3. Four identical charges, each with $q = 5.6 \times 10^{-9}$ C, are placed at the corners of a square with edge length $a = 0.36$ m. The square is centered at the origin and its sides are parallel to the coordinate axes. Take $\Delta s = 0.005$ m and use the program to plot 6 field lines, each starting on a circle of radius 0.10 m centered on the charge at $x = 0.18$ m, $y = 0.18$ m. One line starts parallel to the x axis and the others start at intervals of $\pi/3$ radians around the circle. When a line is within 0.10 m of any charge or more than 0.70 m away from all charges, do not continue it.

One of the lines goes toward the center of the square where, the electric field vanishes. The program does not properly evaluate E_x/E and E_y/E in the limit of vanishing field. Stop plotting when the field becomes less than 10^{-4} times the field at the initial points.

Plot 6 field lines emanating from each of the other charges. This can be done using the data generated for the first charge and a symmetry argument. It is not necessary to calculate new points.

Far away from all charges, the distribution has the same electric field as that of a single charge equal to the net charge in the distribution and located at the origin. The field lines are then radially outward if the net charge does not vanish and is positive. Look at your plot and notice that the lines tend to become more nearly in the radial direction as the distance from the origin increases. Because of the choice of starting points for the lines, they also tend to be arranged with equal angles between adjacent lines. You should see 24 lines, all nearly in a radial direction, with adjacent lines separated by $\pi/12$ radians. Do you?

Now change the sign of the charges at $x = -0.18\,\text{m}$, $y = 0.18\,\text{m}$ and at $x = 0.18\,\text{m}$, $y = -0.18\,\text{m}$ and use the program to plot lines starting at the same points as before. Notice that now the net charge in the distribution is zero and that all lines start and stop at charges within the distribution.

Chapter 24

Gauss's law can be verified directly by using a computer to evaluate the integral $\oint \vec{E}\cdot d\vec{A}$ over a closed surface. Consider the special case of a cube bounded by the six planes $x = 0$, $x = a$, $y = 0$, $y = a$, $z = 0$, and $z = a$, as shown in the figure. Place a single charge q on the line $x = a/2$, $z = a/2$, through the cube center, and carry out the integration one face at a time.

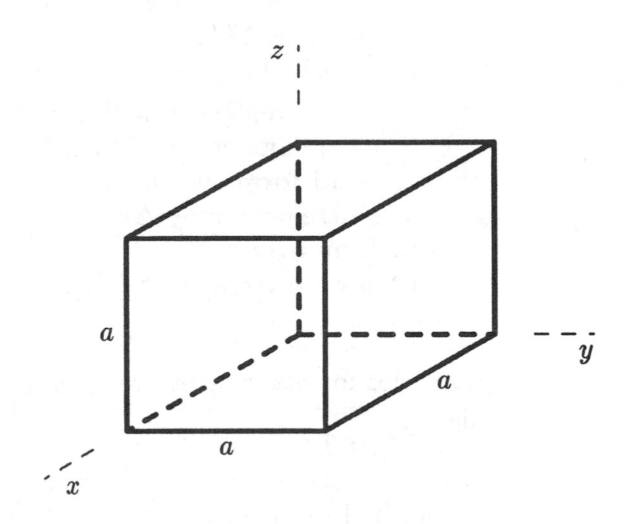

First consider the top face. Divide it into small rectangles of width Δx and length Δy, evaluate the electric field at the center of each rectangle, and calculate $E_z\Delta x\Delta y$. Finally, sum the results. Then, carry out similar calculations for the other faces.

If the y coordinate of the charge is y', then its electric field at x, y, z is given by

$$\vec{E} = \frac{q}{4\pi\epsilon_0}\frac{(x-a/2)\,\hat{\imath} + (y-y')\,\hat{\jmath} + (z-a/2)\,\hat{k}}{[(x-a/2)^2 + (y-y')^2 + (z-a/2)^2]^{3/2}}\,.$$

By symmetry the flux through the top, bottom, front, and back faces are all the same. Calculate the flux through the top face, say, and multiply by 4. On the top face $z = a$ and the quantity to be summed is

$$E_z\Delta x\Delta y = \frac{q}{4\pi\epsilon_0}\frac{a/2}{[(x-a/2)^2 + (y-y')^2 + a^2/4]^{3/2}}\,.$$

For the left face, $y = 0$ and the quantity to be summed is

$$-E_y\Delta x\Delta z = \frac{q}{4\pi\epsilon_0}\frac{y'}{[(x-a/2)^2 + (y')^2 + (z-a/2)^2]^{3/2}}\,.$$

For the right face, $y = a$ and the quantity to be summed is

$$E_y\Delta x\Delta z = \frac{q}{4\pi\epsilon_0}\frac{a-y'}{[(x-a/2)^2 + (a-y')^2 + (z-a/2)^2]^{3/2}}\,.$$

Here is the outline of a program to calculate the flux through the upper face. The x axis and the y axis, both from 0 to a, are each divided into N segments and $\Delta x = \Delta y = a/N$ is computed.

In the program this quantity is called $\Delta\ell$. There are two loops: an outer loop over x and an inner loop over y. The first value of x is $\Delta\ell/2$, at the center of the first segment, and x is incremented by $\Delta\ell$ each time around the loop over x. For each value of x, y starts at $\Delta\ell/2$ and is incremented by $\Delta\ell$ each time around the loop over y. The sum is saved in S. The factor $q(a/2)(\Delta\ell)^2/4\pi\epsilon_0$ appears in every term so it is not included until after the sum is completed.

> input charge and its coordinate: q, y'
> input edge of cube, number of segments: a, N
> calculate segment length: $\Delta\ell = a/N$
> set $S = 0$
> set $x = \Delta\ell/2$
> **begin loop** over x
>   set $A = (x - a/2)^2 + a/4$
>   set $y = \Delta\ell/2$
>   **begin loop** over y
>   replace S with $S = S + \left[A + (y - y')^2\right]^{-3/2}$
>   increment y by $\Delta\ell$
>   **end loop** over y
>   increment x by $\Delta\ell$
> **end loop** over x
> multiply S by $q(a/2)(\Delta\ell)^2/4\pi\epsilon_0$ and display result
> stop

Program lines for other faces are similar. For the left face, the loop lines are:

> set $S = 0$
> set $x = \Delta\ell/2$
> **begin loop** over x
>   set $A = (x - a/2)^2 + (y')^2$
>   set $z = \Delta\ell/2$
>   **begin loop** over z
>   replace S with $S = S + \left[A + (z - a/2)^2\right]^{-3/2}$
>   increment z by $\Delta\ell$
>   **end loop** over z
>   increment x by $\Delta\ell$
> **end loop** over x
> multiply S by $qy'(\Delta\ell)^2/4\pi\epsilon_0$ and display result

For the right face, the loop lines are:

> set $S = 0$
> set $x = \Delta\ell/2$
> **begin loop** over x
>   set $A = (x - a/2)^2 + (a - y')^2$
>   set $z = \Delta\ell/2$
>   **begin loop** over z
>   replace S with $S = S + \left[A + (z - a/2)^2\right]^{-3/2}$
>   increment z by $\Delta\ell$
>   **end loop** over z
>   increment x by $\Delta\ell$

end loop over x
multiply S by $q(a - y')(\Delta\ell)^2/4\pi\epsilon_0$ and display result

PROJECT 1. Use the program to evaluate $\oint \vec{E} \cdot d\vec{A}$ for the surface of a cube with edge $a = 10\,\text{m}$ and with charge $q = 3.7 \times 10^{-9}\,\text{C}$ inside. Use $N = 15$ for three significant figure accuracy. First place the charge at the center of the cube: $x' = 5\,\text{m}$, $y' = 5\,\text{m}$, $z' = 5\,\text{m}$. The flux is the same through each face so you need run the program for only one face, then multiply by 6. Compare the result with q/ϵ_0.

Now place the charge at $x' = 5\,\text{m}$, $y' = 7.5\,\text{m}$, $z' = 5\,\text{m}$. Notice that the value of the flux through each face has changed from the previous situation. Also notice that the flux through the left, right, and top faces differ from each other. Nevertheless, the total flux is the same to within the accuracy of the calculation.

Finally, place the charge at $x' = 5\,\text{m}$, $y' = 12.5\,\text{m}$, $z' = 5\,\text{m}$. It is outside the cube and the total flux through the cube should be zero. The result of the program may not be exactly zero because the flux through each face was computed to only about three significant figures. The total, however, should be significantly less than the flux through any individual face and should be still less if the calculation is done with a larger value of N.

Compare the calculations by filling out the following table with values of the flux:

y'	5 m	7.5 m	12.5 m
top	________	________	________
bottom	________	________	________
right	________	________	________
left	________	________	________
back	________	________	________
front	________	________	________
total	________	________	________

PROJECT 2. If the net charge inside the cube is zero, then according to Gauss' law, the total flux through the surface of the cube is zero. The flux through any particular face, however, does not necessarily vanish. Suppose $q_1 = -3.7 \times 10^{-9}\,\text{C}$ is at $y' = 2.5\,\text{m}$ on the line through the center of the cube and $q_2 = +3.7 \times 10^{-9}\,\text{C}$ is at $y' = 7.5\,\text{m}$ on the same line. Use the program to find the flux through each face of the cube due to each charge separately, then fill out the following table with values of the flux. If you completed the first project, you already have values for q_1.

	q_1	q_2
top	________	________
bottom	________	________
right	________	________
left	________	________
back	________	________
front	________	________
total	________	________

The two total fluxes may not sum to zero because of errors in the calculation. However, it should be considerably less than the flux through any of the faces.

Chapter 25

If charge is distributed along the x axis with linear charge density $\lambda(x')$, then the electric potential at a point in the xy plane is given by

$$V(x,y) = \frac{1}{4\pi\epsilon_0}\int \frac{\lambda(x')}{[(x-x')^2+y^2]^{1/2}}\,dx',$$

where the integral extends over the charge distribution. You can use the Simpson's rule program of Chapter 7 to evaluate the integral. After evaluating the integral, multiply by $1/4\pi\epsilon_0$.

Here is an outline of the program, modified slightly to handle the calculation of an electric potential.

> input limits of integral: x'_i, x'_f
> input number of intervals: N
> replace N with nearest even integer
> calculate interval width: $\Delta x' = (x'_f - x'_i)/N$
> input coordinates of field point: x, y
> initialize quantity to hold sum of values with even labels: $S_e = 0$
> initialize quantity to hold sum of values with odd labels: $S_o = 0$
> set $x' = x'_i$
> **begin loop** over intervals: counter runs from 1 to $N/2$
>   calculate integrand: $I(x') = \lambda(x')\left[(x-x')^2+y^2\right]^{-1/2}$
>   add it sum of values with even labels: replace S_e with $S_e + I(x')$
>   increment x' by $\Delta x'$
>   calculate integrand: $I(x') = \lambda(x')\left[(x-x')^2+y^2\right]^{-1/2}$
>   add it to the sum of values with odd labels: replace S_o with $S_o + I(x')$
>   increment x' by $\Delta x'$
> **end loop**
> calculate integrand at upper and lower limits:
> $I_0 = \lambda(x'_i)\left[(x-x'_i)^2+y^2\right]^{-1/2}$
> $I_N = \lambda(x'_f)\left[(x-x'_f)+y^2\right]^{-1/2}$
> evaluate integral: $(\Delta x/3)(I_N - I_0 + 2S_e + 4S_o)$
> multiply by $1/4\pi\epsilon_0$ and display result
> go back for coordinates of another field point or stop

First use the program to investigate the relationship between the electric potential and the electric field: $E_x = -\partial V/\partial x$, $E_y = -\partial V/\partial y$.

PROJECT 1. Charge is distributed from $x' = 0$ to $x' = 0.10\,\text{m}$ along the x axis with a linear charge density given by $\lambda(x') = 1.83\times10^{-5}\sqrt{x'}\,\text{C/m}$, where x' is in meters. Use the Simpson's rule program to find values for the electric potential at $x = -0.005\,\text{m}$ and at $x = +0.005\,\text{m}$ on the line $y = 0.20\,\text{m}$. Start with $N = 20$ and repeat the calculation with N doubled each time until you get the same results to three significant figures.

Estimate the x component of the electric field at $x = 0$, $y = 0.2\,\text{m}$ by evaluating $-(V_2 - V_1)/\Delta x$, where V_1 is the potential at $x = -0.005\,\text{m}$, V_2 is the potential at $x =$

+0.005 m, and $\Delta x = 0.01$ m. Check your answer by using the program of Chapter 23 to compute the electric field directly. To how many figures do you obtain agreement? Some significance is lost when you subtract the two values of the potential.

Use the Simpson's rule program to calculate the electric potential at $y = 0.195$ m and at $y = 0.205$ m on the y axis. Use the results to estimate the y component of the electric field at $x = 0$, $y = 0.2$ m. Check your result by using the program of Chapter 23.

PROJECT 2. You can use the program to plot equipotential surfaces. In this project you will consider the line charge of the previous project and plot a line in the xy plane along which the electric potential has a given value V. Start with $x = -0.01$ m and use trial and error to find the y coordinate of the point for which the potential has the value V. You might start with $y = .01$ m and increment y by 0.1 m until you find two points with potentials that straddle V, then narrow the gap until the coordinates of the two points at its ends are the same to two significant figures. Increment x by 0.01 m and repeat. Continue until you reach $x = +0.11$ m. Once you have found the first few points a pattern should emerge and later points should be easier to find. Try potentials of 3, 5, and 10 V.

Chapter 28

Many circuit problems involve the solution of simultaneous linear equations. They can be solved on a computer. We describe what is known as the Gauss-Seidel iteration scheme, in which a solution is guessed and the given equations are used to improve the guess.

Suppose there are N equations and the unknowns are i_1, i_2, ..., i_N. For many problems they, are the currents in the various branches of a circuit. Equation number j may be written

$$\sum_{k=1}^{N} A_{jk} i_k = B_j$$

where A_{jk} is the coefficient of unknown i_k in equation j and B_j is a term that contains no unknown. Solve the first equation for i_1 in terms of the other unknowns, the second equation for i_2 in terms of the other unknowns, etc. The result is

$$i_j = \frac{\left[B_j - \sum\limits_{\substack{k=1 \\ k \neq j}}^{N} A_{jk} i_k \right]}{A_{jj}} .$$

The first step is to guess values for i_1, i_2, ..., i_N. These guesses are used in the above equations to calculate new values. The process is then carried out again using the results of the first run. Iteration is continued until two successive runs yield the same results to within an acceptable error. The most current values are used on the right side of the equations as soon as they are calculated.

Care must be taken to arrange the equations so A_{jj} is not zero for any equation in the set. Even so, the results do not converge for some sets of equations. After many iterations successive results may differ greatly and may show no sign of getting closer in value. When this occurs, the original set of equations must be modified by adding (repeatedly, perhaps) some equations to others or subtracting some equations from others and using the resulting equation to replace one of the originals. Such manipulations do not change the solution.

We state without proof that the Gauss-Seidel iteration scheme converges toward the correct solution if, for every equation in the set, the so-called diagonal term (A_{jj} for equation j) is larger in magnitude than the sum of the magnitudes of the other coefficients in the equation. That is,

$$|A_{jj}| > \sum_{\substack{k=1 \\ k \neq j}}^{N} |A_{jk}| .$$

The goal of any modifications made to the original set is to obtain a new set for which this inequality holds.

The first step in the modification process is to put the equations in optimal order. Search for the largest coefficient and arrange the equations so this coefficient becomes a diagonal coefficient. Now search for the largest coefficient that is not in the same equation and does not multiply the same unknown as the first one found, then arrange the equations so this one is also diagonal. Continue until all equations have been considered. For example, suppose you wish to solve the set

$$3i_1 + 2i_2 + 2i_3 = 8$$

$$3i_1 + 4i_2 + 3i_3 = 5$$

$$7i_1 + 5i_2 + 3i_3 = 3 .$$

The largest coefficient is 7. It multiplies i_1 in the third equation, so this equation should be the first. Of the coefficients that multiply other unknowns in other equations, the largest is 4, which multiplies i_2 in the second equation, so this equation should remain the second. The optimal order is

$$7i_1 + 5i_2 + 3i_3 = 3$$

$$3i_1 + 4i_2 + 3i_3 = 5$$

$$3i_1 + 2i_2 + 2i_3 = 8 .$$

None of these equations obey the inequality. Subtract the second from the first and use the result to replace the first. The new set is

$$4i_1 + i_2 = -2$$

$$3i_1 + 4i_2 + 3i_3 = 5$$

$$3i_1 + 2i_2 + 2i_3 = 8 .$$

Now the first equation obeys the inequality but the others do not. Subtract the third from the second and use the result to replace the second. The equations are now

$$4i_1 + i_2 = -2$$

$$2i_2 + i_3 = -3$$

$$3i_1 + 2i_2 + 2i_3 = 8 .$$

To bring the third equation into line, multiply the first by 3, the third by 4, and subtract. Replace the third equation with the result. The new set is

$$4i_1 + i_2 = -2$$

$$2i_2 + i_3 = -3$$

$$5i_2 + 8i_3 = 38 .$$

All three equations now satisfy the inequality and we expect the Gauss-Seidel iteration scheme to work. The scheme may work even if the inequality is not satisfied, so you may want to run through a few iterations before spending time in modifying the equations.

Write a program to solve a set of linear simultaneous equations. Store the coefficients in a subscripted variable $A(j,k)$, the constant terms in the subscripted variable $B(j)$, and the unknowns in the subscripted variable $I(j)$. Take the initial guesses to all be zero. Here's an outline.

input number of equations: N
begin loop over equations; counter j runs from 1 to N
  begin loop over variables; counter k runs from 1 to N
    input coefficients: $A(j,k)$
  end loop over variables
  input constant term: $B(j)$
  set $I(j) = 0$
end loop over equations
* **begin loop** over equations; counter j runs from 1 to N
  set $S = 0$ in preparation for computing sum
  begin loop over variables; counter k runs from 1 to N
    if $k \neq j$ replace S with $S + A(j,k)I(k)$
  end loop over variables
  replace $I(j)$ by $[B(j) - S]/A(j,j)$
end loop over equations
display solution
go back to starred instruction for another iteration or stop

When the solution is displayed you must judge whether convergence has been reached or not. If values of $I(j)$ have not changed much from the last iteration, you will want to stop the program. If they have changed, you will want the program to perform another iteration. You might arrange the display so that results of two or more iterations are on the screen simultaneously.

PROJECT 1. To test the program, use it to solve the following set of 4 simultaneous equations. Obtain three significant figure accuracy.

$$2i_1 - i_2 = 5$$

$$2i_2 - i_3 = 7$$

$$2i_3 - i_4 = 9$$

$$2i_4 - i_1 = 11\,.$$

[ans: $i_1 = 6.47$, $i_2 = 7.93$, $i_3 = 8.87$, $i_4 = 8.73$]

PROJECT 2. Consider the circuit shown below with current arrows and labels. Write down two junction and three loop equations. Modify the set of equations so the convergence conditions are met. Take $\mathcal{E}_1 = 10\,\text{V}$, $R_1 = R_2 = 5\,\Omega$, $R_3 = R_4 = 8\,\Omega$, and $R_5 = 12\,\Omega$. Use the program to find values of the five currents for each of the following values of $\mathcal{E}_2$: 5, 7.5, 10, 12, and 15 V. Assume the given values are exact and find the solutions with an accuracy of at least three significant figures. For each value of $\mathcal{E}_2$, tell if the seats of emf are charging or discharging.

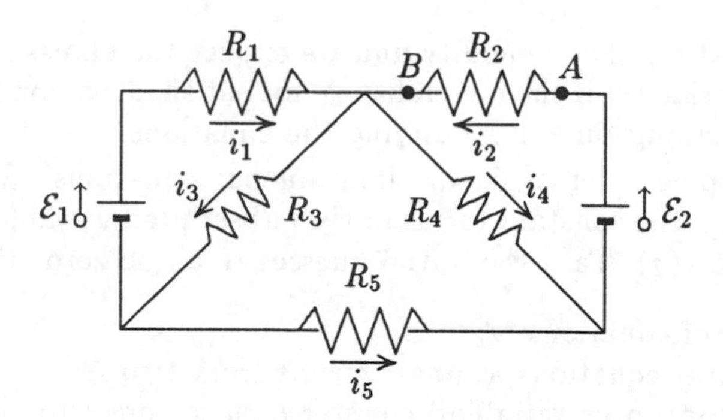

PROJECT 3. Suppose the ends of resistor R_2 in the circuit of the previous project are also connected to an external circuit. Current i_6 enters the external circuit at A and returns to the circuit of the diagram at B.

Take $\mathcal{E}_2 = 15\,\text{V}$ and the other quantities as given in the previous project. For each of the following values of i_6, solve for the values of the other currents: 0, 2, 4, 6, 8, and 10 A. Obtain three significant figure accuracy.

For each of the cases considered, calculate the potential difference ΔV across R_2. Plot ΔV vs. i_6. Notice that it is a straight line. As far as the external circuit is concerned, the circuit shown in the diagram above can be replaced by a seat of emf and a resistor in series. The emf has the value of ΔV for $i_6 = 0$ and the value of the resistance is the slope of the line. These values are $\mathcal{E} =$ ______ V and $R =$ ______ Ω.

We now use the program to investigate the operation of a measuring instrument. In this case the instrument is a Wheatstone bridge and is used to measure resistance. It is typical of many different bridge circuits used for various electrical measurements.

PROJECT 4. The circuit for a Wheatstone bridge is shown below. The symbol G stands for a galvanometer, an instrument that can detect small currents. We suppose the unknown resistor is R_3 and the other resistors are variable. They are set so the galvanometer reads 0 and $i_6 = 0$. Then, $R_3 = R_1 R_4 / R_2$. The resistances R_1, R_2, and R_4 are read and their values are used to calculate R_3.

First, verify that $i_6 = 0$ when $R_3 = R_1 R_4 / R_2$. Take $R_1 = R_2 = 12\,\Omega$, $R_3 = R_4 = 18\,\Omega$, and $R_5 = 3.2\,\Omega$. The resistance of the galvanometer is $2.0\,\Omega$ and $\mathcal{E} = 10\,\text{V}$. Note that the balance condition is met. Write three junction and three loop equations, then use the program to solve for the six currents. Obtain three significant figure accuracy. Don't forget to modify the equations so the convergence conditions are met. You should find that $i_6 = 0$ to within the accuracy of the calculation.

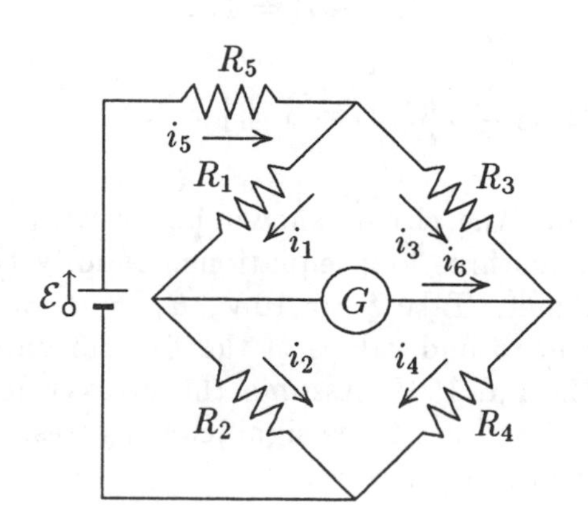

It is usually of some interest to know how sensitive a Wheatstone bridge is. You can test the sensitivity by making an error in the setting of one of the resistors, then seeing if the galvanometer can detect the resulting current i_6.

Take $R_1 = R_2 = 12\,\Omega$, $R_3 = 18\,\Omega$, $R_4 = 19\,\Omega$, and $R_5 = 3.2\,\Omega$. R_g is still $2.0\,\Omega$ and $\mathcal{E}$ is still 10 V. The balance condition is not met, the current in the galvanometer does not vanish, and the balance equation predicts that $R_3 = 19\,\Omega$ instead of the correct value, $18\,\Omega$. Use the program to solve for i_6. The galvanometer must be able to detect a current of this value or less if the bridge is to measure the unknown resistance with an error of less than $1\,\Omega$. Is this a reasonable current for a galvanometer to detect? [ans: -7.67×10^{-3} A; easily detected]

Chapter 30

A computer program can be used to integrate the Biot-Savart equation for the magnetic field of a current loop. In general, the field is given by the integral around the current loop:

$$\vec{B}(\vec{r}) = \frac{\mu_0}{4\pi} i \int \frac{d\vec{r}' \times (\vec{r} - \vec{r}')}{|\vec{r} - \vec{r}'|^3},$$

where i is the current, $\vec{r}$ is the position vector of the point where the field is $\vec{B}$, and $\vec{r}'$ is the position vector of a point on the loop.

Consider circular a loop in the xy plane and use an angular variable of integration. Since the situation has cylindrical symmetry, we can without loss of generality specialize to a point in the yz plane and write $\vec{r} = y\,\hat{\jmath} + z\,\hat{k}$. Take $\vec{r}' = R(\cos\theta\,\hat{\imath} + \sin\theta\,\hat{\jmath})$. Then, $d\vec{r}' = R(-\sin\theta\,\hat{\imath} + \cos\theta\,\hat{\jmath})\,d\theta$. When these substitutions are made, the expressions for the components of the field are

$$B_x = 0,$$

$$B_y = \frac{\mu_0}{4\pi}\frac{iz}{R^2}\int_0^{2\pi} \frac{\sin\theta\,d\theta}{[1 + (y/R)^2 + (z/R)^2 - 2(y/R)\sin\theta]^{3/2}},$$

$$B_z = -\frac{\mu_0}{4\pi}\frac{i}{R}\int_0^{2\pi} \frac{[(y/R)\sin\theta - 1]\,d\theta}{[1 + (y/R)^2 + (z/R)^2 - 2(y/R)\sin\theta]^{3/2}}.$$

Both integrals have the form

$$B_i = \frac{\mu_0 i}{4\pi R}\int_0^{2\pi} f(\theta)\,d\theta.$$

The Simpson's rule program of Chapter 7 can be used to carry out the integrations. Divide the interval in θ from 0 to 2π into N segments, each of length $\Delta\theta = 2\pi/N$. Then, B_i is approximated by $(\Delta\theta/3)(2S_e + 4S_o)$, where S_e is the sum of the values of the integrand at the beginning of segments with even labels and S_o is the sum of the values at the beginning of segments with odd labels. The term $f_N - f_0$ does not appear because the integrands have the same value at $\theta = 0$ and $\theta = 2\pi$. N must be an even integer.

To evaluate both integrals (for B_y and B_z) within the same program, let S_{ye} and S_{yo} collect the sums for the y component and S_{ze} and S_{zo} collect the sums for the z component. Here's an outline.

input number of intervals: N
replace N with nearest even integer
calculate interval width: $\Delta\theta = 2\pi/N$
input coordinates of field point: y, z
initialize quantities to hold sum of values with even labels: $S_{ye} = 0$, $S_{ze} = 0$
initialize quantities to hold sum of values with odd labels: $S_{yo} = 0$, $S_{zo} = 0$
set $\theta = 0$
begin loop over intervals: counter runs from 1 to $N/2$
    calculate $A = \sin\theta$ for future use
    calculate $B = \left[1 + (y/R)^2 + (z/R)^2 - 2(y/R)\sin\theta\right]^{3/2}$ for future use
    update sums over even terms
        replace S_{ye} with $S_{ye} + A/B$
        replace S_{ze} with $S_{ze} + \left[(y/R)A - 1\right]/B$
    increment θ by $\Delta\theta$
    calculate $A = \sin\theta$ for future use
    calculate $B = \left[1 + (y/R)^2 + (z/R)^2 - 2(y/R)\sin\theta\right]^{3/2}$ for future use
    update sums over odd terms
        replace S_{yo} with $S_{yo} + A/B$
        replace S_{zo} with $S_{zo} + \left[(y/R)A - 1\right]/B$
    increment θ by $\Delta\theta$
end loop
evaluate integrals:
    $B_y = (\mu_0 iz/4\pi R^2)(\Delta\theta/3)(2S_{ye} + 4S_{yo})$
    $B_z = -(\mu_0 i/4\pi R(\Delta\theta/3)(2S_{ze} + 4S_{zo})$
display result
go back to input new field coordinates or stop

Running time can be reduced if you add instructions to calculate the new variables $y' = y/R$ and $z' = z/R$ immediately after y and z are read. Then, use y' and z' in succeeding instructions. The equations have been written in a convenient form to do this.

The purpose of the first project is to test the program and to obtain a rough idea of the number of intervals that should be used.

PROJECT 1. If the field point is on the z axis, the integrals can be evaluated analytically, with the result $B_x = 0$, $B_y = 0$, and

$$B_z = \frac{\mu_0 i R^2}{2(R^2 + z^2)^{3/2}}\,.$$

Consider a 1.0-m radius loop carrying a current of 1.0 A and use the program to calculate the field at $z = 0$, 0.50, 1.5, and 2.5 m on the z axis. Since the integrand is constant, you should obtain the correct answers with $N = 2$. Check your answers by evaluating the analytic expression.

Now find the magnetic field at the point $x = 0$, $y = 0.50$ m, $z = 0$. Start with $N = 2$ and on the next trial double N. Continue to double N until the results of two successive trials agree to three significant figures.

Repeat the calculation for the following points along the $x = 0$, $y = 0.50$ m line: $z = 0.10$, 1.0, and 10 m. Note the value of N for which three significant figure accuracy is obtained.

A set of Helmholtz coils consists of two identical loops parallel to each other and carrying the same magnitude of current, in the same direction. In the region between the coils the two magnetic fields tend to be in roughly the same direction and when the distance between the coils is equal to the radius of one of the loops, the field in the region between is particularly uniform. For this reason, Helmholtz coils are often used to produce magnetic fields in the laboratory. In the following project you use the integration program to investigate the uniformity of the magnetic field between two coils.

PROJECT 2. Two circular loops of wire, each with a radius of 1.0 m, are placed parallel to the xy plane, with their centers on the z axis. The center of one is at $z = 0$ while the center of the other is at $z = 2.0$ m. Each carries a current of 1.0 A in the counterclockwise direction when viewed from the positive z axis. Note that the separation is twice the radius.

Use the program to calculate the magnetic field with three significant figure accuracy for field points at $z = 0.60$, 0.80, and 1.0 m, all on the z axis. The last point is at the center of the region between the loops. You will need to run the program twice, once for each loop. The first field point ($z = 0.60$ m) is 0.60 m from the first coil and 1.4 m from the second. Run the program with $z = 0.60$ m, then with $z = -1.4$ m. Vectorially sum the two fields.

Use these calculated fields to test the uniformity of the field along the z axis between the loops. For each of the first two points, subtract the magnitude of the field at the center ($z = 1.0$ m) from the field at the point and divide by the magnitude of the field at the center. If we denote this measure of homogeneity by h, then

$$h(z) = \frac{|B(z) - B_c|}{|B_c|},$$

where B_c is the magnitude of the field at the center of the region.

If h is small, the field does not change much with position and is said to be homogeneous. If h is large, the field is said to be inhomogeneous. For good laboratory magnets, h may be on the order of 10^{-6} or less over several centimeters. The distance between the loops, of course, is usually less than 2 m.

Now consider the same loops, but separated by 1.0 m, the radius of one of them. Use the program to calculate the magnetic field at $z = 0.10$, 0.30, and 0.50 m, all on the z axis. The last point is at the center of the region between the loops. Calculate h for the first two points. Has the field become more or less homogeneous?

PROJECT 3. You can also investigate homogeneity in a transverse direction. Now the field has two non-vanishing components and we define a measure of homogeneity for each:

$$h_z = \frac{|B_z(y) - B_z(y=0)|}{|B_z(y=0)|}$$

$$h_y = \frac{|B_y(y)|}{|B_z(y=0)|}.$$

Both are zero for a perfectly homogeneous field.

Consider the loops of the previous project, with a separation of 2.0 m, and calculate h_y and h_z for $y = 0.20$ and 0.40 m on the line $x = 0$, $z = 1.0$ m.

Now take the separation to be 1.0 m and calculate h_y and h_z for $y = 0.20$ and 0.50 m on the line $x = 0$, $z = 0.50$ m. Has the homogeneity increased or decreased?

If the loops are moved still closer to each other, so their separation is less than the radius, the field becomes less uniform. Consider a separation of 0.50 m and calculate h_y and h_z for $y = 0.20$ and 0.40 m along the line $x = 0$, $z = 0.25$ m.

In the following projects you will use an integration program to investigate Ampere's law. The square shown is in the xy plane, is centered at the origin, and has edges of length a. A wire carrying current i pierces the plane of the square at the point on the x axis with coordinate ℓ. It produces a magnetic field at each point on the perimeter of the square (as well as at other points). According to Ampere's law

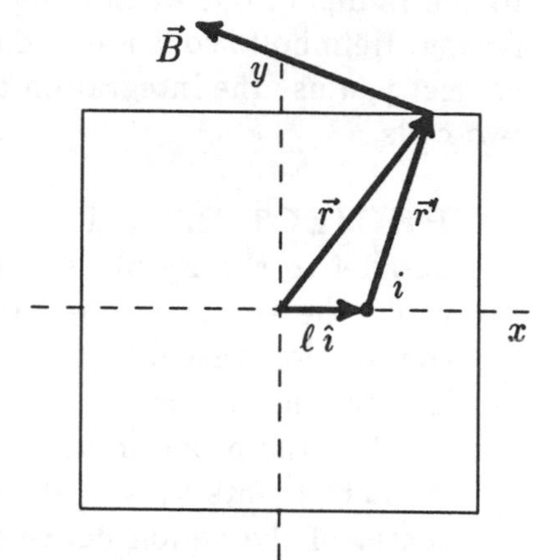

$$\oint \vec{B} \cdot d\vec{s} = \mu_0 i \,,$$

where the integral is around the perimeter and i is the current in the wire. Take the current to be positive if it is out of the page and carry out the integration in the counterclockwise direction around the square.

The magnitude of the magnetic field is given by $B = \mu_0 i/2\pi r'$, where r' is the distance from the wire. The field components are $B_x = -(\mu_0 i/2\pi)y'/(r')^2$ and $B_y = (\mu_0 i/2\pi)x'/(r')^2$. As the diagram shows, $y' = y$, $x' = x - \ell$, and $(r')^2 = (x-\ell)^2 + y^2$ so $B_x = -(\mu_0 i/2\pi)y/[(x-\ell)^2 + y^2]$ and $B_y = (\mu_0 i/2\pi)(x-\ell)/[(x-\ell)^2 + y^2]$.

Consider each of the four sides of the square separately. Across the top $y = a/2$, $d\vec{s} = dx\,\hat{\imath}$ and

$$\int \vec{B} \cdot d\vec{s} = -\frac{\mu_0}{2\pi}\frac{a}{2}\int_{a/2}^{-a/2} \frac{dx}{(x-\ell)^2 + (a/2)^2} \,.$$

The contribution of the bottom is exactly the same. Down the left side $x = a/2$, $d\vec{s} = dy\,\hat{\jmath}$, and

$$\int \vec{B} \cdot d\vec{s} = \frac{\mu_0 i}{2\pi}\left(\frac{a}{2} + \ell\right)\int_{a/2}^{-a/2} \frac{dy}{(\ell + a/2)^2 + y^2} \,.$$

Up the right side $x = a/2$, $d\vec{s} = dy\,\hat{\jmath}$, and

$$\int \vec{B} \cdot d\vec{s} = \frac{\mu_0 i}{2\pi}\left(\frac{a}{2} - \ell\right)\int_{-a/2}^{a/2} \frac{dy}{(-\ell + a/2)^2 + y^2} \,.$$

Each of the integrals can be evaluated using the Simpson's rule program.

PROJECT 4. Suppose the wire carries a current of 1.0 A, out of the page, and the square has sides of length $a = 2.0$ m. Evaluate $\oint \vec{B} \cdot d\vec{s}$, one side at a time, for each of the following positions of the wire: $\ell = 0$, 0.40, 0.80, and 1.2 m. The first three points are inside the square and the fourth is outside. Fill in the table below with values of $\int \vec{B} \cdot d\vec{s}$.

ℓ	0	0.40 m	0.80 m	1.2 m
top	________	________	________	________
bottom	________	________	________	________
right	________	________	________	________
left	________	________	________	________
total	________	________	________	________

For each of the situations, compute $\mu_0 i$, where i is the net current through the square. Compare the result with the totals above.

PROJECT 5. For the square of the previous project, evaluate $\oint \vec{B} \cdot d\vec{s}$ for a current of 2.0 A, out of the page at $\ell = 0$. Evaluate the integral for a current of 1.0 A, out of the page at $\ell = 0.50$ m. Finally, evaluate the integral for a current of 3.0 A, out of the page at $\ell = 0.75$ m. Compare the last answer to the sum of the first two.

Evaluate the integral for the a current of 3.0 A, out of the page at $\ell = 0$. Evaluate the integral for a current of 2.0 A, into the page at $\ell = 0.50$ m. Compare the difference in these results with the value of the integral for a current out of the page at $\ell = 0.40$ m (see the results of Project 4).

Chapter 37

The text deals with the diffraction of light by a single slit for the special case when the viewing screen is far away. Then, the Huygen wavelets emanating from the slit are essentially plane waves. When the viewing screen is close to the slit or when the slit is wide you must take into account the true spherical nature of the wavelets. The diffraction pattern for such wavelets cannot be described analytically but you can use a computer to describe it numerically. As the slit is widened you will be able to see the diffraction pattern change into the geometrical image of the slit.

The diagram shows a plane wave impinging on a single slit from the left. A spherical wave emanates from each point in the slit. One of them is shown. The viewing screen is to the right, a distance x from the slit. Take the origin to be at the center of the slit and let y' be the coordinate of the point within the slit from which the spherical wave emanates. You will calculate the light intensity at the point on the screen a distance y from the center line. According to Huygens' principle the wavelet from y' has the form

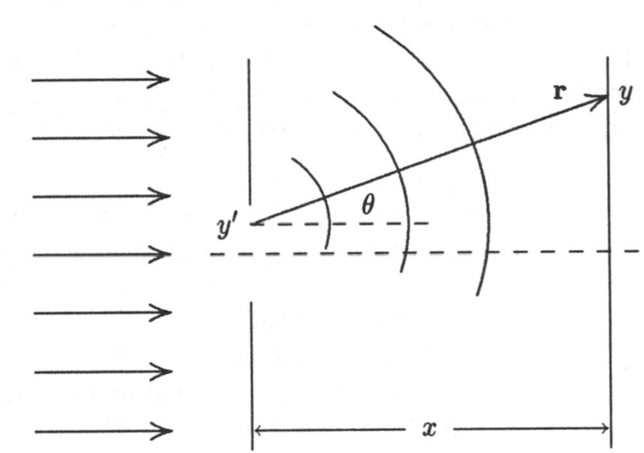

$$E = \frac{B}{r}(1 + \cos\theta)\sin(kr - \omega t)\,,$$

where r is the distance from the wavelet source to the observation point and θ is the angle between the ray from the source to the observation point and the x axis. B is a constant and is chosen so that in the absence of an obstacle the wavelets sum to produce a plane wave with the same amplitude as the original plane wave. Since we are concerned only with changes in the intensity as the observation point moves, we shall choose its value for computational convenience.

The amplitude of the spherical wavelet is not the same as the amplitude of the incident plane wave. Since there are an infinite number of wavelets with sources in the slit, B must be infinitesimal. In addition, the amplitude contains the factor $(1+\cos\theta)$. The wavelet has a larger amplitude in the forward direction (θ near 0) than in the backward direction (θ near 180°). This factor is important if the sum of the wavelets is to reproduce a wave that is traveling in the same direction as the original wave. If the slit is narrow and the viewing screen is far away, then $\cos\theta \approx 1$ for all wavelets that reach the observation point and this factor does not play an important role in determining the intensity pattern. If the slit is wide and the screen is nearby, however, it is important.

The wavelet is spherical and its amplitude decreases as $1/r$. Since the coordinates of the observation point are x and y, $r = [x^2 + (y - y')^2]^{1/2}$ and $\cos\theta = x/r$.

All of the wavelets that originate at points in the slit must now be summed to find the total disturbance at the observation point. Since there is a continuous distribution of wavelet sources in the slit, the sum takes the form of an integral. The amplitude of a wavelet must be infinitesimal and we take $B = dy'/a$, where a is the width of the slit. The total disturbance at the observation point is given by

$$E = \frac{1}{a}\int_{-a/2}^{a/2} \frac{(1+\cos\theta)\sin(kr - \omega t)}{r}\, dy' .$$

The resultant wave has the form $E = E_0 \sin(-\omega t + \alpha)$, where

$$E_0^2 = \frac{1}{a^2}\left[\int_{-a/2}^{a/2} \frac{(1+\cos\theta)\sin(kr)}{r}\, dy'\right]^2 + \frac{1}{a^2}\left[\int_{-a/2}^{a/2} \frac{(1+\cos\theta)\cos(kr)}{r}\, dy'\right]^2$$

The intensity at the observation point is proportional to E_0^2.

The Simpson's rule integration program of Chapter 7 can be modified to evaluate the integrals. S_{1e} and S_{1o} collect the sums of even and odd terms, respectively, for the first integral and S_{2e} and S_{2o} collect the even and odd terms, respectively, for the second integral. The program calculates the intensity at a series of observation points starting at $y = 0$ and ending at $y = y_f$, with an interval of Δy. The intensity pattern is symmetrical about $y = 0$ ($E_0^2(-y) = E_0^2(y)$), so only the upper half need be computed. Here's the outline of a program.

input slit width: a
input number of segments: N
replace N with the nearest even integer
calculate wavelet source interval: $\Delta y' = a/N$
input wavelength: λ
calculate $k = 2\pi/\lambda$
input observation interval and final observation coordinate: Δy, y_f
input distance from slit to viewing screen: x
set $y = 0$
begin loop over observation coordinate
  set $S_{1e} = 0$, $S_{1o} = 0$, $S_{2e} = 0$, $S_{2o} = 0$, $y' = 0$
  begin loop over wavelet sources; counter runs from 1 to $N/2$
    calculate r: $r = \left[x^2 + (y - y')^2\right]^{1/2}$
    calculate $(1 + \cos\theta)/r$: $A = (1 + x/r)/r$
    update sums of even terms:
      replace S_{1e} with $S_{1e} + A\sin(kr)$
      replace S_{2e} with $S_{2e} + A\cos(kr)$
    increment y' by $\Delta y'$
    calculate r: $r = \left[x^2 + (y - y')^2\right]^{1/2}$
    calculate $(1 + \cos\theta)/r$: $A = (1 + x/r)/r$
    update sums of odd terms:
      replace S_{1o} with $S_{1o} + A\sin(kr)$
      replace S_{2o} with $S_{2o} + A\cos(kr)$
    increment y' by $\Delta y'$

```
        end loop over wavelet sources
        calculate integrands at upper and lower limits: I_1N, I_10, I_2N, I_20
        calculate intensity:
           E_0^2 = (Δy'/3a)^2[(I_1N − I_10 + 2S_1e + 4S_1o)^2
                 +(I_2N − I_20 + 2S_2e + 4S_2o)^2]
        display or plot intensity
        if y ≥ y_f exit loop over observation coordinate
        increment y by Δy
    end loop over observation coordinate
    stop
```

You might write instructions to plot the intensity pattern directly on the monitor screen or you might plot it by hand using data generated by the program.

First use the program to investigate the pattern when the viewing screen is far from the slits. This project also acts as a check on the program. The plot of the intensity as a function of viewing coordinate should agree with the diagram in the text.

PROJECT 1. A 1.0×10^{-4}-m wide slit is illuminated by a plane wave with a wavelength of 5.0×10^{-7} m and the intensity pattern is viewed on a screen 1.0 m away. Use the numerical integration program with $N = 16$ to plot E_0^2 every 1.0×10^{-3} m from $y = -15 \times 10^{-3}$ m to $y = +15 \times 10^{-3}$ m. The intensity for negative y is exactly the same as that for positive y, so you need to run the program for positive y only. Indicate on the plot the geometric image of the slit. It extends from $y = -5.0 \times 10^{-5}$ m to $y = +5.0 \times 10^{-5}$ m.

The pattern is dominated by a broad, intense, central region, with a maximum at $y = 0$. This is followed on either side by a series of less intense bright fringes, separated by regions where the intensity is low. The fringes, of course, come about as a result of the interference of the Huygen wavelets. At $y = 0$, the wavelets are all nearly in phase with each other and add constructively. As y increases, the various wavelets travel different distances to the observation point and arrive with different phases. At a minimum of the intensity, the phases are such that the sum of the wavelets is small.

Notice that the intensity pattern spreads well beyond the region of the geometric image. In fact, the central bright region alone is roughly 100 times as wide as the geometric image. The appearance of secondary maxima makes the pattern even wider. This is diffraction.

The pattern spreads as the slit width is narrowed. Suppose a 7.0×10^{-5}-m wide slit is illuminated by a plane wave with a wavelength of 5.0×10^{-7} m and the intensity pattern is viewed on a screen 1.0 m away. Use the program to plot the intensity as a function of the observation coordinate y. Use the graph to find the coordinates of the minima of intensity closest to the central bright area. Compare the values of these coordinates to those when the slit width is 1.0×10^{-4} m.

An increase in wavelength also produces a broadening of the intensity pattern. Suppose a 1.0×10^{-4}-m wide slit is illuminated by a plane wave with a wavelength of 6.5×10^{-7} m and the intensity pattern is viewed on a screen 1.0 m away. Use the program to plot the intensity as a function of the observation coordinate y. Find the coordinates of the minima of intensity closest to the central bright area. Compare with the coordinates of these minima when the wavelength is 5.0×10^{-7} m and the slit width is the same (1.0×10^{-4} m).

As the slit is widened or the viewing screen is brought closer to the slit, the fringe system narrows. Eventually the central maximum of intensity occupies approximately the region of the geometric image and very little light reaches regions of the geometric shadow. There is still some fringing near the edges of the geometric image, however. Something else happens as the slit is widened: fringes appear in the region of the geometric image. The intensities at minima within the image region are not zero so the fringes are not as noticeable as the fringes of a narrow slit.

The following project is designed to show you how the intensity pattern changes as the slit widens. To carry it out requires considerable running time. To shorten the time the number of integration intervals has been selected so the calculation is accurate to only two significant figures. If a fast machine is available and higher accuracy is desired, you might double the number of intervals in each case.

PROJECT 2. A plane wave with a wavelength of 5.0×10^{-7} m illuminates a 7.0×10^{-5}-m wide slit and the intensity pattern is viewed on a screen 0.020 m away. Use the program to plot the intensity as a function of the coordinate y of the observation point. Take $N = 16$ and plot points every 2.0×10^{-5} m from $y = -3.6\times10^{-4}$ m to $y = +3.6\times10^{-4}$ m. Locate the edges of the geometric image on the graph.

Qualitatively the pattern is quite similar to the patterns obtained in the previous project. The bright central region extends a considerable distance beyond the geometric image and this region is followed by a series of fringes. The central maximum is not quite as bright and the dark regions between secondary maxima are slightly broader than for greater slit-to-screen distances.

The slit is now widened to 1.4×10^{-6} m. The wavelength and slit-to-screen distance remain the same. Use the program, with $N = 32$, to plot the intensity every 1.0×10^{-5} m from $y = -2.2\times10^{-4}$ m to $y = +2.2\times10^{-4}$ m. Locate the edges of the geometric image on the graph.

Notice that the pattern is much more narrow than before. The central bright region is now within the geometric image and the pattern has a shoulder near the edge of the image. This shoulder is a remnant of the first minimum of the pattern for a narrow slit.

The slit is now widened to 2.8×10^{-4} m. Use the program, with $N = 50$, to plot the intensity every 1.0×10^{-5} m from $y = -2.0\times10^{-4}$ m to $y = +2.0\times10^{-4}$ m. Locate the edges of the geometric image on the graph.

The geometric image is now discernible in the intensity pattern. There are fringes deep within the image but the intensity does not become zero anywhere in that region. There is a gray central area where the intensity is about half that at the maximum but this region merges into a bright region and then, as the edge of geometric image is approached, the intensity falls off rapidly. There is fringing in the neighborhood of the image edge but the illuminated region does not extend very far beyond the edge.

Chapter 38

First write a program that carries out the Lorentz transformation for a succession of values of the relative velocity v of the reference frames. Input the coordinate x and time t of the event, as measured in frame S. Then, input the first value of v to be considered, the last, and the interval Δv. Positive values of v mean S' is moving in the positive x direction; negative values mean it is moving in the negative x direction. Use $x' = \gamma(x - vt)$ and $t' = \gamma(t - vx/c^2)$, where $\gamma = 1/\sqrt{1 - v^2/c^2}$ and c is the speed of light, to calculate the coordinate and time in the primed frame. Here's an outline.

input coordinate and time in S: x, t
input first velocity, last velocity, and increment: v_i, v_f, Δv
set $v = v_i$
begin loop over intervals
    exit loop if $|c - v|/c < 10^{-5}$
    calculate γ: $\gamma = 1/\sqrt{1 - v^2/c^2}$

```
        calculate x': x' = γ(x − vt)
        calculate t': t' = γ(t − vx/c²)
        display v, x, t, x', t'
        increment v by Δv
    end loop if v > v_f
    stop
```

The computer will overflow if it tries to calculate γ for v nearly equal to c so these values are rejected. If $v_f = v_i$, then results for only one velocity are produced.

First use the program to investigate simultaneity and length measurements.

PROJECT 1. Two events occur at time $t = 0$ in reference frame S, one at the origin and the other at $x = 5.0 \times 10^7$ m. Plot the distance $|\Delta x'|$ between the events as measured in frame S', as a function of the velocity v of that frame relative to frame S. Consider values from $-0.95c$ to $+0.95c$.

For the range considered, which value of v results in the greatest value of $|\Delta x'|$? Which results in the smallest value? [ans: $\pm 0.95c$; 0]

Plot the time interval $|\Delta t'|$ between the events as measured in S', as a function of v. For the range considered, which value of v results in the greatest value of $|\Delta t'|$? Which results in the smallest value? [ans: $\pm 0.95c$, 0]

For what range of v does the second event occur before the first ($\Delta t' < 0$)? For what range does it occur after the first ($\Delta t' > 0$)? [ans: $v > 0$; $v < 0$]

You may think of the values of $|\Delta x'|$ as the results of a series of length measurements, with the object being measured traveling at the various velocities considered. Its length is S is measured by simultaneously making marks at the front and back ends on the x axis and measuring the distance between them. Since the marks are made simultaneously, $\Delta t = 0$ and $\Delta x' = \gamma \Delta x$, where $\Delta x'$ is the length in the rest frame of the object. Note that $|\Delta x'| > |\Delta x|$ and the discrepancy becomes greater as the speed becomes greater.

PROJECT 2. Two events at separated places that are simultaneous in one frame are not simultaneous in any other frame that is moving with respect to the first. The closer together the two events are in space, however, the smaller is the time interval between them in S'. Verify this statement by repeating the calculations of the last project, but take $x = 2.5 \times 10^7$ m. Compare the results you obtain with those of the last project for the same value of v.

If two events occur at the same coordinate, then the time between the events is the proper time interval. The time between the events, as measured in another frame, is longer than the proper time by the factor γ. This is the phenomenon of time dilation.

PROJECT 3. Two events occur at $x = 0$ in reference frame S, one at $t = 0$ and the other at $t = 8.0 \times 10^{-5}$ s. Use the program to plot the time interval $|\Delta t'|$ between the events as measured in S', as a function of the velocity v of that frame relative to S. Consider values of v from $-0.95c$ to $+0.95c$.

For the range considered, which value of v results in the greatest value of $|\Delta t'|$? Which results in the smallest value? [ans: $\pm 0.95c$, 0]

The proper time interval for these events is measured in frame S, where both events occur at the same place. Notice that the time interval, as measured in any other frame, is longer.

Plot the distance $|\Delta x'|$ in frame S' as a function of the velocity of that frame. For what range of v is $\Delta x'$ negative? For what range is it positive? [ans: $v < 0$; $v > 0$]

The two events considered may not occur at the same coordinate in either S or S'. Then, neither Δt nor $\Delta t'$ is the proper time interval between them and these two quantities are not related to each other by a factor of γ. If $\Delta x/\Delta t < c$, then a frame traveling at less than the speed of light exists for which the two events occur at the same coordinate. The time interval, as measured in that frame, *is* the proper time interval between the events.

PROJECT 4. One event occurs at $x = 0$, $t = 0$ and another occurs at $x = 200\,\text{m}$, $t = 9.5 \times 10^{-7}\,\text{s}$. Use the program and a trial and error technique to find the velocity of a frame for which the events occur at the same place. Check your answer by a direction calculation: $\Delta x' = 0$ means $\gamma(\Delta x - v\Delta t) = 0$, or $v = \Delta x/\Delta t$. [ans: $2.11 \times 10^8\,\text{m/s}$ $(0.702c)$]

The next project gives a nice demonstration of time dilation. Each observer signals the other, giving the time read by his clocks. Each finds that the other's clocks run slowly. You may wish to revise the program so it will calculate x' and t' for various values of x and t, all for the same value of the relative velocity v. Here's an outline.

input velocity of S' relative to S: v
calculate γ: $\gamma = 1/\sqrt{1 - v^2/c^2}$
input coordinate and time in S: x, t
calculate x': $x' = \gamma(x - vt)$
calculate t': $t' = \gamma(t - vx/c^2)$
display v, x, t, x', t'
another calculation?
    if yes go back to third line
    if no stop

PROJECT 5. A rocket starts on Earth and travels away at $0.97c$. Every hour for the first five hours an Earth-bound transmitter at the launch pad sends a radio signal to the rocket. These signals are electromagnetic and travel at the speed of light. Assume the rocket starts at time $t = 0$ and the first signal is sent at $t_{s1} = 3600\,\text{s}$. Signal n is sent at $t_{sn} = 3600n$, where $n = 1, 2, 3, 4, 5$. Take the launch pad to be at the origin of Earth's rest frame S and the rocket to be at the origin of its rest frame S'.

Fill in the first 3 columns of the following table (t_{sn} is the time signal n is sent, t_{rn} is the time signal n is received, x_{rn} is the coordinate of the rocket when signal n is received, all in the rest frame of Earth. You will need to show that signal n is received at time $t_{rn} = t_{sn}c/(c - v)$, where v is the speed of the rocket. This is easy since the coordinate of the rocket is given by $x = vt$ and the coordinate of signal n is given by $x = c(t - t_{sn})$. The signal is received when these are equal. The distance from Earth to the rocket when it receives signal n is given by $x_{rn} = vt_{rn}$.

n	t_{sn}	t_{rn}	x_{rn}	t'_{rn}	t'_{sn}
1					
2					
3					
4					
5					

Now use the program to find the time t'_{rn} when signal n is received, as measured by a clock on board the rocket. Fill in the fourth column of the table.

An observer on the rocket can calculate the time, according to his clock, when each signal was sent. The signal starts from the coordinate of the launch pad at time t'_{sn}. This is $x' = -vt'_{sn}$, where the negative sign appears because in the frame of the rocket, Earth is moving in the negative x direction. The coordinate of the signal is given by $x' = -vt'_{sn} + c(t' - t'_{sn})$. This must be zero when $t' = t'_{rn}$ (the rocket receives the signal at the origin of S'. So $t'_{sn} = t'_{rn}c/(c+v)$. Fill in the last column of the table. According to clock on the rocket is the Earth clock slow or fast? ________

Now suppose the rocket emits a signal every hour, according to on-board clocks. Let t'_{sn} be the time signal n is sent from the rocket (at $x' = 0$), let t'_{rn} be the time it is received on Earth, and let x'_{rn} be the coordinate of Earth when the signal is received. Fill out the following table for those signals.

n	t'_{sn}	t'_{rn}	x'_{rn}	t_{rn}	t_{sn}
1	________	________	________	________	________
2	________	________	________	________	________
3	________	________	________	________	________
4	________	________	________	________	________
5	________	________	________	________	________

According to Earth clocks is the clock on the rocket slow or fast? ________

The first program above can be used to investigate the velocity of a particle as measured by observers moving at various velocities. Suppose the particle has constant velocity u along the x axis of S. If reference frame S is placed so its origin is at the position of the particle at $t = 0$, then the coordinate of the particle is given by $x = ut$. Pick a value for t and calculate x. Now use the program to find x' and t', the coordinate and time in another frame. Finally use $u' = x'/t'$ to compute the velocity in the primed frame.

PROJECT 6. A particle moves along the x axis of reference frame S with a velocity of $0.10c$. Plot the velocity in frame S' as a function of the velocity v of that frame. Consider values from $-0.95c$ to $+0.95c$. Notice that for v close to zero the Galilean transformation is nearly correct. That is, u' is nearly $u - v$. But for v close to the speed of light u' is also close to the speed of light.

If the particle speed is c in frame S, then it is c in every frame. Take $u = c$ and find the particle velocity for values of v in the range from $-0.95c$ to $+0.95c$. The answer should be c for every value of v.

Now consider a particle moving along the y axis of S. The y component of its velocity in S' is not the same as the y component in S because $\Delta t' \neq \Delta t$. In addition, the velocity in S' has an x component equal to $-v$. Test these assertions. Take the y component of the velocity in S to be $u_y = 0.10c$ and the x component to be $u_x = 0$. Modify the program to find the components of the velocity and the speed in S' as function of the velocity of that frame. Let $x = 0$ and $y = u_y t$. Use $u'_x = x'/t'$ and $u'_y = y'/t'$ to calculate the components of the particle velocity in S'. Consider values of v from $-0.95c$ to $+0.95c$.

Notice that the y' component becomes smaller as the speed of S' approaches the speed of light. This is because $\Delta t'$ becomes larger. In the limit as $v \to c$, $u'_y \to 0$ and $u'_x \to c$. The particle moves along the x' axis at the speed of light.

As the speed of a particle with mass increases toward the speed of light, its kinetic energy increases and, for speeds near the speed of light, the increase is dramatic. According to the defining equations, $\mathbf{p} = m\mathbf{u}/(1 - u^2/c^2)^{1/2}$ and $E = mc^2/(1 - u^2/c^2)^{1/2}$, both the energy and momentum become infinite as u approaches c. The ratio of the magnitude of the momentum to the energy, however, does not blow up but approaches $1/c$ in the limit as the speed approaches the speed of light. When pc is much greater than mc^2 then $E \approx pc$.

The first program of this chapter can be modified to plot the energy and the magnitude of the momentum as functions of the particle velocity. Consider a particle that moves along the x axis. Here's an outline.

> input mass, first velocity, last velocity, and increment: m, u_i, u_f, Δu
> set $u = u_i$
> **begin loop** over intervals
>   exit loop if $|c - u|/c < 10^{-5}$
>   calculate constant α: $\alpha = \sqrt{1 - u^2/c^2}$
>   calculate p: $p = mu/\alpha$
>   calculate E: $E = mc^2/\alpha$
>   display u, p, E, p/E
>   increment u by Δu
> **end loop** if $u > u_f$
> stop

PROJECT 7. A 6.5×10^{-29}-kg particle travels along the x axis. Use the program to make separate plots of the energy, momentum, and ratio p/E as function of the particle speed. Plot points every $0.05c$ from $u = 0$ to $u = 0.95c$. Mark the value of the rest energy on the energy graph.

At slow speeds the energy is nearly the rest energy. Added to this is the kinetic energy, which increases as the square of the speed. Close to the speed of light the energy increases more rapidly with the speed of the particle and becomes infinite at $u = c$. The momentum starts at zero and increases linearly with the speed of the particle. At relativistic speeds it increases more rapidly and becomes infinite at $u = c$. The ratio p/E is small near $u = 0$ since p is small. It increases since E is nearly constant and p increases. In the relativistic region, p/E approaches $1/c$ as a limiting value.

Newton's second law is valid in the form $\mathbf{F} = d\mathbf{p}/dt$, but emphatically not in the form $\mathbf{F} = m\mathbf{a}$. If the force is given as a function of time, then $\mathbf{p} = \mathbf{p}_0 + \int_0^t \mathbf{F}\, dt$ can be used to find the momentum at time t. The velocity $\mathbf{u}$ of the particle is given in terms of its momentum by $\mathbf{u} = \mathbf{p}c/(m^2c^2 + p^2)^{1/2}$ and the position vector is given by $\mathbf{r} = \mathbf{r}_0 + \int_0^t \mathbf{u}(t)\, dt = \mathbf{r}_0 + \int_0^t [\mathbf{p}c/(m^2c^2 + p^2)^{1/2}]\, dt$.

Consider motion along the x axis and suppose the momentum is given as a function of time. Write a program, essentially a modification of the Simpson's rule program of Chapter 7, to compute the velocity and position as functions of time. Take the initial position to be at the origin. Let Δt_p be the display interval, let N be the number of integration intervals used in each display interval, and let t_f be the final time. Here's an outline.

> input mass and initial velocity: m, u_0
> input display interval, number of integration intervals, final time: Δt_p, N, t_f
> replace N with nearest even integer
> calculate integration interval: $\Delta t = \Delta t_p/N$
> initialize quantity to hold sum of values with even labels: $S_e = 0$
> initialize quantity to hold sum of values with odd labels: $S_o = 0$
> set $t = 0$
> calculate momentum at time $t = 0$: p_0
> calculate velocity at time $t = 0$: $u_0 = p_0c/(m^2c^2 + p_0^2)^{1/2}$

```
begin loop over display intervals
    begin loop over integration intervals: counter runs from 1 to N/2
        calculate momentum at time t: p
        calculate velocity at time t: u = pc/(m^2c^2 + p^2)^{1/2}
        add velocity to sum of values with even labels: replace S_e with S_e + u
        increment t by Δt
        calculate momentum at time t: p
        calculate velocity at time t: u = pc/(m^2c^2 + p^2)^{1/2}
        add velocity to sum of values with odd labels: replace S_o with S_o + u
        increment t by Δt
    end loop over integration intervals
    calculate momentum at time t: p_N
    calculate velocity at time t: u_N = p_N c/(m^2c^2 + p_N^2)^{1/2}
    evaluate integral: x = (Δt/3)(u_N − u_0 + 2S_e + 4S_o)
    display t, u_N, x
if t ≥ t_f end loop over display intervals
stop
```

PROJECT 8. A 7.6×10^{-25}-kg particle starts from rest and is acted on by a constant force of $F = 4.1 \times 10^{-18}$ N, in the positive x direction. Use the program to plot the velocity u as a function of time from $t = 0$ to $t = 200$ s. Use $\Delta t_p = 5$ s and $N = 30$. Use $p = Ft$ to calculate the momentum.

Although the momentum increases uniformly, the velocity does not. At first, when $p \ll mc$, $u \approx p/m$ and the velocity does increase uniformly. But later, when p is a significant fraction of mc or larger, u changes much more slowly. In fact, it approaches the speed of light as a limit. It cannot increase beyond that limit no matter how large the momentum becomes. To verify that the acceleration decreases as the speed increases, use your data to estimate the average acceleration in the first and last 5 s intervals. [ans: $5 \times 10^6\,\text{m/s}^2$; $1 \times 10^5\,\text{m/s}^2$]

On the same graph plot the velocity as a function of time using the classical kinematic relation $u(t) = (F/m)t$, where F is the force. Estimate the time for which the velocity deviates from the classical approximation by 10%. [ans: 25 s]

Now plot the coordinate x for the same time interval. Notice that at first the curve has the shape predicted by classical kinematics: $x = \frac{1}{2}(F/m)t^2$ but later it becomes nearly a straight line with slope equal to c.

No matter how large the force becomes the qualitative results are the same: it cannot accelerate the particle beyond the speed of light. Take the force to be $4.1 \times 10^{-20}t^2$ N, where the time t is in seconds. Use the program to plot the velocity u as a function of time from $t = 0$ to $t = 50$ s.